Antisense Technology

The Practical Approach Series

SERIES EDITOR

B. D. HAMES
Department of Biochemistry and Molecular Biology
University of Leeds, Leeds LS2 9JT, UK

★ **indicates new and forthcoming titles**

Affinity Chromatography

★ Affinity Separations

Anaerobic Microbiology

Animal Cell Culture
(2nd edition)

Animal Virus Pathogenesis

Antibodies I and II

★ Antibody Engineering

★ Antisense Technology

★ Applied Microbial
Physiology

Basic Cell Culture

Behavioural Neuroscience

Bioenergetics

Biological Data Analysis

Biomechanics—Materials

Biomechanics—Structures and
Systems

Biosensors

Carbohydrate Analysis
(2nd edition)

Cell–Cell Interactions

The Cell Cycle

Cell Growth and Apoptosis

Cellular Calcium

Cellular Interactions in
Development

Cellular Neurobiology

Clinical Immunology

★ Complement

Crystallization of Nucleic
Acids and Proteins

Cytokines (2nd edition)

The Cytoskeleton

Diagnostic Molecular Pathology
I and II

★ DNA and Protein Sequence
Analysis

DNA Cloning 1: Core
Techniques (2nd edition)

DNA Cloning 2: Expression
Systems (2nd edition)

★ DNA Cloning 3: Complex
Genomes (2nd edition)

★ DNA Cloning 4: Mammalian
Systems (2nd edition)

Electron Microscopy in
Biology

Electron Microscopy in
Molecular Biology

Electrophysiology

Antisense Technology

A Practical Approach

Edited by

CONRAD LICHTENSTEIN

School of Biological Sciences, Queen Mary and Westfield College,
University of London, London E1 4NS, UK

and

WOLFGANG NELLEN

Abteilung Genetik, FB 19, Universität Kassel, Heinrich-Plett-Strasse 40,
D-34132 Kassel, Germany

OXFORD UNIVERSITY PRESS
Oxford New York Tokyo

Oxford University Press, Great Clarendon Street, Oxford OX2 6DP

Oxford   New York

Athens   Auckland   Bangkok   Bogota   Bombay   Buenos Aires
Calcutta   Cape Town   Dar es Salaam   Delhi   Florence   Hong Kong
Istanbul   Karachi   Kuala Lumpur   Madras   Madrid   Melbourne
Mexico City   Nairobi   Paris   Singapore   Taipei   Tokyo   Toronto   Warsaw

and associated companies in
Berlin   Ibadan

Oxford is a trade mark of Oxford University Press

Published in the United States
by Oxford University Press Inc., New York

Users of books in the Practical Approach Series are advised that prudent
laboratory safety procedures should be followed at all times. Oxford
University Press makes no representation, express or implied, in respect of
the accuracy of the material set forth in books in this series and cannot
accept any legal responsibility or liability for any errors or omissions
that may be made.

A catalogue record for this book is available from the British Library

Library of Congress Cataloging in Publication Data
Antisense technology: a practical approach/edited by Conrad
Lichtenstein and Wolfgang Nellen.
(Practical approach series; 185)
Includes bibliographical references and index.
1. Antisense nucleotides—Laboratory manuals.   2. Antisense
peptides—Laboratory manuals.   I. Lichtenstein, Conrad.
II. Nellen, Wolfgang.   III. Series.
QP623.5.A58A586   1997   572.8—DC21   97–23434

ISBN 0 19 963584 6 (Hbk)
ISBN 0 19 963583 8 (Pbk)

Typeset by Footnote Graphics, Warminster, Wilts
Printed in Great Britain by Information Press, Ltd, Eynsham, Oxon.

Preface

Antisense nucleic acids, complementary to a target, usually mRNA, can inhibit expression of the specific target gene. Antisense RNA was first applied experimentally in eukaryotes in 1983; that same year, it was also discovered to be a natural form of gene regulation in prokaryotes. More recently, examples of natural antisense mechanisms in eukaryotes are emerging. Although the details of the molecular mechanisms of antisense-mediated gene silencing in eukaryotes remain to be elucidated, this technology has proved to be a very powerful tool to study gene function. As it is often very difficult to predict whether a given antisense nucleic acid will succeed in blocking expression of its target gene, the approaches taken have been largely empirical. In this book we have brought together a variety of experimental systems and experience, drawn from the many diverse approaches taken.

Can we predict which antisense:target combinations are likely to be the most effective *in vivo* by first performing *in vitro* studies? A crucial determinant in the effectiveness of the antisense approach is that intermolecular interactions should be favoured over intramolecular ones. We begin with a chapter by Detlev Riesner on nucleic acid structure which reviews computer programs to predict RNA secondary structure and provides protocols to map such structure and to measure the stability of RNA:RNA duplexes using temperature gradient gel electrophoresis (TGGE). Stephen Munroe extends this approach with protocols for RNA:RNA annealing assays. This allows an analysis of the role of proteins in catalysing such annealing or in destabilizing RNA:RNA duplexes. Also provided are protocols to study the effect of antisense RNA on splicing and on translation *in vitro*. He also includes a useful guide to handling RNA to minimize degradation. Noémi Lukács gives some future perspectives by presenting a novel approach to detecting dsRNAs both *in vivo* (to allow localization) and *in vitro* using dsRNA-specific antibodies, the latter complementing the TGGE approach.

There is now a very broad repertoire of antisense nucleic acids and their derivatives available to silence gene expression. These include antisense oligodeoxyribonucleotides and Jean-Jacques Toulmé and colleagues discuss the merits of such DNA-based oligos and their various nuclease-resistant derivatives. They present strategies for the design of such oligos, protocols for their synthesis and purification, analysis of activity in cell-free extracts, and methods to deliver them to living cells. Alluded to in this chapter are the non-specific effects of DNA-based oligos. Cy Stein and Arthur Krieg expand upon this to review the range of non-sequence specific, even toxic, effects of such oligos including their stimulation of the immune system.

Antisense RNA can either be used as a direct reagent, or following expression from a transgene. For the former approach, Ute Weber and Hans

Gross present protocols for purification of the bacteriophage T7 RNA polymerase and the use of this for the *in vitro* synthesis of antisense RNAs. For the latter, Martin Hildebrandt reviews the applications of expression of anti-sense RNA in a lower eukaryotic model system, the slime mould *Dictyostelium discoideum* and Anja Kuipers and colleagues do the same for plants. In addition to their protocols they provide tables of successful and unsuccessful experiments which will help the researcher to design antisense experiments on the basis of accumulated experience to increase their success rate.

Hammerhead ribozymes are autocatalytic satellite RNAs that replicate via a helper RNA virus in plants. Martin Tabler and Georg Sczakiel discuss the general exploitation of these to function as specific endoribonucleases against any desired target sequence. They call them catalytic antisense RNA because, by such cleavage, they can accelerate the destruction of their mRNA target. They provide protocols for the design and synthesis of catalytic antisense RNAs, how to choose the best secondary-structure-free region of mRNA target, and *in vitro* assays of activity. Another approach for destruction of the mRNA target, this time with a DNA-based oligo, is presented by Robert Silverman and colleagues. They have developed a novel magic bullet to achieve antisense-mediated degradation of the mRNA target. Here the antisense oligo is coupled to a tetradenylate moiety. This chimaera then attracts a specific (naturally, interferon-induced) ribonuclease, RNase L, to destroy the mRNA target. Protocols are provided to chemically synthesize these chimaeras and to analyse their activity both in cell-free systems and using purified RNaseL.

Michael Strauss provides a chapter reviewing the medical applications of antisense technology and protocols for their use in cell culture and whole animals.

It is sometimes also desirable to replace the expression of a target gene by a mutated or related one. Jeffrey Holt describes an antisense rescue approach where an antisense gene targeted against the 5′ untranslated region of a target is co-transfected with another antisense-resistant gene (where this region is deleted). Antisense RNA is typically thought of as a precise reverse genetic tool to apply against a specific gene. Timothy Spann and colleagues describe a novel strategy they call shotgun antisense mutagenesis to generated a library of antisense cDNAs. An advantage of this approach is the ease of identifying the antisense gene in a particular transformant that is responsible for the phenotypic change observed.

We also have a chapter with a checklist to assess the success of an anti-sense experiment and provide protocols to determine whether the observed change in phenotype, if found, is indeed antisense-mediated, and if not, at what stage the experiment failed.

London   C. L.
Kassel   W. N.
June 1997

Contents

Contents

5. Catalytic antisense RNA based on hammerhead ribozymes 93

Martin Tabler and Georg Sczakiel

6. 2–5A-antisense chimeras for targeted degradation of RNA 127

R. H. Silverman, A. Maran, R. K. Maitra, C. F. Waller, K. Lesiak, S. Khamnei, G. Li, W. Xiao, and P. F. Torrence

7. *In vitro* applications of antisense RNA and DNA 157

Stephen H. Munroe

8. Antisense applications in *Dictyostelium*: a lower eukaryotic model system   173

Martin Hildebrandt

9. Applications of antisense technology in plants   191

Anja G. J. Kuipers, Evert Jacobsen, and Richard G. F. Visser

Contents

Contents

14. Detection of sense:antisense duplexes by structure-specific anti-RNA antibodies 281

Noémi Lukács

A1. List of Suppliers 297

Index 301

Contributors

DEBRA A. BROCK
Howard Hughes Medical Institute, Department of Biochemistry and Cell Biology, Rice University, Houston, TX 77251-1892, USA.

CHRISTIAN CAZENAVE
INSERM U 386, Laboratoire de Biophysique Moléculaire, Université Victor Segalen, 146, rue Léo Saignat, 33076 Bordeaux Cedex, France.

RICHARD H. GOMER
Howard Hughes Medical Institute, Department of Biochemistry and Cell Biology, Rice University, Houston, TX 77251-1892, USA.

HANS J. GROSS
Institut für Biochemie, Universität Würzburg, Biozentrum, Am Hubland, D-97074 Würzburg, Germany.

MARTIN HILDEBRANDT
Abteilung Genetik, FB19, Universität Kassel, Heinrich-Plett-Strasse 40, D-34132, Kassel, Germany.

JEFFREY T. HOLT
Department of Cell Biology, Vanderbilt University Medical School, Nashville, TN 37232, USA.

EVERT JACOBSEN
Department of Plant Breeding, Agricultural University, Wageningen, PO Box 386, NL-6700 AJ Wageningen, The Netherlands.

S. KHAMNEI
Section on Biomedical Chemistry, Laboratory of Medicinal Chemistry, National Institute of Diabetes and Digestive and Kidney Diseases, National Institutes of Health, Bethesda, MD 20892, USA.

ARTHUR KRIEG
Department of Internal Medicine, University of Iowa, 540 EMRB, Iowa City, IA 52242, USA.

ANJA G. J. KUIPERS
Department of Plant Breeding, Agricultural University, Wageningen, PO Box 386, NL-6700 AJ Wageningen, The Netherlands.

K. LESIAK
Section on Biomedical Chemistry, Laboratory of Medicinal Chemistry, National Institute of Diabetes and Digestive and Kidney Diseases, National Institutes of Health, Bethesda, MD 20892, USA.

Contributors

G. LI
Section on Biomedical Chemistry, Laboratory of Medicinal Chemistry, National Institute of Diabetes and Digestive and Kidney Diseases, National Institutes of Health, Bethesda, MD 20892, USA.

CONRAD LICHTENSTEIN
School of Biological Sciences, Queen Mary and Westfield College, University of London, London E1 4NS, UK.

NOÉMI LUKÁCS
Hungarian Academy of Sciences, Biological Research Center, Institute of Plant Biology, H-6701 Szeged, Hungary.

R. K. MAITRA
The Cleveland Clinic Foundation, 9500 Euclid Avenue, Cleveland, Ohio 44195-5178, USA.

A. MARAN
The Cleveland Clinic Foundation, 9500 Euclid Avenue, Cleveland, Ohio 44195-5178, USA.

SERGE MOREAU
INSERM U 386, Laboratoire de Biophysique Moléculaire, Université Victor Segalen, 146, rue Léo Saignat, 33076 Bordeaux Cedex, France.

STEPHEN H. MUNROE
PO Box 1881, Marquette University, Department of Biology, Wehr Life Science Building, Milwaukee, WI 53201, USA.

WOLFGANG NELLEN
Abteilung Genetik, FB19, Universität Kassel, Heinrich-Plett-Strasse 40, D-34132 Kassel, Germany.

DETLEV RIESNER
Institut für Physikalische Biologie, Heinrich-Heine-Universität Dusseldorf, Universität Strasse 1, D-40225, Düsseldorf, Germany.

GEORG SCZAKIEL
Deutsches Krebsforschungzentrum, Im Neuenheimer Feld 242, D-6900, Heidelberg, Germany.

R. H. SILVERMAN
Department of Cancer Biology, NN1, The Cleveland Clinic Foundation, 9500 Euclid Avenue, Cleveland, Ohio 44195-5178, USA.

TIMOTHY P. SPANN
Department of Cell and Molecular Biology, Northwestern University Medical School, Chicago, Illinois, USA.

Contributors

C. A. STEIN
Department of Medicine, Columbia University, College of Physicians and Surgeons, 630 W. 168 St., New York, NY 10032, USA

MICHAEL STRAUSS
Department of Molecular Cell Biology, Humbold-Universität, Max Debrueck Centre for Molecular Medicine, Robert-Roessle-Str. 10, D-13125 Berlin-Buch, Germany.

MARTIN TABLER
Foundation for Research and Technology-Hellas, Institute of Molecular Biology and Biotechnology, PO Box 1527, GR-711 10 Heraklion, Crete, Greece.

P. F. TORRENCE
Section on Biomedical Chemistry, Laboratory of Medicinal Chemistry, National Institute of Diabetes and Digestive and Kidney Diseases, National Institutes of Health, Bethesda, MD 20892, USA.

JEAN-JACQUES TOULMÉ
INSERM U 386, Laboratoire de Biophysique Moléculaire, Université Victor Segalen, 146, rue Léo Saignat, 33076 Bordeaux Cedex, France.

RICHARD G. F. VISSER
Department of Plant Breeding, Agricultural University, Wageningen, PO Box 386, NL-6700 AJ Wageningen, The Netherlands.

C. F. WALLER
Department of Hematology and Oncology, Freiburg University Medical Center, Hugstetter Strasse 55, D-79106 Freiburg, Germany.

UTE WEBER
Institut für Biochemie, Universität Würzburg, Biozentrum, Am Hubland, D-97074 Würzburg, Germany.

W. XIAO
Section on Biomedical Chemistry, Laboratory of Medicinal Chemistry, National Institute of Diabetes and Digestive and Kidney Diseases, National Institutes of Health, Bethesda, MD 20892, USA.

Abbreviations

APS	ammonium persulfate
CaMV	cauliflower mosiac virus
CAT	chloramphenicol acetyltransferase
cDNA	complementary DNA
CHS	chalcone synthase
CIP	calf intestinal alkaline phosphatase
CPG	controlled pore glass
DEPC	diethyl pyrocarbonate
DMAP	4-dimethylaminopyridine
DMS	dimethyl sulfate
DMT	dimethoxytrityl
DNA	deoxyribonucleic acid
DNase	deoxyribonuclease
dNTP	mixture of dATP, dGTP, dCTP, dTTP
DOC	deoxycholate
dsRNA	double-stranded RNA
DTT	dithiothreitol
ELISA	enzyme linked immunosorbent assay
GBSS	granule-bound starch synthase
IFN	interferon
IPTG	isopropyl-β-D-thiogalactopyranoside
LDL	low density lipoprotein
LPS	lipopolysaccharide
MAR	matrix attachment region
MMTV	mouse mammary tumour virus
MP	methylphosphonate
nt	nucleotide
PAGE	polyacrylamide gel electrophoresis
PBS	phosphate-buffered saline
PCR	polymerase chain reaction
PG	polygalacturonase
PKC	protein kinase C
PKR	protein kinase R
PMSF	phenylmethylsulfonyl fluoride
PO	phosphodiester
PS	phosphorothioate
RRL	rabbit reticulocyte lysate
RT	reverse transcriptase
SDS	sodium dodecyl sulfate
TBE	Tris, borate, EDTA

T_c	critical temperature
T-DNA	transferred DNA
TEAA	triethylammonium acetate
TEMED	tetramethylethylenediamine
TGGE	temperature-gradient gel electrophoresis
TGMV	tomato golden mosiac virus
TLC	thin-layer chromatography
T_m	melting temperature
UV	ultraviolet
WGE	wheat germ extract

1

Nucleic acid structures

DETLEV RIESNER

1. Introduction

From the wide discipline of nucleic acid structure only selected topics have a direct bearing on antisense technology. Thus nucleic acid structure will not be discussed in this chapter as an independent field of research but as an essential factor for a particular field of application. The restrictions for selecting the topics of interest concern the structural elements as well as the solution conditions and the thermodynamic conditions. I shall mainly address RNA intramolecular structure and RNA:RNA double-strand formation. A few examples of DNA oligonucleotides directed against RNA targets will also be considered. Antisense technology is predominantly an *in vivo* technology, whereas studies on nucleic acid structure have been carried out in the past mainly *in vitro*. Therefore, *in vitro* conditions either have to be adapted to reflect the *in vivo* situation or the data have to be extrapolated. This requirement restricts the applicability of several experimental methods. Not only the solution conditions but also the thermodynamic approach has to resemble the requirements of antisense technology. Antisense RNA as well as target RNA act in structures which are formed right after synthesis in the cell. These structures, which might also exist as mixtures of conformers, deviate in many cases from the equilibrium structure. This complicates the experimental as well as the theoretical approach. Having these restrictions in mind, an outline will be given on the structures, on methods to predict them, and on experimental approaches for structure analysis.

2. Structure of RNA:RNA and RNA:DNA double-strands and intramolecular structure of RNA

2.1 Intermolecular double-strand formation

Antisense:target interaction is in most cases the interaction of two RNA strands. This is true for all cases of naturally occurring control by antisense RNA as well as of most cases of antisense technology. For some applications, DNA oligonucleotides have been applied as antisense nucleic acids which then form RNA:DNA hybrids (see Chapter 3). Naturally occurring antisense RNAs (reviewed in ref. 1) vary from < 50 to > 1000 nucleotides in length, artificial from about 20 to several hundred. The smallest size is limited by the thermodynamic stability and varies with the G:C content (see below). The structure of the antisense:target RNA complexes is that of a homogeneous RNA double-strand, i.e. the A-form, right-handed double helix (2). Also RNA:DNA hybrids assume the structure of an A-form double helix, although slightly different from the RNA:RNA double helix (2). The details of the double helices, when regarded at the level of atomic coordinates, also depend upon the sequence, but need not be considered for most cases of antisense technology. The structure of a double-helical stem might also depend on the number of base pairs. This determines the details of tRNA structure with stems of four to seven pairs but not for double-strands of more than 20 base pairs.

Antisense:target complexes are not restricted to the formation of homo-geneous double-strands as described above. A well known example is the ribozyme action, where the catalytically active site (hammerhead- or hairpin-type) (3) is not a homogeneous double-helical region but consists of a highly structured single-strand which is bound by two double-helical stretches to the target RNA. Antisense:target interactions might also be designed to simulate structures, which are formed naturally as intramolecular structures and serve as recognition sites for regulatory factors, structure-specific nucleases, etc.

2.2 Intramolecular RNA structure

Intramolecular RNA structure has to be considered for antisense technology, as the double-stranded complex is not formed from two structureless single-strands, but both antisense RNA and target RNA might be present as highly structured molecules. This raises two questions:

(a) Do antisense:target RNA complexes form at all?

(b) What are the kinetics and mechanism of double-strand formation?

Both problems are a matter of primary interactions between antisense and target RNA and of competition between intra- and intermolecular inter-actions. For a detailed description of RNA structural elements and their biological function, refer to the review article by Wyatt and Tinoco (4) and Chastain and Tinoco (10).

The effects of stem-loop structures on complex formation have been studied in detail in naturally occurring antisense interactions. The primary interaction occurs between a few single-stranded bases in loops; vividly this step is called the 'kissing reaction' (1, 5). After the primary, still labile complex has been formed, double-strand formation proceeds into the stem region up to com-pletion. It is known from studies on the binding of two tRNA anticodon loops (6) with complementary bases, that particular loop structures present bases in a conformation which is single-stranded but preformed for double-strand formation, and therefore the primary complex is sufficiently stable to promote further double-strand formation at the expense of the intramolecular structure. Consequently the stems of the antisense and target RNAs are required to present the loops for the kissing reaction and to increase the stability of the individual RNA molecules.

Naturally occurring antisense systems provide a useful model in designing antisense experiments. At present, the detailed mechanisms of double-strand formation, e.g. the kissing reaction, can hardly be predicted. Furthermore, the sequence of the target RNA is fixed and only the position and the length of a segment within the sequence might be optimized for antisense interaction. I will outline which structural features can be taken into account in a routine analysis and what kind of pitfalls might exist in general.

If complex formation is not mediated by a particular intramolecular structure, one would expect any intramolecular base pairs to compete with intermolecular complex formation. Intramolecular structure can be considered on two levels. On the first level the structure is described by the well known secondary structure or base pairing scheme. It includes, as structural elements double-helical stems, hairpin loops, internal loops, bulged loops, and bifurcation loops. They are represented in *Figure 1* (left panels). In the absence of further information one has to assume that the bases in the loops have a high degree of freedom and flexibility. Programs which predict these secondary structures with good accuracy are discussed below.

These programs yield not only the structure of lowest free energy, but also suboptimal or metastable structures. The latter two are not of mere theoretical interest, but might have an advantage in fast folding after transcription. The problem of forming suboptimal structures and coexisting structures is discussed in Section 3. *Figure 2* shows a sequence which forms a single stem-loop as the structure of lowest free energy (left panel). However, if one considers the process of structure formation during synthesis, the first hairpin formed (right panel) is sufficiently stable to allow the formation of the second hairpin. Since the activation barrier between the metastable and the stable structure is high, the former one may be maintained.

A new level of viewing RNA structures became evident from investigations applying experimental methods with atomic resolution. The simple assumption of flexible loops does not hold in all cases; in several examples looped segments are involved in higher ordered structures. Without discussing each structure in detail the best known examples are redrawn from the original literature in *Figure 1* (right panels). In addition to loops, structural elements like pseudoknots and triple-strands are typical components of RNA tertiary structure and, by definition, were not contained in secondary structure prediction. In a few cases, like tetra-loops and pseudoknots, it is possible to predict these higher ordered structures, but in most cases it is not, and one has to consider all these potential structures and their possible role in affecting antisense:target interactions.

Hairpin

 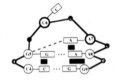 

Internal

Bulge

 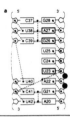 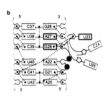

Bifurcation

Pseudoknot

Triple strand

4

Figure 1. On the left side, schematic representations of secondary structure elements of RNA are shown. On the right side, examples from the literature are selected, for which the three-dimensional structure of the looped regions is known from NMR studies. The hairpin structures are the so-called tetra-loops (7, 8); the internal loop is the E. loop from eukaryotic 5S RNA (9, 10); the bulged loop is the TAT region of HIV RNA, which is (a) uncomplexed or (b) bound to argininamide (11). The pseudoknot is represented by the orginal model (12) confirmed later by NMR (13). The triple-stranded structure is a domain of the *Tetrahymena* group I intron determined by NMR (14) or by model building (15). Reproduced from the literature with permission.

Finally, structures like those in *Figure 1* represent, whenever known, functional signals. To give a few examples: hairpins are involved in mRNA regulation including stability, frame shifting, and binding of specific proteins, the internal loop in *Figure 1* is the recognition site for TFIIIA in eukaryotic 5S RNA, also bulged loops serve as binding sites for proteins and can change conformation upon protein binding. This is exemplified by the binding of the transcriptional activator protein TAT to the TAR region of HIV RNA. Pseudoknots have been found in viral RNA, mRNA, and ribosomal RNA. Among other functions, they might induce pausing of translation and frame shifting. The loops in pseudoknots are quite large in some cases and lead to long-range interactions. The triple-strand is a small part of the tertiary structure of the *Tetrahymena* group I intron. It is obvious that these interactions define relative orientations of helices which are distant in the sequence but which need close proximity for proper function. This list of biological functions is far from complete (for review see ref. 4). It should be emphasized that these structures are of particular interest for manipulation by antisense technology, however, their accessibility for interfering antisense RNA is difficult to predict.

3. Prediction of intramolecular structures and double-strand formation

Calculations to predict the intramolecular and intermolecular structures of antisense and target RNA may be carried out for two reasons:

(a) They provide clues on possible structures present in the separate RNA

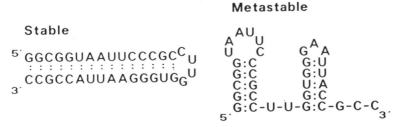

Figure 2. Stable and metastable secondary structure of the same sequence.

molecules which may affect double-strand formation. These predictions are incomplete to the extent of structures which cannot be calculated by simple rules (see *Figure 1*, right panel).

(b) They provide a reliable estimate of double-strand formation for the case of thermodynamic equilibrium. In most cases this equilibrium will be far on the side of the double-strands, and only in cases of particularly strong intramolecular structures and very low concentrations, the equilibrium might be less favourable for antisense:target complexes. In practice, the problem is more to reach the state of equilibrium than the situation after equilibrium has been established.

3.1 Estimation of the equilibrium between intermolecular and intramolecular structures

Let us assume that the intramolecular structures of both antisense RNA and target RNA are known from applying computer programs as described in Section 3.2. The situation is presented schematically in *Figure 3*. The structure of the antisense RNA has to be disrupted completely and that of the target RNA partially, to form a bimolecular complex consisting of a double-stranded section and the remaining intramolecular structure of the target. To convert most of the target RNA molecules into the complexed form, antisense RNA has to be present in excess molar concentration.

Protocol 1 provides equations for calculation of the ratio of complexed and free RNA. While *Protocol 1A* gives the equation for 37°C, *Protocol 1B*

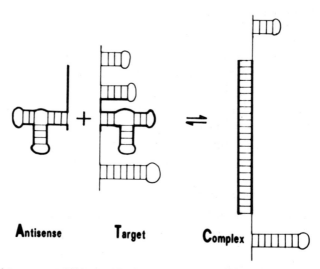

Figure 3. Antisense:target RNA complex formation is shown. The heavy contours represent segments which form the homogeneous double helix in the complex. Representation of the complex is slightly out of scale.

provides the equation for the more general case where secondary structure models for antisense and target RNA are known (from experiments or computer predictions) but the values for $\Delta G°_A$ and $\Delta G°_T$ are not and have to be calculated.

All calculations refer to 1 M ionic strength. This is reasonably close to the cellular conditions, since the lower ionic strength in the cell is balanced by the higher concentration of divalent cations like Mg^{2+}.

Protocol 1. Calculation of the ratio of complexed and free target RNA

Definitions

- [C], molar concentration of complexed target RNA
- [T], molar concentration of free target RNA
- c_0, molar concentration of excess antisense RNA
- $\Delta G°$, standard free energy for secondary structure formation
- $\Delta H°$, standard enthalpy for secondary structure formation
- $\Delta S°$, standard entropy for secondary structure formation
- $\Delta G°_A$, $\Delta G°$ for secondary structure formation of antisense RNA A (see *Figure 3*)
- $\Delta G°_C$, $\Delta G°$ for double helix formation in the complex C (see *Figure 3*), without the second order reaction

- $\Delta G°_T$, $\Delta G°$ for formation of that part of secondary structure of target RNA which is disrupted during complex formation (see *Figure 3*)
- β, nucleation parameter for second order reaction, assumed value: 10^{-3} M^{-1}
- R, gas constant (8.314 J/mol K)
- T_{th}, temperature of the calculation
- T_{ex}, temperature of the experiment[1]
- J_{th}, ionic strength of the calculation (1 M)
- J_{ex}, ionic strength (~ concentration of Na ions) of the experiment
- $f_{G:C}$, G:C content of the RNA
- $l_{A:U} = 20°C$
- $l_{G:C} = 8.4°C$ (17)

(All values of the temperature *T* in formulas [1], [2], and [3] are in K)

A. *Calculation for T = 37°C*

1. Take the values for $\Delta G°_A$, and $\Delta G°_T$ at 37°C from the programs for calculating the intramolecular structures (see Section 3.2). If these are not available, proceed as described under part B.

2. Calculate $\Delta G°_C$ as the sum of $\Delta G°$ values at 37°C of all stacks of neighboured base pairs as listed in *Table 1*.

3. Calculate the ratio of complexed and free target RNA:
$$\ln ([C] / [T]) = \ln c_0\beta + (\Delta G°_C - \Delta G°_A - \Delta G°_T) / RT \qquad [1]$$

B. *Calculation for any temperature T*

1. Calculate $\Delta G°_C(T)$ from the values of $\Delta H°_{stack}$ and $\Delta S°_{stack}$, which are given for all neighbouring base pairs in *Table 1*, where:
$$\Delta G°_C(T) = \Sigma (\Delta H°_{stack} - T \times \Delta S°_{stack}) \qquad [2]$$

2. Select from *Table 2* all $\Delta S°$ values for hairpin loops, internal loops, and bulged loops which are present in the secondary structure models. As a first approximation loops do not contribute to $\Delta H°$, but only to $\Delta S°$.

7

Protocol 1. *Continued*

3. For bifurcation, calculate $\Delta S°$ as that of an internal loop with the loop size p, where:

$$p = \text{all unpaired nucleotides} + 4 \qquad [3]$$

Count each additional helix in the loop as four single-stranded bases.

4. Calculate $\Delta G°_A(T)$ or $\Delta G°_T(T)$ where:

$$\Delta G°_{A(\text{or }T)} (T) = \Sigma \ (\Delta H°_{\text{stack}} - T \times \Delta S°_{\text{stack}}) + T\Sigma \ \Delta S°_{\text{loop}} \qquad [4]$$

$$\text{all stacks} \qquad\qquad \text{all loops}$$

5. Insert values of $\Delta G°_C(T)$, $\Delta G°_A(T)$, and $\Delta G°_T(T)$ into *Equation 1*.

C. *Correction for different ionic conditions and different concentrations of urea*

1. To compare experimental results (Section 4) with calculations, shift the temperature scale according to *Equation 5* which is taken from ref. 16.

$$(T_{\text{th}} - T_{\text{ex}}) / \log \ (J_{\text{th}} / J_{\text{ex}}) = f_{\text{G:C}} \cdot I_{\text{G:C}} + (1 - f_{\text{G:C}}) \cdot I_{\text{A:U}} \qquad [5]$$

2. Shift the temperature (e.g. of double-strand dissociation) for each M urea in the sample down by 2.9°C (16).

3.2 Programs to predict intramolecular secondary structures

Most of the computer programs which are currently available for RNA struc-ture prediction are based on the algorithm described first by Nussinov *et al.* (24) and Zuker and Stiegler (25). In this section three programs are recommended to predict RNA structures relevant for antisense technology. These are the programs *MFold* (26), which is part of the GCG package available in most molecular biology laboratories, and the program *LinAll* (27, 28), which is adapted for calculation of the temperature-dependence of RNA structure and is available from Dr Gerhard Steger upon request (available on the home page of the Department of Biophysics, Heinrich-Heine-Universität, Düsseldorf: http://www.biophys.uni-duesseldorf.de). A program of comparable quality is *RNAfold* from Fontana *et al.* (29) which is based on the algorithm of McCaskill (30).

MFold is described in detail in the handbook of the GCG package (31). It predicts optimal and suboptimal secondary structures using the energy mini-mization method of Zuker (26) and the stability parameters of Turner and colleagues (32). If, however, other stability parameters are used, other struc-tures might be determined as 'optimals'. Therefore, the 'correct' folding may not be the 'optimal' folding determined by the program, and one should view several optimal and suboptimal structures within a slight variation of the stability parameters. From comparing the different structures one can see which parts of the structure remain the same upon slight changes of the parameters and thus might be the most reliable.

Table 1 Parameters for calculation of $\Delta G°(T)$ of a double-stranded segment according to Equation 2.[a]

		3' A:U	3' U:A	3' G:C	3' C:G
$-\Delta G°_{37°C}$ kJ/mol	5' A:U	4.5	4.4	7.7	9.8
	5' U:A	5.5	4.5	9.1	11.1
	5' G:C	11.1	9.8	13.8	15.7
	5' C:G	9.1	7.7	9.1	13.8
$-\Delta H°/$ kJ/mol	5' A:U	27.6	23.9	31.8	42.7
	5' U:A	33.9	27.6	44.0	55.7
	5' G:C	55.7	42.7	51.1	59.5
	5' C:G	44.0	31.8	33.5	51.1
$-\Delta S°/$ kJ/mol K	5' A:U	0.077	0.0649	0.0804	0.01097
	5' U:A	0.0946	0.0770	0.1164	0.1486
	5' G:C	0.1486	0.1097	0.1243	0.1461
	5' C:G	0.1164	0.0804	0.0812	0.1243

[a] Each stack of two base pairs consists of a 5' and a 3' base pair. The values are from Turner and his colleagues (18).

Table 2 $\Delta S°$ values for loops in intramolecular structures as collected from the literature (119–22)[a]

Loop size	ΔS (kJ/mol K) Hairpin	Internal[b]	Bulge
1	–	–	0.0394
2	–	0.0014	0.0549
3	0.1185	0.0129	0.0649
4	0.0825	0.0225	0.0703
5	0.0569	0.0299	0.0724
6	0.0599	0.0351	0.0745
7	0.0624	0.0392	0.0766
8	0.0641	0.0421	0.0787
9	0.0660	0.0449	0.0807

[a] For larger loop sizes see ref. 23.
[b] For loopsize of internal loops count the unpaired nucleotide in both strands.

MFold can be applied to linear and circular sequences. Up to 850 bases can be included in one run of *MFold*. They have to be located within the first 10 000 bases of a given sequence. Because of the algorithm used, the CPU time increases with the third power of the number of bases included. In order to search for structures which are restricted to the target site of an antisense interaction, or to find a target site without stable structures, it might be advantageous to calculate the structure of appropriate sequence windows (~ 400 nt) and to shift the window in consecutive runs through the sequence of interest (see also Chapter 5). If the individual fragments are overlapping (by ~ 200 nt) this is an effective way to find local structures but does not allow one to find long-distance interactions.

In addition to the parameters of *Tables 1* and *2*, base stacking adjacent to double-helical stems in loops and bonus energies for recognized 'tetra-loops' are included. All parameters might be varied, to test for such an influence on the structure. As long as the structure at 37°C is required, $\Delta G°$ values at that temperature are needed. With values of $\Delta H°$ and $\Delta S°$, the temperature can be chosen between 0°C and 100°C. Three types of folding constraints can be introduced, which might take into account local structures known for example from experimental studies (Section 4):

(a) You can insist that all optimal and suboptimal structures contain one or several specified helices. This is done by specifying for each helix two segments in the sequence, or only one segment being double-stranded, leaving the counter-segment variable.

(b) In the same way two segments or only one segment can be prevented from being in a specified or partially specified helix, respectively.

(c) Regions of the RNA sequence can be excluded from the folding, when for example the secondary structure is known for that region. In that case the base pair which closes off that region is specified. If the effect on the structure of, e.g. splicing out a segment is being studied, adjacent bases before and behind that segment are ligated in the program.

As with *MFold*, the program *LinAll* allows calculation of optimal and sub-optimal secondary structures of linear and circular sequences at various temperatures. Formation of certain base pairs may be prevented and certain subsequences may be excluded from base pairing. From an ensemble of structures covering the temperature range of interest, the partition function of the RNA is approximated which allows one to simulate optical denaturation curves and mobilities in temperature-gradient gel electrophoresis (TGGE). *RNAfold* predicts the optimal secondary structure at a given temperature and calculates the partition function. Additional programs (*RNA melt* and *RNA heat*) simulate optical denaturation curves and heat capacity curves, respectively.

All programs provide similar representations of the predicted structures. *PlotFold* displays the optimal and suboptimal secondary structures for an RNA molecule predicted by *MFold*. It is also contained in the GCG package and allows the user to choose from among seven different representations of the structures. Three of the representations are used most often: *Energy Dotplot*, *Squiggles*, and *Text Output*.

Energy Dotplot (*Figure 4A*) displays all base pairs of the optimal structure and of all suboptimal structures, which fall into a specified increment of total free energy. As with the early Tinoco plot (33), it takes the form of a two-dimensional graph where both axes represent the same RNA sequence. Each point in the graph indicates a base pair between the ribonucleotides whose positions in the sequence are the coordinates of that point of the graph. An example using the version of the *LinAll* program is depicted in *Figure 4*. As with *RNAfold*, the probability of formation of a base pair is proportional to the area of the corresponding dot. A colour plot allows for differentiation between base pairs belonging to the optimal structure or to suboptimal structures of different maximum free energy by using different colours. For each position in the sequence it can easily be tested, how many different base pairings are possible within the energy increment for suboptimal structures. Drawing vertical and horizontal lines from any point in the sequence to the diagonal, allows one to count how many dots are hit by the line (see *Figure 4A*). Most reliable base pairs are represented by single hits with the colour of the optimal structure.

With *LinAll* such a dot plot may be presented in a three-dimensional graph in which the height of a peak at position i, j is proportional to the probability of base pair i:j (*Figure 4B*).

Squiggles Plot yields the well known two-dimensional representation of a distinct secondary structure (*Figure 4F*). The scales for stems and loops are different: larger loops appear too small. With *LinAll* (*Figure 4C–E*) and *RNAfold* similar representations are possible which avoid the inappropriate scaling of loops. In *Figure 4E* and *F*, the same structure is represented in both ways.

The text output representation of the RNA structure is similar to *Squiggles* but written as a text, no graphics device is needed. The same output is possible with *LinAll*. *RNAfold* allows one to depict both, the optimal structure and the structure distribution, by a linear string of characters which code for the probability of the structural elements.

4. Experimental determination of secondary structure

From the large number of experimental methods to determine nucleic acid structure (reviewed in refs 34 and 35), two most suitable for antisense technology will be presented. The first method is RNA structure chemical mapping, which can be performed in nearly every laboratory. The method yields a model for the RNA structure as it is present under the conditions of the experiment. The second method is the analysis by temperature-gradient gel electrophoresis (TGGE), which can elucidate the situation right after transcription and shows whether more than one structure of the same RNA species is present.

4.1 RNA structure mapping by DMS modification

RNA structure mapping by dimethyl sulfate (DMS) modification is a handy routine procedure (36). It identifies all adenosines (A) and cytosines (C)

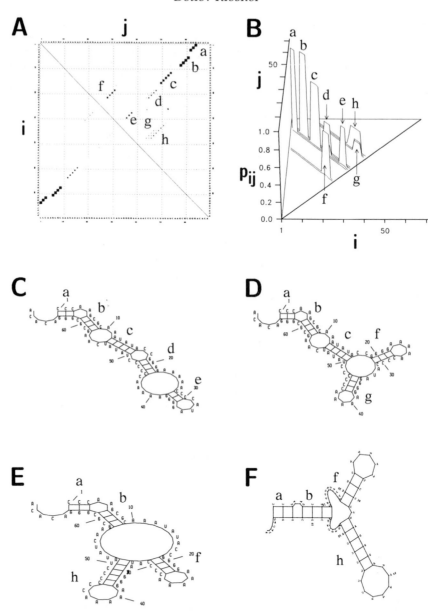

which are not protected by base pairing or tertiary structure interactions. The method can be applied in all laboratories, where sequence analysis with the primer extension method is routine. As an advantage over methods, in which protection against enzymatic attack is analysed, chemical probing by DMS can be applied under *in vivo* conditions, even in intact cells.

Figure 4. This shows representations of RNA secondary structures as obtained from computer programs. Small letters designate identical helices in all representations. (A) *Energy Dotplot*, in which each dot represents a base pair present in structures within an increment of free energy; in the *LinAll* or *RNAfold* version (shown in A) the size of the dot is proportional (above the diagonal) or with the third root (below diagonal) of the thermodynamic probability of the base. (B) $P_{i,j}$ plot with the matrix of base pairs as in the dot plot but with the thermodynamic probability $P_{i,j}$ on the vertical axis. Bars represent stems which are either present in all structures ($P_{i,j} = 1$) or in coexisting structures. (C–E) Base pairing schemes of the optimal structure (C) and two suboptimal structure (D, E) of the same sequence in the representation of *LinAll*. (F) Same structure as in (E) but represented by *Squiggles*; note that the looped regions are out of scale.

In *Figure 5*, the Watson–Crick base pairs are depicted together with the sites of methylation by DMS treatment: *N*1 of A, *N*3 of C, and *N*7 of G. It is obvious that the sites in A and C are protected by canonical base pairing, whereas base paired as well as single-stranded G may be methylated except where the *N*7 position is protected by strong stacking or tertiary structure interactions. If protein binding of the corresponding segments has to be taken into account (e.g. under *in vivo* conditions), additional protection occurs. Conditions of methylation have to be chosen such that each position is modified only in a few per cent of the molecules. The methylated nucleotides A and C are stop sites in the consecutive primer extended reverse transcription; methylated G does not appear as a specific stop site but increases the background of non-specific stopping. When the ladder of a dideoxy sequencing reaction with the method is compared with the positions of stop sites, the methylated nucleotides in the sequence can be determined. An example of the gel analysis and the resulting secondary structure is shown in *Figure 6*. The detailed protocol modified from ref. 37, and Section C from ref. 45 is given in *Protocol 2*.

Figure 5. This shows sites of chemical modification by dimethyl sulfate (DMS): *N*1 of adenosine (A), *N*3 of cytosine (C), and *N*7 of guanosine (G) are modified. *N*1 of A and *N*3 of C are protected against modification by base pairing, *N*7 of G is not.

Protocol 2. RNA structure mapping with DMS *in vitro* and *in vivo*

Equipment and reagents

- Buffer I: 200 mM Hepes–NaOH pH 8.0, 50 mM KCl, 10 mM $MgCl_2$, 0.1 mM EDTA
- DMS (dimethyl sulfate): > 99% pure (Aldrich)
- Buffer II: 1 M NaCl, 500 mM Tris–HCl pH 8.0
- Buffer III: 250 mM Tris–HCl pH 8.3, 375 mM KCl, 15 mM $MgCl_2$
- Primer extension mixture (all components from Gibco BRL): 4 µl buffer III, 2 µl 0.1 M DTT (dithiothreitol), 1 µl mixed dNTP stock solution (10 mM each dATP, dGTP, dCTP, dTTP pH 7.0), 1 µl distilled H_2O, 1 µl (200 U of Superscript™ II)
- TE buffer: 10 mM Tris–HCl pH 8.0, 1 mM EDTA
- Loading buffer: 80% deionized formamide, 0.1 mg/ml xylene cyanol FF, 0.1 mg/ml bromophenol blue, 0.1 mM EDTA pH 8

- Guanidine thiocyante reagent: 5 M guanidine thiocyanate (59 g/100 ml), 25 mM Tris pH 8.0, 2% Sarkosyl, 10 mM EDTA, check pH between 7 and 8, filter; before use, add β-mercaptoethanol to 0.1 mM (0.7 ml/100 ml) and diethyldithiocarbamate to 3–4 mg/ml
- RNase-free DNase (DNaseQ, Promega)
- Buffer IV: 50 mM Tris–HCl pH 8.0, 10 mM EDTA, 100 mM NaCl, 0.2% SDS, 3 mg/ml diethyldithiocarbamate (added fresh)
- Centrifuge: 4°C with 15 ml tubes and 13 000 *g*
- Gel electrophoresis instrument for sequencing gels
- Water-bath
- Vacuum desiccator
- Heat block

A. *DMS treatment in vitro*

1. Dissolve 50–100 µg of total cellular RNA or 5–50 µg of *in vitro* transcripts in 200 µl buffer I.

2. Incubate for 5 min at room temperature.

3. Add 1 µl DMS (final concentration 0.5%).

4. Incubate one sample for 10 min, another for 20 min, and a third for 30 min at room temperature (22–25°C), to find the best conditions.

5. Set-up a control reaction by omitting DMS.

6. Terminate the reaction on ice.

7. Precipitate RNA by adding 1 µl containing ~ 14 µg carrier tRNA, 30 µl of 3 M sodium acetate, 1 ml of cold 96% (v/v) ethanol, and chill on dry ice for 10 min.

8. Centrifuge the precipitate for 30 min at 13 000 *g*.

9. Repeat the precipitation by dissolving in 0.3 ml and adding 1 ml of cold 96% (v/v) ethanol.

10. Rinse the pellet with 100 µl of cold 70% (v/v) ethanol.

11. Dry at room temperature.

12. Redissolve the RNA in 20 µl TE buffer.

B. *DMS treatment in vivo (plants)*

1. Harvest leaves immediately before DMS treatment.

2. Cover 5 g leaf material with 20 ml buffer I, add 100 µl DMS, and 0.1% (v/v) Triton X-100.

3. Incubate for 10, 20, 30 min at room temperature (22–25°C) under vacuum to find optimal conditions.

4. Release vacuum abruptly several times for infiltration.

5. Terminate incubation by transferring the leaves into ice–water (*c.* 250 ml).

6. Wash for 10 min.

7. Dry the leaves and freeze in liquid nitrogen before RNA preparation.

C. *Preparation of RNA from plant tissue (after DMS modification)*

1. Grind frozen tissue in at least 2 ml/g tissue guanidine thiocyanate reagent until thawed to a homogeneous mixture.

2. Pellet cell debris for 20 min at 8000 *g*, add to supernatant 0.03 vol. 3 M NaAc pH 5.0, and 0.75 vol. EtOH, and precipitate at –20°C for > 15 min.

3. Centrifuge for 20 min at 8000 *g*, and resuspend pellet as completely as possible in 1 ml buffer IV/mg of starting tissue.

4. Extract with 2 vol. phenol:chloroform:isoamyl alcohol and then with 2 vol. chloroform:isoamyl alcohol.

5. Add NaCl to 200 mM and 0.6 vol. isopropanol. Precipitate at –20°C for at least 15 min.

6. Pellet for 10 min at 13 000 *g*, wash pellet in 70% EtOH, and spin again.

7. Resuspend pellet in 0.6 ml TE in a centrifuge tube and add $MgCl_2$ to 10 mM.

8. Add RNase-free DNase to 20 μl/ml, and incubate at room temperature for 15 min.

9. Add SDS to 0.2% and EDTA to 10 mM. Extract with phenol: chloroform, add NaAc to 0.3 M, and 2 vol. EtOH. Precipitate on ice at least 15 min.

10. Pellet RNA for 10 min at 13 000 *g* and resuspend final pellet in 0.1 ml TE/mg starting tissue. Store at –70°C.

D. *Analysis of modification sites using primer extension*

1. Label 10 picomoles of a specific primer at the 5′ end with [^{32}P] according to Maniatis *et al.* (38). (^{32}P-labelled primers cause fewer non-specific terminations as compared to labelling by incorporating [^{32}P] into the extension products.)

2. To anneal the labelled primer to the DMS modified RNA, add to 2 μl of buffer II 50 ng modified precipitated transcript or 10 μg modified total RNA, ~ 5 × 10^5 c.p.m. from 10 picomoles labelled primer, and add distilled water to final volume of 20 μl.

15

Protocol 2. *Continued*

3. Heat the hybridization mixture at 85 °C for 3 min in a heat block and cool to room temperature over a period of about 3 h.

4. Precipitate the annealing mixture with ethanol and let the pellet dry at room temperature.

5. To carry out the primer extension, add the primer extension mixture to the precipitated RNA.

6. Vortex gently, collect the mixture by brief centrifugation, and incubate for 1 h at 48 °C.

7. Precipitate the reaction with ethanol, dry the pellet at room temperature, and redissolve in gel loading buffer.

8. Load the samples on a sequencing gel and run the gel electrophoresis together with a sequencing reaction as a control.

9. Identify modified A and C and check for single-stranded regions in the secondary structure model (*Figure 6*).

4.2 Analysis of structural transitions and complex formation of RNA by temperature-gradient gel electrophoresis (TGGE)

The analysis of complex formation by the method of TGGE has the advantage over other methods that it allows one not only to separate complexed from uncomplexed RNA, but also to differentiate between different conformations of RNA. It was applied first by Hecker *et al.* (39) to analyse complex formation between plus- and minus-strands of viroid sequence. The principle and a number of applications of TGGE have been outlined in several original and review articles (40–43). TGGE is described in *Protocol 3* which is very similar to an earlier protocol (42). The major features are explained best with the example in *Figure 7*. The RNA sample is applied to the broad slot at the top of a normal polyacrylamide slab gel; thus it migrates in the electric field as a broad band from top to bottom. Perpendicular to the electric field, a linear temperature-gradient is established in the gel by an ancillary device. Thus molecules at the left side run at low, but constant temperature, molecules in the middle part at intermediate temperatures, etc. At temperatures where a structural conformation occurs, the electrophoretic mobility of the RNA changes, and the band assumes the shape of a transition curve, from lower to higher mobility or vice versa.

TGGE has been applied successfully to RNA in the size range of 100 nt to several thousand (41). It yields most accurate results for RNA sizes between $\sim$ 150 nt and $\sim$ 700 nt, which is the characteristic range of polyacrylamide gel electrophoresis. The concentration of polyacrylamide in the gel, electric field,

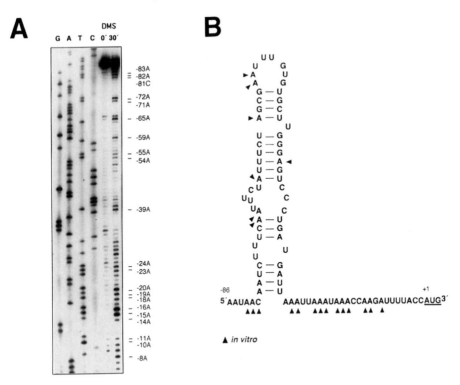

Figure 6. This shows an example for structural analysis, as derived from DMS modification, of the 5′ non-translated region of the spinach chloroplast psbA mRNA. DMS modifies single-stranded adenosine and (with lower efficiency) cytosine residues. (A) Primer extension analysis of the modification pattern; G, A, T, C sequencing reaction using the cloned psbA DNA and a phosphorylated primer. 0′: total RNA from spinach without DMS treatment; 30′: total RNA from spinach treated with 0.5% DMS, 23 °C, 30 min. (B) Secondary structure model as predicted by *LinAll* and compared with the modification sites. The figure is reproduced with permission from P. Klaff (46).

buffer conditions, and temperature range have to be optimized for each situation as detailed in reviews (41, 42). The protocol below describes standard conditions adapted to the application in antisense technology.

The advantages of TGGE over conventional physical methods to study conformational transitions of RNA are obvious. Minute amounts of RNA, depending upon the method of detection, are sufficient for an analysis. If silver staining is used for detection, 10 ng RNA per band are sufficient. If radioactive transcripts are analysed, bands can be visualized on an autoradiogram of the gel or a membrane after blotting. If mixtures of RNA, even natural RNA, are analysed, the RNA can be transferred from the gel to a membrane by a Northern blot, and a specific RNA can be detected with the corresponding hybridization probe. Double-stranded RNAs can be stained specifically by

pHa 106

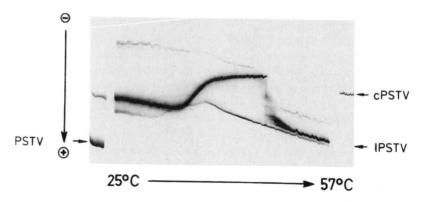

Figure 7. This shows TGGE of the transcript Ha106 from cDNA of a viroid sequence. Marker slots (left and right side of the broad sample slot) contain natural circular (c) and linear (l) potato spindle tuber viroid (PSTVd). Transcripts were analysed by TGGE without any treatment after transcription. A single RNA species assumes several different conformations as represented by different transitions curves. The two faster running bands represent monomolecular conformers, the weak bands, which run slower, are from bimolecular complexes which dissociate at the same temperature as the discontinuous transition of the main conformer. At high temperature all bands represent the identical RNA strand without base pairing; the different positions are from the first step in electrophoresis when the run was carried out for 10 min without temperature-gradient. Details of TGGE are described in *Protocol 3*. Figure modified from ref. 44.

dsRNA-specific antibodies as described in Chapter 14. Standard procedures are used for staining the RNA.

Different RNA species or different RNA conformations can be analysed in the same solution. An example is given in *Figure 7*, where a single RNA species assumes four different conformations. Therefore, antisense RNA and target RNA and complexes thereof can be detected, not only as single bands as in normal gel electrophoresis, but can be identified and characterized by their typical conformational changes. This advantage is outlined in an example of antisense:target RNA interaction in an *in vitro* transcription experiment simulating an *in vivo* situation: a constitutively transcribed antisense RNA is targeted against a newly synthesized viral or mRNA. To simulate this situation the target RNA is transcribed by T7 polymerase from a plasmid vector in the presence of the antisense RNA. Transcription is then stopped by adding 5 mM EDTA, the assay solution is adapted to TGGE conditions by dilution with loading buffer, and the analysis is carried out without any further treatment. Thus, conformations and complexes, exactly as generated in the transcription assay, are analysed. In *Figure 8* the interaction of a short

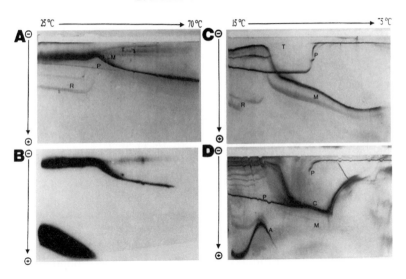

Figure 8. This shows TGGE of complex formation between antisense and target RNA. Panels A and B show the analysis of complexes of short antisense RNA (antiL) and plus-strand dimeric transcripts of PSTVd after pre-transcription of antisense RNA. Panel A represents a silver stained gel, panel B an autoradiogram from the radiolabelled antisense RNA. The fast migrating band in the autoradiogram is most probably from complexes of antisense RNA with early terminated transcripts. Panel C shows the minus-strand dimeric transcript, and panel D the minus-strand dimeric transcript after pre-transciption of VL_+ antisense RNA. The bands are assigned to three different structures (R, T, M); T is the tri-helical structure, M the multi-hairpin structure, and R the extended rod-like structure similar to that of native viroid. P is the transcription vector. For details of TGGE see *Protocol 3*. Figure reproduced with permission from ref. 43.

antisense RNA antiL (17 nt) and a long antisense RNA VL_+ (183 nt) with dimeric transcripts of viroid (PSTVd) sequences is shown (43). On the gel in *Figure 8A* in which the complex formation with antiL is analysed, several transition curves are visible. Those designated as R, M, and T are similar to these obtained by Hecker *et al.* (39) analysing the dimeric transcript alone. The transition curve M (multi-hairpin) is characteristic for the situation right after transcription; this structure is not in thermodynamic equilibrium but slowly undergoes transitions into structures represented by curves T (tri-helical) and R (rod-like, similar to the native structure of viroids). In addition to the three curves R, M, T, bands between those of T and M are detected, which originate from antisense RNA:transcript complexes. Band P is the plasmid transcription vector which was not removed from the sample. The complexes could be identified unequivocally from autoradiograms (*Figure 8B*) of the identical experiment. It is shown that the radiolabelled antisense RNA migrates within the transition curve of the M structure and within the additional weaker bands. From *Figure 8B* it can be seen that the antisense:target complexes are stable up to a temperature over 65 °C.

Complex formation of the longer antisense RNA VL$_+$ and oligomeric minus-strand transcripts is depicted in *Figure 8C* and *D*. Dimeric minus-strand was transcribed from the plasmid first without antisense RNA present and analysed by TGGE (*Figure 8C*). In accordance with an earlier study on isolated transcripts (39) and similarly to the results on plus-strand transcripts, three different structures can be detected. The predominant structure, which is characteristic for the situation directly after transcription, is represented by curve M, and other structures T and R are present in low concentration only. The band of lower intensity but parallel to M might represent an early termination.

After pre-transcription of the antisense RNA, which was carried out exactly as in the experiment of *Figure 8A* and *B*, the influence of antisense RNA binding is evident (*Figure 8D*). The uncomplexed structure M nearly vanished, whereas most of the molecules, i.e. more than 95%, migrate as complex C with antisense RNA. This attribution is obvious from the slower migration as compared to the uncomplexed state, from the similar mono molecular transition in the range of 30°C to 35°C, and from the typical transition of homogeneous double-strands between 55°C and 65°C. The latter consists of a partial denaturation of the double-stranded regions showing up as a drastic retardation and a discontinuous transition to a state of faster mobility at still higher temperature, which is visible only as diffuse bands of the dissociated antisense RNA and dimeric transcript. Binding to the other structures might also occur but does not contribute significantly. Since complex formation shifted the curves significantly, radiolabelling as in the case of *Figure 8B* was

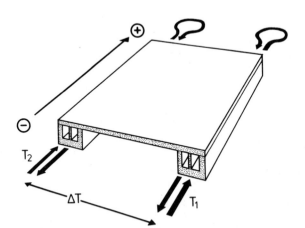

Figure 9. The figure shows a temperature-gradient plate for electrophoresis with a gradient of temperature *T* perpendicular to the direction of migration (− to + for nucleic acids). The metal plate (210 × 190 × 30 mm) is electrically insulated against the gel; each double channel for cooling (210 × 30 × 30 mm) is attached to a thermostating bath of temperature T1 or T2 respectively. Figure reproduced with permission from ref. 40.

not necessary. As the major result from TGGE analysis it can be concluded that complex formation of antisense RNA VL$_+$ and dimeric minus-strand intermediates occurs with a yield close to 100% under conditions which simulate the *in vivo* situation. In contrast to this complete complex formation, only 30% of the transcripts were found in the complexed form, if completely transcribed antisense RNA and dimeric minus-strand transcript were incubated and analysed by TGGE (data not shown).

These experiments demonstrate that the intramolecular structure of target RNA, in this case viroid RNA, and the process of structure formation after synthesis has to be taken into account for antisense:target interaction. Since it is nearly impossible at present to predict these properties from calculations, TGGE is a valuable tool to analyse interactions under close to cellular conditions.

Protocol 3.   Temperature-gradient gel electrophoresis of RNA

Equipment and reagents

- The TGGE gel apparatus: a conventional horizontal gel electrophoresis apparatus to which a constant temperature-gradient is applied. The gradient is established by a metal plate connected to two thermostating baths with different temperatures on the opposite edges of the plate (*Figure 9*). An instrument consists of a specially designed buffer tank and an aluminium plate electrically insulated to greater than 500 V by a sintered layer of epoxide. The thermostating bath for the lower temperature should be a cryostat (e.g. Haake F3-C or Julabo F20-HC) with sufficient cooling power (at least 200 W at 20°C). For the higher temperature a small circulator, such as Haake F3-S or Julabo UC-5, is sufficient. To ensure ease of operation and control during electrophoresis, both thermostating baths should be equipped with digital temperature displays. The baths are connected to the edges of the gradient plate by large diameter water tubes shielded in neoprene tubes. To ensure accurate control of temperature the tubes should be as short as possible and the water pumped by strong pumps with a flow rate above 10 litres/min. Air bubbles in tubes or channels of the gradient plate must be avoided. Good electrical contact between the gel and the electrodes is made through electrode contact cloths (well soaked) that are insulated in polyethylene foil in order to prevent drying out at a higher temperature, except where in direct contact with the gel. A number of gel compositions and buffers may be used in the TGGE system. The instrument described is no longer available commercially; an equivalent operated, however, by Peltier-elements instead of cooling liquid is available from Biometra (37079 Göttingen, Germany).

- Gel support films and gel casting moulds (available from Biometra)

- 10 × TBE buffer: 890 mM Tris (109 g/litre), 890 mM boric acid (55 g/litre), and 25 mM EDTA (9.3 g/litre) pH 8.3, adjust the buffer with glacial acid to pH 8.3 and autoclave (for electrophoresis dilute 50-fold)

- Electrophoresis buffer: 0.2 × TBE

- 2 × sample loading buffer: 400 μl of 10 × TBE buffer, 120 μl of bromophenol blue stock solution (10 mg/ml), 120 μl of a xylene cyanol FF stock solution (10 mg/ml), fill up to 10 ml with autoclaved water

- PAA gel stock: 30% acrylamide: 1% *N',N'*-methylene *bis*-acrylamide solution

- Gel solution for one TGGE (5%): add 8.3 ml of PAA gel stock (30:1) to 1 ml of 10 × TBE buffer, fill up to 48 ml with distilled water—start polymerization with 50 μl TEMED and 2 ml of 1% ammonium persulfate

- Buffer A (for silver staining): 100 ml of ethanol, 5 ml of acetic acid, made up with distilled water to 1 litre

- Buffer B: dissolve 1 g of AgNO$_3$ in 1 litre of distilled water (1 litre buffer B can be reused five to ten times)

- Buffer C: dissolve 15 g of NaOH, and 0.1 g of NaBH$_4$ in 1 litre of distilled water, and then add 4 ml of formaldehyde (stock 37% in water) (this buffer must be freshly prepared immediately before use)

- Buffer D: dissolve 7.5 g of Na$_2$CO$_3$ in 1 litre water

Protocol 3. *Continued*

A. *Gel casting*

TGGE gels are polyacrylamide gels covalently bound to gel support films. A slot-forming plate with a single long slot is used for TGGE with the temperature-gradient perpendicular to the electric field.

1. Prior to gel casting clean the plates with distilled water, avoiding organic solvents. Put a gel support film with the hydrophobic side against the plain glass plate. Cover the hydrophilic side of the gel support film with the attached sheet of paper. Fix the gel support film by rubbing the film against the glass plate with the aid of a straight edge of a ruler or another suitable device. Remove the protective paper.

2. Layer the slot-forming plate onto the gel support film. The slot formers are oriented toward the bottom of the gel mould. Fix the gel mould with clamps.

3. Mix the gel solution and add the TEMED and ammonium persulfate to start polymerization. Transfer the gel solution into the gel mould. Avoid air bubbles.

4. Polymerize for at least 60 min and then remove the gel from the glass plates. To prevent leakage the gel mould may be placed into a horizontal position.

B. *Running TGGE*

1. Fill each buffer tank with 1 litre of electrophoresis buffer. Boil the synthetic electrode wicks in distilled water, wring, and put them into the buffer tanks. Soak the wicks with electrophoresis buffer. Mount the gel with the gel support film facing the plate. Avoid trapping air bubbles by 'rolling' the gel onto the plate. The long slot of the perpendicular TGGE gel is oriented towards the cathode (black). Set-up the buffer bridges consisting of a double layer of electrode wicks. Place one side of each wick on the gel while the other side is submerged in the buffer inside the tank. Ensure that the cathode wick does not approach the slot closer than 5 mm, otherwise your sample might be absorbed by the wick.

2. Mix 400 μl of sample with 400 μl of 2 × sample loading buffer and load the mix into the slot. Cover the gel with the glass plate. Close the safety lid and start electrophoresis at 300 V (constant voltage). Stop after 10 min.

3. Cover the gel with protective plastic wrap (Saran Wrap™) to prevent evaporation. Establish the temperature-gradient in the gel by switching on the thermostat and the cryostat. The standard temperature-gradient suitable for most RNAs is 20°C at the left end of the TGGE plate and 60°C at the right edge. When the heating and cooling

devices have reached the desired temperatures, wait a further 10 min for equilibration.

4. Once equilibrated, run the gel at 300 V for 1 h. Stop electrophoresis. Take the gel out of the TGGE system and remove the protective foil.

5. For silver staining of DNA and RNA in polyacrylamide gels (especially TGGE gels), all incubations are done by shaking the gel gently while submerged in approx. 300 ml of the appropriate solution. Put the gel into a plastic tray. The gel support film is oriented towards the bottom of the tray. Cover the gel with buffer A and incubate for 3 min. Discard the buffer. Repeat the incubation with buffer A. Incubate in buffer B for 10 min. Recover this buffer because 1 litre of buffer B may be reused for five to ten gels. Wash twice with distilled water for 10 sec. Incubate the gel in buffer C for 20 min. Discard the solution. Incubate in buffer D for 5–10 min. Discard the solution.

The analysis of complex formation by TGGE facilitates the identification of different bands in the gel as antisense:target complexes. After the identification of the bands in TGGE, conventional electrophoresis can be applied at the appropriate constant temperature.

Acknowledgements

I am indebted to Drs Petra Klaff and Gerhard Steger and Ms Astrid Schröder for stimulating discussions and for providing protocols and figures. I thank Ms Heidi Gruber for help in preparing the manuscript.

References

1. Simons, R. W. (1988). *Gene*, **72**, 35.
2. Saenger, W. (1984). *Principles of nucleic acid structure.* Springer–Verlag, New York.
3. Cech, T. T. (1987). *Science*, **236**, 1532.
4. Wyatt, J. R. and Tinoco, I. Jr. (1993). In *The RNA world* (ed. R. F. Gesteland and J. F. Atkins), pp. 465–96. Cold Spring Harbor Laboratory Press.
5. Persson, G., Wagner, E. G. H., and Norström, K. (1990). *EMBO J.*, **9**, 3767.
6. Grosjean, H., Söll, D. G., and Corthers, D. M. (1976). *J. Mol. Biol.*, **103**, 499.
7. Heus, H. and Pardi, A. (1991). *Science*, **253**, 191.
8. Varani, G., Cheong, C., and Tinoco, I. Jr. (1991). *Biochemistry*, **30**, 3280.
9. Varani, G., Wimberly, B., and Tinoco, I. Jr. (1989). *Biochemistry*, **28**, 7760.
10. Chastain, M. and Tinoco, I. Jr. (1991). *Prog. Nucleic Acids Res. Mol. Biol.*, **41**, 131.
11. Puglisi , J. D., Tan, R., Calnan, B. J., Frankel, A. D., and Williamson, J. R. (1992). *Science*, **257**, 76.
12. Pleij, C. W. A., Rietveld, K., and Bosch, L. (1985). *Nucleic Acids Res.*, **13**, 1717.
13. Puglisi, J. D., Wyatt, J. R., and Tinoco, I. Jr. (1990). *J. Mol. Biol.*, **214**, 437.
14. Chastain, M. and Tinoco, I. Jr. (1992). *Biochemistry*, **31**, 12733.
15. Michel, F. and Westhof, E. (1990). *J. Mol. Biol.*, **216**, 585.

16. Steger, G. (1994). *Nucleic Acids Res.*, **22**, 2760.
17. Steger, G., Müller, H., and Riesner, D. (1980). *Biochim. Biophys. Acta*, **606**, 274.
18. Freier, S. M., Kierzek, R., Jaeger, J. A., Sugimoto, N., Caruthers, M. H., Neilson, T., *et al.* (1986). *Proc. Natl. Acad. Sci. USA*, **83**, 9373.
19. Fink, T. R. and Crothers, D. M. (1972). *J. Mol. Biol.*, **66**, 1.
20. Gralla, J. and Crothers, D. M. (1973). *J. Mol. Biol.*, **73**, 497.
21. Gralle, J. and Crothers, D. M. (1973). *J. Mol. Biol.*, **78**, 301.
22. Fink, T. R. and Krakauer, H. (1975). *Biopolymers*, **14**, 433.
23. Riesner, D. and Steger, G. (1990). In *Landolt-Börnstein, Group VII Biophysics*, Vol. 1, *Nucleic acids*, Subvolume d, *Physical data II, Theoretical investigations* (ed. W. Saenger), pp. 194–243. Springer–Verlag, Berlin.
24. Nussinov, R., Pieczenik, G., Griggs, J. R., and Kleitman, D. J. (1978). *SIAM J. Appl. Math.*, **35**, 68.
25. Zuker, M. and Stiegler, P. (1981). *Nucleic Acids Res.*, **9**, 133.
26. Zuker, M. (1989). In *Methods in enzymology* (ed. J. E. Dahlberg and J. N. Abelson), Vol. 180, pp. 262–88.
27. Steger, G., Hofmann, H., Förtsch, J., Gross, H. J., Randles, J. W., Sänger, H. L., *et al.* (1984). *J. Biomol. Struct. Dyn.*, **2**, 543.
28. Schmitz, M. and Steger, G. (1992). *Comp. Appl. Biosci.*, **8**, 389.
29. Fontana, W., Konings, D. A., Stadler, P. F., and Schuster, P. (1993). *Biopolymers*, **33**, 1389.
30. McCaskill, J. S. M. (1990). *Biopolymers*, **29**, 1105.
31. Devereux, J., Haeberli, P., and Smithies, O. (1984). *Nucleic Acids Res.*, **12**, 387.
32. Jaeger, J. A., Turner, D. H., and Zuker, M. (1990). In *Methods in enzymology* (ed. R. F. Doolittle), Vol. 183, pp. 281–306.
33. Tinoco, I. Jr., Uhlenbeck, O. C., and Levine, M. D. (1971). *Nature*, **230**, 363.
34. Jaeger, J. A., Santalucia, J. Jr., and Tinoco, I. Jr. (1993). *Annu. Rev. Biochem.*, **62**, 255.
35. Lilley, D. M. J., Heumann, H., and Such, D. (ed.) (1992). *Structural tools for the analysis of protein–nucleic acid complexes*. Birkhäuser, Basel.
36. Ehresmann, Ch., Baudin, F., Mougel, M., Romby, P., Ebel, J.-P., and Ehresmann, B. (1987). *Nucleic Acids Res.*, **15**, 9109.
37. Senecoff, J. F. and Meagher, R. B. (1993). In *Methods in enzymology* (ed. E. A. Zimmer, T. J. White, R. L. Cann, and A. C. Wilson), Vol. 224, pp. 357–72.
38. Maniatis, T., Fritch, E. F., and Sambrook, J. (1982). *Molecular cloning: a laboratory manual.* Cold Spring Harbor Laboratory Press.
39. Hecker, R., Zhi-min Wang, Steger, G., and Riesner, D. (1988). *Gene*, **72**, 59.
40. Rosenbaum, V. and Riesner, D. (1987). *Biophys. Chem.*, **26**, 235.
41. Riesner, D., Henco, K., and Steger, G. (1991) In *Advances in electrophoresis* (ed. A. Chrambach, M. J. Dunn, and B. J. Radola), Vol. 4, pp. 169–250. VCH Verlagsgesellschaft, Weinheim.
42. Henco, K., Harders, J., Wiese, U., and Riesner, D. (1994). In *Methods in molecular biology* (ed. J. M. Walker), pp. 211–28. Humana Press, Clifton, New Nersey.
43. Matousek, J., Schröder, A. R. W., Trnena, L., Reimers, M., Baumstark,T., Dedic, P., *et al.* (1994). *Biol. Chem. Hoppe-Seyler*, **375**, 765.
44. Steger, G., Baumstark, T., Mörchen, M., Tabler, M., Tsagris, M., Sänger, H. L., *et al.* (1992). *J. Mol. Biol.*, **227**, 719.
45. Barkan, A. (1989). *The Plant Cell*, **1**, 437.
46. Klaff, P. and Gruissem, W. (1995). *Photosyn. Res.*, **46**, 235–48.

2

Evaluation of antisense effects

WOLFGANG NELLEN and CONRAD LICHTENSTEIN

1. Introduction

Antisense-mediated gene silencing by expression of antisense RNA in transfected cells has become a complementing alternative to gene disruption or
gene 'knock-outs' by homologous recombination. Though powerful and often
successful, there are unknown mechanisms which may sometimes hamper the
effectiveness of the experiment. As with experiments with antisense oligonucleotides used to transiently interfere with gene expression, there may also
be unexpected side-effects leading to the wrong interpretation of the results
(see Chapter 11).

Antisense transformation is mostly used:

(a) If no other means to disrupt gene expression are available (e.g. in plants).

(b) If a complete gene knock-out may cause a lethal phenotype while cells
with residual, small amounts of the gene product may still survive.

(c) To specifically abolish expression of a given gene in a certain tissue, or
cell type, or during a specific developmental stage.

(d) If a family of closely related genes is to be silenced (e.g. spliced leader in
Trypanosomes).

(e) If one antisense transcription unit is to be targeted against, e.g. different,
related viruses.

In most cases, a certain, or at least a change in, phenotype is expected, but
an altered phenotype does not necessarily mean that the antisense experiment
worked! Conversely, if there is no manifestation of the expected phenotype,
this does not necessarily mean that antisense did not work! There is re-
dundancy in eukaryotic genomes and various, apparently unrelated genes
may take over the function of the silenced gene. Alternatively, small residual
amounts of the gene product may be sufficient to fulfil functions so that,
under laboratory conditions, a mutant phenotype is not detected.

Therefore, it is necessary to carefully analyse the cell line or organism
to understand and evaluate the results. Once it is demonstrated that gene

expression is indeed reduced or abolished, one can think about more sophisticated methods in the search for more subtle phenotypes.

In this chapter, the methods we provide are a mixture of absolutely necessary controls and of additional experiments for those who really want to know what happened. As stated above, the use of antisense oligos to silence gene expression can yield numerous side-effects; this is fully discussed in Chapter 11. So far, no such problems have been seen with antisense transcripts; nevertheless, this does not mean that they do not occur. Thus, it is essential to provide proof that any effect is specific for the targeted gene.

Therefore a checklist should be performed to assess the success of the antisense experiment.

(a) Is the observed change in phenotype due to a *bona fide* antisense effect?

    i. Is the endogenous target gene still intact? Perform Southern blots.

    ii. Is the antisense construct expressed? Perform Northern blots and/or nuclear run-ons.

    iii. Is expression of the endogenous target gene affected? Perform Northerns/nuclear run-ons to detect mRNA and/or Westerns to detect the gene product.

    iv. Is the endogenous target gene silenced by methylation? Perform DNA methylation assays.

    v. Has integration of the antisense gene affected expression of adjacent genes? Compare phenotypes of transformants with different integration sites; examine length of antisense transcript to prove no transcription run-through of adjacent genes has occurred.

(b) Why has the expected change in phenotype not occurred?

    i. Is the antisense construct expressed in sufficient amounts and appropriate lengths? Perform run-on assays, Northern blots.

    ii. What is the cellular localization of sense and antisense transcripts? Perform strand-specific *in situ* hybridization, as described in another volume in this series (1).

    iii. Is lack of antisense activity caused by poor interaction with the target mRNA? As discussed in Chapter 1, intramolecular secondary structure may preclude efficient interactions and this can be overcome by using a different part of the target gene for the antisense construct.

Protocols to address the questions posed in the checklist are given below.

2. Protein expression studies

One of the most obvious first tests is to show that the levels of the gene product (in the case of a coding RNA) are indeed reduced. If an antibody is available, standard techniques of Western blotting can be used. In other cases, enzyme assays, or other tests for the gene product of interest may be performed.

For *Dictyostelium* and probably also for other cells and organisms which can be grown on or applied to nitrocellulose filters, an easy and rapid test can be run by performing a protein colony blot. Even though this is only semi-quantitative, it may help to process many samples simultaneously. A protocol for this procedure is given in Chapter 8.

3. Analysis of steady state RNA levels

In most, but not all, cases of successful antisense experiments, the transcripts of the targeted gene are reduced in quantity or they are completely lost. It is therefore useful to examine steady state RNA levels in a Northern blot. This will also confirm if the antisense construct introduced into the cells is efficiently transcribed and can accumulate. We have found that in some cases antisense transcripts are rather unstable and so are not always easy to detect. Although a random primed probe or a nick translated fragment should detect both, sense and antisense RNA, we strongly recommend the use of strand-specific *in vitro* transcripts as hybridization probes; protocols for the production of these are described in Chapter 4. This does not only unambiguously allow for distinguishing between the two RNAs, but is also more sensitive. The sense or antisense transcripts may not always correspond in size to the wild-type mRNA or to the length expected from the vector construct, respectively; sometimes specific degradation products are observed.

RNA blots will also help to detect potential problems resulting from fortuitous transcription in either orientation. They could come from cryptic promoters in the vector construct or by read-through from genomic promoters close to the integration site of the input antisense construct.

If many samples have to be processed, it is also possible to perform a semi-quantitative RNA colony blot with strand-specific probes (*Protocol 1*). This procedure has been successfully applied for *Dictyostelium* and yeast. The major limitations are in the ease of efficient lysis of cells and, more importantly, the absolute abundance of the transcripts of interest (2).

Protocol 1. RNA colony blot (adapted from refs 2 and 3)

Equipment and reagents

- Nylon membrane (e.g. Pall Biodyne) or microtitre plates with a peel-off membrane at the bottom (e.g. 'Silent Monitor' plates)
- Lysis solution: 4 M guanidinium thiocyanate, 25 mM sodium citrate pH 7, 0.5% Sarcosyl, 0.1 M 2-mercaptoethanol
- Standard hybridization solution: 50% formamide, 3 × SSC, 10 mM EDTA pH 7.2, 60 mM phosphate buffer pH 7, 0.2% SDS, 4 × Denhardt solution, 200 µg/ml yeast tRNA

Protocol 1. *Continued*

Method

1. Transfer cells to, or grow on, a nylon membrane, or grow cells in microtitre plates with a peel-off membrane at the bottom.

2. Place filters (cells 'sunny-side' up) onto a drop of lysis solution on a piece of 'Saran Wrap' for 5 min. About 15 μl per cm^2 of solution is sufficient to have the filter well soaked.

3. Transfer filters to a drop of 20 × SSC for 5 min, then air dry, and bake for 1 h at 80 °C under vacuum.

4. Perform hybridization in standard hybridization solution plus [α-^{32}P]UTP labelled *in vitro* transcripts overnight at 55 °C.

5. Wash filters in 2 × SSC, 0.1% SDS several times, and then expose to X-ray film.

Since colony blots provide only an estimate of RNA quantities, and since clones or colonies contain approximately equal cell numbers, a control blot using a housekeeping gene as probe is not absolutely required.

4. Analysis of RNA transcription rates

4.1 Nuclear run-ons

To determine whether the effect of the antisense transcript in these clones is at the level of transcription or at the level of RNA stability, run-on assays with strand-specific probes can be performed. This also allows an examination of whether the antisense construct is transcribed at a sufficient level.

The first step in the run-on assay is the preparation of transcriptionally competent nuclei. The ease of preparation of nuclei very much depends on the cell type used. We therefore can only suggest methods which have been successfully used in our laboratory on *Dictyostelium* (*Protocol 2*), for plants we recommend a protocol by Somssich (4).

Protocol 2. Preparation of nuclei (adapted from ref. 5)

Equipment and reagents

- Lysis buffer for parts A and B: 50 mM Hepes pH 7.5, 40 mM MgCl$_2$, 20 mM KCl, 0.15 mM spermidine, 5% sucrose, 14 mM 2-mercaptoethanol
- Lysis buffer for part C: 20 mM Tris pH 7.5, 5 mM MgCl$_2$, 1% Trasylol (aprotinin, Bayer)
- 10% NP-40
- 10% Percoll
- 1% Triton X-100
- Nuclei storage buffer: 40 mM Tris pH 8, 10 mM MgCl$_2$, 1 mM EDTA, 50% glycerol, 14 mM 2-mercaptoethanol
- Polycarbonate filter (Nuclepore, pore size 0.5 mm)

All steps are carried out in the cold.

A. *Standard detergent method*

1. Resuspend 10^8–10^9 cells in 1–10 ml lysis buffer.

2. On ice, add 10% NP-40 dropwise with repeated shaking and vortexing. Stop addition of NP-40 as soon as the suspension becomes clear. Do not add more than a total of 1% NP-40 because nuclear membranes could then be damaged.

3. Immediately spin at 4000 r.p.m. for 5 min, carefully discard supernatant, and resuspend the pelleted nuclei in lysis buffer without NP-40.

4. Spin again at 1000 r.p.m. to remove unlysed cells and large particles. Carefully transfer supernatant with a pipette to a fresh tube and spin again at 4000 r.p.m. At this step you may add 10% Percoll to the suspension to increase the density and to remove smaller particles.

5. Resuspend the pelleted nuclei at a concentration of 10^9/ml in nuclei storage buffer and immediately freeze as 20 μl aliquots in liquid nitrogen.

6. Store at –70°C; nuclei are usually good for several months.

B. *Mechanical cell disruption*

NP-40 may damage the nuclei, especially when high concentrations or longer incubation times are needed to lyse the cells. Under these circumstances, mechanical disruption of the cells may be advantageous.

1. Resuspend cells in lysis buffer (as in part A) at a concentration of 5×10^7/ml and press through a polycarbonate filter. The filtrate is collected in a beaker on ice. In our hands, two to three passages are required to completely lyse the cells.

2. Centrifuge the suspension at low speed (1000 r.p.m.) to pellet unbroken cells.

3. Spin the supernatant at high speed (4000 r.p.m.) to pellet the nuclei.

4. Resuspend the nuclei in fresh buffer with 10% Percoll and spin again at 4000 r.p.m.

5. Aliquot the nuclei and store as above.

C. *Gradient detergent method*[a]

An alternative and very gentle procedure to lyse cells is the gradient method (6).

1. Resuspend the cells in lysis buffer C at a concentration of 5×10^7 to 10^8 cells/ml.

2. Prepare a step gradient in a 15 ml Falcon tube. The bottom layer of 5 ml contains lysis buffer with 440 mM sucrose, the top layer of 3 ml contains lysis buffer with 320 mM sucrose and 1% Triton X-100.

Protocol 2. *Continued*

3. Carefully layer the cell suspension onto the gradient and centrifuge the tube at 3000 r.p.m. for 5 min. Cells lyse when passing through the upper step and nuclei are collected in the lower solution which contains no detergent.

4. Wash the nuclei once with lysis buffer, resuspend in storage buffer, and store as described above.

[a] Nuclei pellets are usually white, big, soft, and fluffy. Nuclei still contain a fair amount of cytoskeleton but are good for run-on transcription.

The nuclei are used to prepare run-on transcripts as described in *Protocol 3*.

Protocol 3. Preparation of run-on transcripts

Equipment and reagents

- 5 × transcription buffer: 200 mM Tris pH 7.9, 1.2 M KCl, 50 mM $MgCl_2$, 25% glycerol, 0.5 mM DTT
- ATP, GTP, CTP (4 mM)
- [^{32}P]UTP
- RNasin
- RNaseA

Method

1. Set-up the run-on transcription in a total volume of 100 μl with 20 μl of the nuclei preparation (2×10^7 nuclei), 20 μl 5 × transcription buffer, 5 μl each ATP, GTP, CTP (4 mM), 1 μl RNasin, and 10 μl (100 μCi) [^{32}P]UTP. Incubate at 22 °C for 30 min.

2. Stop reaction by addition of 10 μl 20% SDS, 10 μl 0.2 mM EDTA, and extract with phenol:chloroform.

3. To purify transcripts from unincorporated radioactivity, remove the aqueous phase and pass it over a Sephadex G50 spin column.

4. For hybridization with the strand-specific targets, use the smallest volume possible (approx. 500 μl) in a seal-a-meal plastic bag cut to the size of the filter containing the RNA samples. To compensate for the volume of the run-on transcript probe, use a 1.5 × concentrated hybridization solution and fill up with water to the appropriate volume.

5. Hybridize for at least 16 h, wash three times in 2 × SSC, 0.1% SDS, a last wash may be done in 2 × SSC containing 0.1 μg RNaseA/ml (to reduce background non-specific ssRNA); however, this is usually not necessary.

6. Air dry filters and expose to X-ray film.

4.2 Targets for strand-specific run-on assays

For analysis of the mRNA and antisense RNA abundance in the run-on transcripts, it is important to have *in vitro* prepared positive and negative controls to compare them to. We suggest the following:

(a) Make cold *in vitro* transcripts in both orientations from your gene of interest using the standard protocols for SP6 or T7 RNA polymerase reactions (Chapter 4). Transcribe sufficient material to get at least 1–2 µg of RNA.

(b) Make transcripts of a marker gene (e.g. actin or even better some gene from your collection which is transcribed *in vivo* at similar levels to your gene of interest) as a positive control. In addition we recommend making transcripts of a DNA fragment which is not produced *in vivo* (e.g. CAT) as a negative control.

(c) You may apply the four transcripts (sense, antisense, negative, and positive control) to a nylon membrane with a dot or slot blot device. We prefer, however, to run them on a Northern gel and blot the RNA to the membrane. Including ethidium bromide in the sample allows confirmation that equal amounts have been blotted. This way you also separate residual template DNA from the transcripts and obtain a clear band which contains only full-length RNA. Transcription artefacts (e.g. from non-specific initiation of the polymerase at the ends of the linearized template) are also separated from the specific transcript.

(d) From the filters, cut out an area, as small as possible, containing the set of bands, bake or cross-link by UV, pre-hybridize, and hybridize with the nuclear run-on transcripts as described in *Protocol 3*.

5. Polysome assays for antisense activity

In some cases it has been observed that both sense and antisense RNAs are present in the cells. Nevertheless, there are other means than mRNA degradation which exert antisense function. We and others have observed that translation may become impaired by antisense transcripts (7, 8). One way to test this is the examination of polysome association of the mRNA of interest: if it is associated with large polysomes (as found in wild-type cells) then there is obviously no effect of the antisense transcripts. If, however, the mRNA is found in the monosome peak, or with smaller polysomes, it is most likely that there is an antisense-mediated inhibition of translation.

Protocol 4 describes a rapid easy method to prepare polysomes. Polysomes of cells expressing antisense RNA and wild-type cells should be prepared concurrently and run on parallel gradients to allow a direct comparison of the mRNA position.

Protocol 4. Rapid polysome preparation

Equipment and reagents

- Polysome lysis buffer (PLB): 50 mM Hepes pH 7.5, 40 mM MgCl$_2$, 20 mM KCl, 5% RNase-free sucrose (the buffer is auto-claved before addition of the filter sterilized sucrose)
- 20% NP-40 autoclaved stock solution: 20% NP-40 may form two phases after autoclaving, allow to cool then shake
- Proteinase K (2 mg/ml)

Method

1. Harvest cells and wash in an appropriate buffer but without EDTA.
2. Resuspend cells at a concentration of 2 × 10^8/ml in PLB.
3. On ice, add 20% NP-40 to a final concentration of 1.5%. Leave on ice for 5 min and shake repeatedly; the cell suspension becomes clear when the cells are lysed.
4. Spin the sample for 3 min at 10 000 r.p.m. Carefully remove the super-natant with a sterile Pasteur pipette or a blue tip, and load immediately onto a cold, preformed 7–50% sucrose gradient (sucrose in PLB). Do not try to load all of the supernatant and absolutely avoid taking any of the pellet. Tubes for a Beckman SW41 swinging bucket rotor can be loaded with up to 500 µl (corresponding to 5 × 10^7 cell equivalents), tubes for an SW27 rotor with up to 1 ml of the sample.
5. Run the gradient at 25 000 r.p.m. for 3 h at 4°C.
6. Fractionate the gradient either through a UV monitor or simply by carefully taking fractions from the top with a blue tip. It is most con-venient to collect 400 µl fractions into Eppendorf tubes pre-loaded with 50 µl of a solution containing 1% SDS and 2 mg/ml proteinase K.
7. After 20 min digestion at 37°C, precipitate the RNA with ethanol, wash once with 70% ethanol, and then load onto agarose gels containing formaldehyde. Fractions close to the bottom of the gradient (high sucrose concentration) should be redissolved in water and precipitated again with ethanol if the first pellet is large, transparent, and viscous.
8. Add ethidium bromide to the gel loading buffer to allow for a rapid estimate on the position of large and small ribosomal subunits, mono-somes, and polysomes in the gradient even while the gel is running.
9. Blot the gel to a nylon membrane and hybridize with an *in vitro* tran-script which recognizes the sense mRNA.[a]

[a]The same filter can be used as an easy control by probing for antisense transcripts. These RNAs should definitely not be associated with polysomes. If they are (which has not been observed to our knowledge), it could explain a non-functional antisense experiment: presumably, loading the complementary RNA with ribosomes would keep it from interacting with the target.

Lack of both detectable targeted RNA and gene product is not, however, unambiguous evidence of an effective antisense experiment. We have observed that sometimes antisense constructs result in unusually efficient gene disruption by homologous recombination. We found a truncated mRNA and a truncated protein which nevertheless fully served its function. Gene disruption by homologous recombination can be easily ruled out by genomic Southern blotting to show that the genomic structure of the gene is unaltered. Homology-dependent gene silencing can also occur (see below).

6. RNA localization studies

Is the asRNA:mRNA duplex unable to leave the nucleus? In some cases it has been observed that sense:antisense hybrids are formed in the nucleus and prevent nucleo-cytoplasmic transport (e.g. ref. 9).

As a way to determine if transport to the cytoplasm is inhibited, we suggest making crude preparations of nuclei and cytoplasm as described in *Protocol 5*. Samples of wild-type control cells should be prepared in parallel to compare the ratio of nuclear and cytoplasmic RNA and analysed by electrophoresis on an agarose gel containing formaldehyde.

Protocol 5. Preparation of nuclear and cytoplasmic RNAs

Equipment and reagents
- See *Protocols 3* and *4*
- LiCl

A. *Preparation of cytoplasmic RNA*

1. Make a crude preparation of cytoplasm according to the protocol for polysome preparations (*Protocol 4*).

2. Rapidly supplement with SDS to a final concentration of 0.5%, and proteinase K to a final concentration of 0.1 mg/ml, and incubate at 37°C for 20 min.

3. Extract with phenol, precipitate with ethanol and 0.1 vol. of LiCl, wash once with 70% ethanol, and resuspend in sterile water.

B. *Preparation of nuclear RNA*

1. Make a crude preparation of nuclei as described in *Protocol 3*.

2. Extract RNA from nuclei as in part A, step 3.

3. To eliminate the DNA, perform a final selective RNA precipitation with 4 M LiCl (1–5 h at –20°C). Wash twice with 70% ethanol.

---

7. Antisense-independent gene silencing

Transformation with antisense constructs may also interfere with gene expression by influences on the state (structure) of the endogenous gene copy. It has been observed that in response to the genomic integration of a sense or an antisense construct, the resident gene can be methylated and so transcriptionally inactivated (10).

The easiest way to determine DNA methylation is by comparative restriction digests with a methylation-sensitive enzyme on DNA isolated from transformants and from the corresponding wild-type strain. The samples are separated on an agarose gel, blotted to a nylon membrane, and hybridized with a labelled probe of the gene of interest. Larger hybridizing fragments in the antisense transformant demonstrate that at least parts of the gene are methylated due to antisense effects. Since only the sequences of the respective restriction site are examined in this experiment, it is recommended to use a variety of enzymes to cover a more representative portion of the gene. Tables of C methylation-sensitive restriction enzymes are provided by most suppliers.

However, if no change is seen in the restriction pattern between the wild-type and the transformant, this does not exclude methylation. The analysis is limited by the availability of suitable isoschizomers and by the fact that restriction sites are mostly symmetrical. In recent years it has become clear that methylation is not restricted to CpG sequences.

A more general (but also more laborious) analysis is the sequencing of PCR fragments from genomic DNA following treatment with bisulfite. Bisulfite converts unmethylated cytosines to uridines thus resulting in A:T base pairs transitions in the PCR product. In contrast, me^{5}cytosines are not changed. The reaction occurs in the following steps:

(a) Sulfonation of the 5 position of cytosine at low pH.

(b) Hydrolytic deamination at the 1 position catalysed by bisulfite, also carried out at acidic pH.

(c) Alkaline desulfonation resulting in uracil.

The reaction is highly single-strand specific (*Protocol 6*).

Protocol 6. Determination of C methylation by the bisulfite method (11–13)

Equipment and reagents
- Sodium bisulfite reagent: 1.5 M sodium bisulfite, 0.5 mM hydrochinone pH 5.5
- Nitrogen gas
- Promega Magic DNA Clean-Up System (optional)

Method

1. Digest at least 2 µg of genomic DNA with a restriction enzyme to generate a suitable fragment of the region of interest. Isolate the appropriate size class of fragments from an agarose gel to enrich for the DNA of interest. This step may be omitted, restriction digestion is nevertheless recommended to reduce the viscosity of the sample. Alternatively, DNA may also be sheared by passage through a narrow gauge needle.

2. Denature DNA at 95°C for 5 min in a volume of 100 µl.

3. Add 1.2 ml of freshly made up sodium bisulfite reagent. Carry out reaction under nitrogen at 50°C for 20 h. Especially for highly G + C-rich regions in the DNA, repeated heat denaturation during the reaction may be advantageous.

4. (a) Dialyse samples at 4°C against 1 litre of 5 mM NaOAc, 5 mM hydrochinone pH 5.2, followed by 0.5 mM NaOAc pH 5.2, and then distilled water. (Extensive time of dialysis is required—at least 8 h in each solution!)

   (b) Alternatively, remove bisulfite by passing the sample over a desalting column (Promega Magic DNA Clean-Up System).

5. Desiccate the dialysed sample in vacuum almost to dryness and resuspend in 100 ml 10 mM Tris, 0.1 mM EDTA pH 7.5.

6. Add NH_4OH to a final concentration of 0.1 M (desulfonation step), incubate at room temperature or at 37°C for 30 min.

7. Add NH_4OAc (pH 7) to a final concentration of 3 M, precipitate with 2.5 vol. of ethanol, resuspend in 100 ml 10 mM Tris, 0.1 mM EDTA pH 7.5. Sample may be stored at −20°C for up to two months.

8. To avoid PCR amplification of incompletely denatured DNA fragments, samples may be treated with *MnI*.

9. Carry out PCR with primers described in Section 7.1 under standard conditions. Control reactions with both unmethylated, and methylated plasmid DNA (methylated with *HaeIII* methylase) should also be set-up.

7.1 Design of primers

7.1.1 Strand-specific primers

Strand-specific primers can be designed because the bisulfite reaction converts unmethylated C to U by deamination. This is reflected in a G to A exchange in the primers and will confer strand-specificity. Obviously, one has to be sufficiently certain about the methylation pattern of the priming site but this is mostly not the case. Primer sequences should therefore be chosen following the suggestions overleaf.

(a) Use a priming site of approximately 30–35 nucleotides with multiple, equally distributed Cs, this confers strand-specificity, at the same time, however, there is uncertainty about the methylation status of these Cs. Primers are designed assuming complete conversion of Cs to Us (except for CpG dinucleotides, see below). One has to be aware that these primers will select for DNA which is not methylated in the priming site!

(b) Avoid CpG dinucleotides in the priming site, these are the major (but not the only!) potential methylation sites.

(c) If CpG dinucleotides cannot be avoided, degenerate the primer at this position incorporating an A and a G to pair with cytosine or uracil.

(d) There should be no internal complementarity in the primers.

(e) Complementarity between pairs of primers should be reduced to a minimum.

The effect of bisulfite treatment of DNA containing unmethylated C residues is exemplified by the arbitrary sequence given below of the reverse primer for the upper (A) strand and the corresponding forward primer for the lower (B) strand:

```
Genomic A strand         5' GACTACCTGAGCTCAG 3'
Bisulfite treated        5' GAUTAUUTGAGUTUTG 3'
1st PCR product from 3'  3' CTAATAAACTCAAAAC 5'
Reverse primer A         5' GATTATTTGAGTTTTG 3'
Genomic B strand         3' CTGATGGACTCGAGTC 5'
Bisulfite treated        3' UTGATGGAUTUGAGTU 5'
Forward primer B         5' AACTACCTAAACTCAA 3'
```

As can be seen, the two primers, though directed against the same segment of the DNA are very different and specific for the respective strand. The second primers (forward A and reverse B) are designed in the same way assuming full conversion of Cs to Us.

Restriction enzyme recognition sites added to the 5' ends of the primers allow for convenient cloning of the PCR fragments into suitable vectors for sequencing. Multiple clones have to be sequenced to obtain an overview of the methylation pattern in the DNA because methylation may be different in individual molecules.

Unmethylated Cs will be represented by As in one strand and Ts in the other strand when both orientations of the PCR product are sequenced. Methylated Cs, in contrast, will appear as usual as Cs in one and Gs in the other strand.

7.1.2 Non-strand-specific primers

Non-strand-specific primers should be designed to be complementary to priming sites lacking Cs or should be degenerate in positions opposite to a C.

In this case a single primer set is sufficient because both strands and PCR products from both strands will be recognized, for example:

```
Genomic A strand              5' GACTACCTGAGCTCAG 3'
Bisulfite treated             5' GAUTAUUTGAGUTUAG 3'
1st PCR product from 3'       3' CTAATAAACTCAAATC 5'
Genomic B strand              3' CTGATGGACTCGAGTC 5'
Bisulfite treated             3' UTGATGGAUTUGAGTU 5'
so giving a degenerate primer 5' RAYTAYYTRARYTYAR 3'
```

Thus this degenerate primer can pair with both methylated and unmethylated genomic B strand template and 1st PCR product from the genomic A strand template.

The high degeneracy of the non-strand-specific primer shown in this example (only given for comparison with the strand-specific primer) should be avoided by choosing a priming site with a minimum of C residues.

In contrast to the strand-specific primers, clones derived from the PCR products are a mixture of sequences derived from both strands.

Example (assuming the C residues in both strands are not methylated):

```
A strand          5' ATACGTAA 3'
B strand          3' TATGCATT 5'
A-derived         5' ATATGTAA 3'
PCR clone         3' TATACATT 5'
B-derived         5' ATACATAA 3'
PCR clone         3' TATGTATT 5'
```

If the DNA is hemi-methylated or different genomic molecules have different methylation patterns, the interpretation of the sequence patterns may become more complicated.

Controls are essential in this assay. The completeness of the bisulfite reaction is monitored by reactions carried out with the sequence of interest on an *E. coli* plasmid preferentially grown in a dcm⁻ strain. This should result in an empty G lane in the sequencing reaction since all cytosines have been converted to uridine.

For comparison, the same plasmid methylated by *Hae*III methylase *in vitro* should show no sequence alterations after bisulfite treatment.

Acknowledgements

We thank M. Wassenegger and Peter Meyer for help with setting-up the protocol for the bisulfite method. The authors are supported by an ARC grant, W. N. is further supported by a grant by the Deutsche Forschungsgemeinschaft (Ne285/4) and C. P. L. by a grant from the BBSRC.

References

1. Wilkinson, D. G. (ed.) (1992). *In situ hybridisation: a practical approach.* IRL Press, Oxford.
2. Maniak, M., Saur, U., and Nellen, W. (1989). *Anal. Biochem.*, **176**, 78.
3. Chomczynski, P. and Sacchi, N. (1987). *Anal. Biochem.*, **162**, 156.
4. Somssich, I. E. (1994). In *Plant molecular biology manual E1* (ed. S. B. Gelvin and R. A. Schilperoort), 2nd edn, pp. 1–11. Kluwer Academic Publishers, Dordrecht.
5. Nellen, W., Datta, S., Crowley, T., Reymond, C., Sivertsen, A., Mann, S., *et al.* (1987). *Methods Cell Biol.*, **28**, 67.
6. May, T., Blusch, J., Sachse, A., and Nellen, W. (1991). *Mech. Dev.*, **33**, 147.
7. Hildebrandt, M. and Nellen, W. (1996). In *Mechanisms and applications of gene silencing* (ed. D. Grierson, G. Lycette, and G. Tucker), pp. 15–20. Nottingham University Press, Nottingham, UK.
8. Melton, D. A. (1985). *Proc. Natl. Acad. Sci. USA*, **82**, 144.
9. Kim, S. K. and Wold, B. J. (1985). *Cell*, **42**, 129.
10. Wassenegger, M., Heimes, S., Riedel, L., and Saenger, H. L. (1994). *Cell*, **76**, 1.
11. Frommer, M., McDonald, L. E., Millar, D. S., Collis, C. M., Watt, F., Grigg, G. W., *et al.* (1992). *Proc. Natl. Acad. Sci. USA*, **89**, 1827.
12. Meyer, P., Niedenhof, I., and ten Lohuis, M. (1994). *EMBO J.*, **13**, 2084.
13. Clark, J. S., Harrison, J., Paul, C. L., and Frommer, M. (1994). *Nucleic Acids Res.*, **22**, 2990.
14. Lee, R. C., Feinbaum, R. L., and Ambros, V. (1993). *Cell*, **75**, 843.
15. Wightman, B., Ha, I., and Ruvkun, G. (1993). *Cell*, **75**, 855.

3

Selecting, preparing, and handling antisense oligodeoxyribonucleotides

JEAN-JACQUES TOULMÉ, CHRISTIAN CAZENAVE, and
SERGE MOREAU

1. Introduction

Combined developments in DNA sequence analysis and in oligonucleotide (oligo) chemistry make the antisense oligo approach a very attractive one to regulate gene expression. Knowing the primary sequence of the gene one wishes to turn down allows the design of a complementary antisense oligo sequence which, upon binding to its cognate RNA, may selectively prevent the function of the target gene. Since this very simple idea was first proposed by a Russian team in the late 1960s (1), numerous examples of successful inhibition of gene expression by antisense oligos have been published in the scientific literature. This ranges from pioneering work in cell-free extracts and cultured cells (2, 3) to recent *in vivo* experiments (4, 5). Even therapeutic trials in human beings are under way for AIDS and cancer (6). See Chapter 10 for medical applications.

Although the naive concept of hybrid-arrested transmission of genetic information via complementary sequences is still valid, we are now far from the initial idea. Inhibition of protein synthesis does not only result from the hybrid-arrested translation, i.e. from the competition between antisense oligo and translation machinery for mRNA binding. Studies in cell-free extracts and *Xenopus* oocytes have shown that RNase H-mediated degradation of the RNA target largely contributes to the antisense effect. RNase H enzymes, found in both prokaryotic and eukaryotic cells, can cleave the RNA strand of RNA:DNA hybrids. Therefore, they recognize antisense oligodeoxyribonucleotide:mRNA complexes as substrate and cleave the RNA, so irreversibly preventing reading of the information borne by the RNA.

In addition, the range of target sites that we can consider for oligos has been considerably widened since the early 1970s. Besides mRNA, pre-RNA, or viral RNA, one can take advantage of the formation of a triple-stranded structure from particular double-stranded DNA regions, thus allowing inter-ference with transcription. Although still in its infancy this strategy is worth considering. Many problems are shared by antisense and triplex-forming oligonucleotides (e.g. nuclease-sensitivity, cell uptake, toxicity). Specific constraints for this approach are briefly discussed in Section 4.4.

It seems clear that naked linear phosphodiester beta oligodeoxy- or oligo-ribonucleotides (unmodified DNA or RNA) are of limited interest as inhibitory agents in complex biological systems. Such oligos are susceptible to nuclease attack and are taken up by cells with limited efficiency. Antisense oligos must also deal with RNA secondary structures, compete with proteins for binding to the target, be delivered to the appropriate cell compartment, and bind selectively to the sense sequence. In addition, for *in vivo* use, one needs non-toxic molecules. Lastly, molecules which are easy to synthesize are preferable, especially for therapeutic use.

At first sight the antisense strategy is appealing because it rests on the **rational** design of a **specific** inhibitor of gene expression. As the antisense sequence is derived from the sense one, base pairing between the two strands should ensure the selective association of the partners. However, the numerous parameters that an antisense molecule should fulfil suggest that the design of an efficient oligo will not be an easy task. This chapter focuses on the selec-tion of antisense DNA oligos aimed at inhibiting translation of mRNA. How-ever, most of the information presented here is relevant for any target. Several recent reviews (7, 8) and books (9–11) have been published on anti-sense oligos in the last few years to which the reader should refer for a more in-depth view of the field.

One of the first questions asked by colleagues who would like to eliminate the expression of their favourite gene is which analogue to use amongst the numerous chemically modified oligos that have been described. Most of these oligos are, however of limited interest to non-specialists. We restrict our chapter to molecules available from specialized suppliers or which can be home-made on an oligo synthesizer. Phosphorothioate, methylphosphonate, 2'-O-methyl oligos are analogues which display resistance to nucleases and are therefore able to exert their inhibitory effect for a longer period of time. We also describe the synthesis of terminally modified oligos, as such conju-gates proved to be of special interest for improved binding efficiency (inter-calating agents) or for increased uptake (hydrophobic ligands). The reader's attention is drawn to a key parameter: affinity of the oligo for target and non-target ligands. This governs both the efficiency and the specificity of the pro-cess. The lifetime of the molecule and the mechanism of action (RNase H-dependent or independent) is also important for the success of the experi-ment. In the last part of the chapter we describe biological applications of antisense technology that we have used in our laboratory.

2. Chemical synthesis of oligodeoxynucleotides

All moieties of the polymer have been considered for modification: the inter-nucleoside bridge, the sugar residue, the bases, and the 5' or 3' ends (see refs 8 and 12–14 for reviews). In this section we focus on the main modifications which have proved to be of value for the antisense approach.

Today, unmodified oligos are prepared almost exclusively by solid phase phosphoramidite techniques (15). The modification and functionalization of oligos are therefore mainly based on this approach. We will first develop the salient features of unmodified oligodeoxyribonucleotide synthesis using phos-phoramidite chemistry and then present some strategies to introduce classical modifications (phosphorothioate and 2'-*O*-methyl derivatives) on a commer-cial synthesizer, as well as the attachment of ligands at the 5' end of oligos. The chemical synthesis of oligos involves the formation of a phosphodiester bridge between two nucleoside units. The polymerization reaction between phosphorus (PIII) derivatives of nucleosides (the building blocks) takes advantage of solid state chemistry.

2.1 Main features of solid phase phosphoramidite chemistry

First developed for peptide synthesis, solid phase synthesis was readily extended to nucleic acid synthesis. The first monomer is attached to an inert solid support and extension results from the addition of the second monomer in solution. When the reaction is complete, reagents are eliminated and the dimer is retained on the solid support. The cycle is repeated until the desired oligomer is obtained. At the final step the product is released from the solid support. Such a synthesis offers multiple advantages:

(a) Chemical reactions can be driven to completion by the use of a large excess of reagents relative to the polymer-bound chain.

(b) Excess reagents are washed away after the reaction, avoiding tedious purification after each monomer addition.

(c) The process is amenable to automation.

The key step of the synthetic scheme is the coupling reaction, which intro-duces the phospho bridge between two successive nucleoside units. The development of trivalent derivatives of phosphorus compounds allowed high rate and high yield in the coupling step (typically more than 98%) to be achieved. After successive optimizations the best compromise was reached by using 3' phosphoramidite derivatives of nucleosides.

The nucleophilic 5' hydroxyl group of one unit reacts with the electrophilic phosphorous centre of the ensuing monomer, activated by tetrazole, a mild acidic compound. This activation is one of the most important advantages of the method. It allows the building units (the phosphoramidites) to be easily handled and stored. All other nucleophilic species have to be eliminated from the reaction medium. This holds for the 5' OH group of the phosphoramidite, the exocyclic amino groups of bases, and OH on phosphorous. These reactive species must therefore be temporarily blocked. Moreover water must be eliminated. Contamination by traces of water in the synthesizer lines, solvents, or amidites is one of the major reasons for failure. For this reason it is highly desirable to synthesize a successive series of oligos to avoid preparation of oligos from reagents kept on the machine for a while.

The 5' hydroxyl function is protected as 4,4' dimethoxytrityl (DMT) ether; this protecting group can be introduced regioselectively at the 5' position and can be removed under mild acidic conditions. The exocyclic amino groups of adenine, guanine, and cytosine are protected as amides that are cleaved under basic conditions after the synthesis of the DNA oligo. The 2-cyanoethyl group on the phosphorous is also easily removed under mild basic conditions. An oxidation step is introduced to convert the phosphite into a pentavalent compound.

2.2 The synthesis cycle

Solid phase phosphoramidite chemistry is routinely implemented on auto-matic synthesizers. The chemical synthesis of DNA is carried out in the 3' to 5' direction, taking advantage of the higher reactivity of the primary 5' hydroxyl group compared to the secondary 3' one. The 3' end of the first monomer is linked to the solid support through a succinyl spacer which allows the easy recovery of the 3' OH group at the end of the synthesis after a mild basic treat-ment. The cycle of the synthesis consists of three main steps (*Figure 1*):

(a) Deprotection of the 5' OH position by treatment with dichloroacetic or trichloroacetic acid in dichloromethane.

(b) Coupling by simultaneous addition of a large excess of a phosphoramidite building unit and of activator (tetrazole) in anhydrous acetonitrile.

(c) Oxidation of the phosphite triester to phosphate triester using an aqueous iodine reagent.

A capping reaction is generally performed to acetylate the 5' OH group which failed to react. This step is introduced either at the end of the coupling reaction or after the oxidation step.

Average coupling yields of 98.5–99% are reached on automatic machines. The cycle time ranges from 5–8 min depending on the protocol and scale of synthesis used; 0.2 μmole or 1 μmole scales are currently used for antisense oligo synthesis, at least for *in vitro* experiments. The solid support consists of porous glass beads of 500 Å average pore size for most of synthesis. However for monomers exceeding 35 nucleotides in length, we recommend the use of 1000 Å pore size solid support.

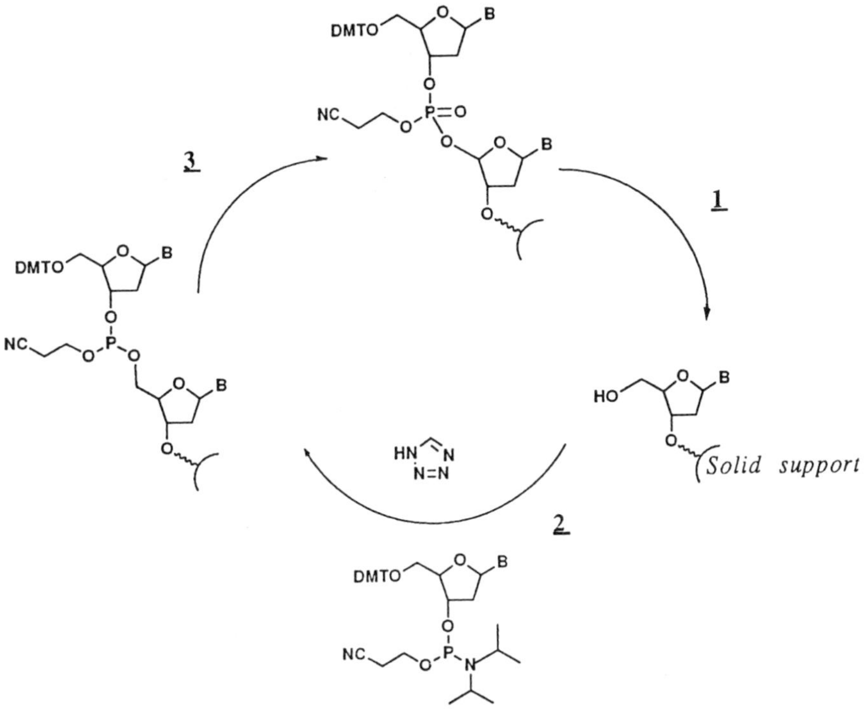

Figure 1. The three main steps of the synthesis cycle. (1) 5' position deprotection by acidic treatment (dichloroacetic acid in dichloromethane). (2) Coupling reaction with the tetrazole as activator: the free 5' OH group is condensed with the phosphoramidite. (3) The oxidation step: the P(III) phosphoramidite derivative is converted to P(V) phosphotriester compound through the iodine:water mixture.

2.3 Deprotection and purification

Post-synthetic treatment of the oligo includes cleavage from the solid support and elimination of base, phosphate, and ribose protections. A combined strategy can be used which takes advantage of the DMT group on the full-length oligo to allow an easy separation from the aborted sequences by reverse-phase HPLC chromatography. Here the DMT group is kept on the last added monomer. The deprotection is thus carried out in two steps.

The solid phase linked oligo is treated with concentrated ammonium hydroxide. The succinyl link is easily cleaved after one hour of incubation at room temperature. Nucleobase protecting groups and the phosphate protecting group require a prolonged incubation time (16 h at 55°C). This procedure is described in *Protocol 1*.

Reverse-phase HPLC chromatograhy of 5' DMT oligo is based on the hydrophobic character of the DMT group. The interaction of this aromatic moiety with the non-polar bound phase of the column packing allows the specific retention of DMT-containing species. Elution is obtained by increasing the amount of the less polar component of the mobile phase (e.g. acetonitrile). This procedure is described in *Protocol 2*. After preparative reverse-phase HPLC, the DMT group is removed by an acidic treatment (acetic acid:water, 80:20) for one hour at room temperature.

We recently used a new hydrophobic column based on a polystyrene support (R3, Perseptive Biosystem). The crude ammonia solution of oligos is directly injected as the column and trityl-off failure sequences are eluted by a 0–12% acetonitrile gradient. The treatment of the oligos on the column with 2% trifluoroacetic acid leads to detrytilation. The detrytilated oligos are then eluted by increasing the acetonitrile concentrations (up to 60%). We thus obtained purified oligos in one step (30 min).

Protocol 1.   Oligonucleotide cleavage and deprotection

Equipment and reagents
- Concentrated ammonium hydroxide (27%)
- Triethylamine
- 5 ml Luer tip syringes
- Screw-capped vials

Method

1. After the synthesis, remove the column from the synthesizer. Fill a syringe with 3 ml of concentrated aqueous ammonia and attach it to one end of the column. Fit the other end with an empty syringe. Depress the syringe plungers to force the solution back and forth through the column. Repeat two or three times to ensure that the solid phase is totally immersed in the ammonia solution.

2. Leave at room temperature and pump the solution back and forth through the column from time to time.

3. After 2 h draw all the solution into one of the syringes. Carefully remove the full syringe from the column and deposit the solution in a screw-capped vial.

4. Heat the tightly screwed vial for 16 h at 55°C.

5. Cool the vial and transfer the solution into a flask suitable for rotatory evaporation. Add 30 µl of triethylamine to prevent detrytilation, if the oligo is to be purified by reverse-phase HPLC.

Protocol 2.   Purification of phosphorothioate oligonucleotides by reverse-phase HPLC

Equipment and reagents

- Column: a C-18 bonded phase on silica with spherical particles (5 µm) and 300 Å pore size,   with additional capping of free silanol group (e.g. Nucleosil 100–5 C-18, 100 mm × 4 mm)[a]

- Chromatographic buffer A: 0.1 M triethyl-ammonium acetate pH 7
- Chromatographic buffer B: 0.1 M triethyl-ammonium acetate in 80% acetonitrile
- Sample
- 80% glacial acetic acid in water
- Ethyl acetate
- Rotary evaporator

A. Analytical reverse-phase HPLC

1. Dissolve the crude sample (from 0.2 or 1 μmole scale), obtained after ammonium hydroxide treatment, in 300 μl of buffer A.

2. Set the UV monitor to 260 nm and the flow rate to 1 ml/min.

3. Use the following gradient: 0–50% buffer B over 50 min, then 50–100% buffer B over 5 min.

4. Inject a small portion (e.g. 5 μl) of the crude sample (from a total volume of 300 μl).

5. Monitor the separation by absorbance at 260 nm.

B. Preparative reverse-phase HPLC

1. Set the UV monitor to 290 nm and the flow rate to 1ml/min.

2. Use the same gradient as in part A, step 3.

3. Inject 300 μl of the sample.

4. Monitor the A_{290} and collect the central part of the absorbance peak, discarding the leading and trailing edges.

5. Evaporate the collected fraction using a rotary evaporator *in vacuo*.

6. Add 1 ml of 80% acetic acid and incubate for 1 h at room temperature.

7. Evaporate the acetic acid solution *in vacuo* using a rotary evaporator.

8. Dissolve the residue in 1 ml H_2O, extract DMT-OH twice with an equal volume of ethyl acetate. Reconcentrate the sample *in vacuo* to dryness.

9. Suspend the residue in a known volume of water and measure the absorbance at 260 nm on a 1/100 dilution.[b] The extinction coefficient of the single-stranded oligo is calculated taking into account nearest-neighbour interactions (16) according to the following equation:

$$\varepsilon_{DpEpFpG...KpL} = 2(\varepsilon_{DpE} + \varepsilon_{EpF} + \varepsilon_{FpG} + ...\varepsilon_{KpL}) - (\varepsilon_E + \varepsilon_F + ...\varepsilon_K)$$

[a] This column allows injection of oligos from 1 μmole scale synthesis. The same column is used either for analytical purposes or for preparative runs (0.2 or 1 μmole scale of synthesis).
[b] Before quantification by UV one or two precipitation steps can be performed in order to eliminate residual salts from the HPLC buffer. Suspend the dry residue in 100 μl of water, add 1 ml of *n*-butanol, vortex for 15 sec, and centrifuge for 1 min. Discard the supernatant.

3. Synthesis of modified oligonucleotides

3.1 Phosphorothioates

Phosphorothioate (PS) analogues of oligos have sulfur in place of one of the non-bridging oxygen atoms bonded to phosphorous (*Figure 2*). This substitution

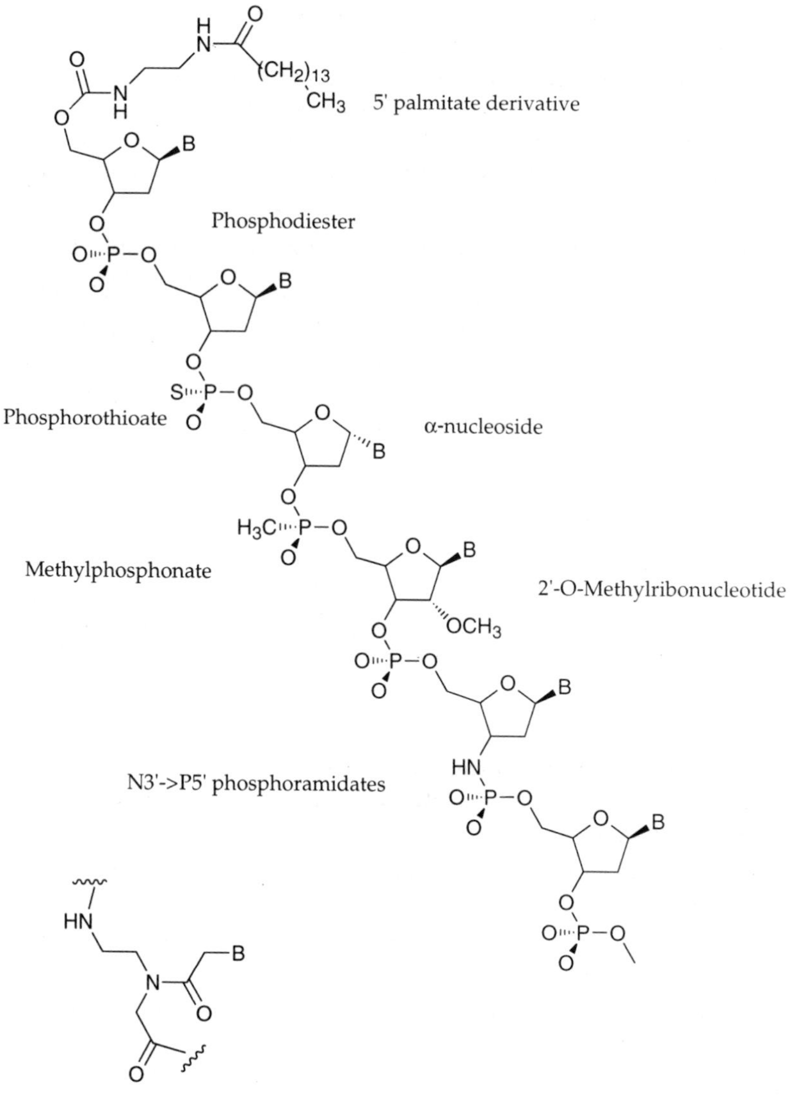

Figure 2. Chemical structure of some oligonucleotide analogues.

introduces an asymmetric centre at the phosphorous atom. PS are therefore mixtures of diastereoisomers. They are probably the most widely used analogues of oligos for antisense application (17). This is related to their fairly good nuclease-resistance and to their ability to elicit RNase H activity.

PS oligomers can be synthesized by a simple adaptation of the standard synthesis cycle by substituting the oxidation step by a sulfurization one by a number of methods. The two main reagents used are tetraethylthiuram (TETD) (18) and 'Beaucage's reagent' (3H-1,2-benzodithiol-3one-1,1-dioxide) (19). *Protocol 3* describes a method which gives short reaction times and high efficiency of incorporation of a sulfur atom, as demonstrated by ^{31}P NMR. The capping step in this case is performed after the sulfurization reaction.

Protocol 3. Synthesis of phosphorothioate oligonucleotides

The sulfurization step is performed by delivering 240 μl of a 0.05 M solution of Beaucage's reagent for 60 sec onto the column. This sulfurization step is introduced after the coupling step. The capping step must be performed at the end. Carefully check this point on the program implemented on your synthesizer. If available, use an auxilliary flask to deliver the reagent.

Equipment and reagents

- Dichlorodimethylsilane
- Dichloromethane
- Oven
- Desiccator
- Methanol

- Oligo synthesizer fitted with standard reagents
- Beaucage's reagent: 3H-1,2-benzodithiol-3one-1,1-dioxide (Glen Research, Pharmacia, or Sigma)
- Dry acetonitrile

Method

1. Add dichlorodimethylsilane (12 ml) to the auxilliary flask containing dichloromethane (200 ml). After 5 min remove the solution and rinse with methanol.

2. Dry the auxilliary flask in an oven overnight (110°C), allow to cool in a desiccator.

3. Dissolve 0.5 g (2.5 mmol) of Beaucage's reagent in dry acetonitrile (50 ml) and transfer to the flask.

4. Fit the bottle to the corresponding port on the synthesizer. Prime the reagent twice.

We routinely use *Protocol 2* in our laboratory to purify both PS oligos and unmodified oligos. A typical chromatogram is given in *Figure 3* for a 17-mer phosphorothioate. This analysis of a crude reaction mixture indicates the presence of several DMT-containing species and some non-DMT failure sequences at shorter retention times.

3.2 Methylphosphonates

Methylphosphonate (MP) oligos (*Figure 2*) contain a non-ionic internucleosidic linkage. One non-bridging oxygen atom is substituted by a methyl group

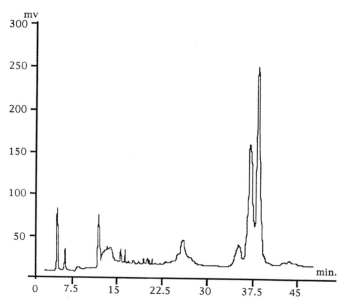

Figure 3. Analytical HPLC chromatogram of a crude phosphorothioate oligomer (5′ CTG ATA CTT TAGC). The synthesis has been performed at the 0.2 μm scale according to *Protocol 1*. The chromatography was run as described in *Protocol 2* with a linear gradient of buffer B (0.1 M TEAA pH 7.0 in 80% acetonitrile) in buffer A (0.1 M TEAA pH 7.0). The 5′ ODMT oligo (16-mer) eluted as a doublet centred at 37.5 min. This is due to a mixture of configurations at the phosphorous atom adjacent to the 5′ ODMT group. Detritylated oligos are detected at 25 min.

(20). As already mentioned for PS oligos this substitution leads to a mixture of diastereoisomers; this partly accounts for the low stability of MP RNA hybrids. MP are presently of limited use in antisense applications although they are resistant to nucleases. However, chimeric oligos associating phosphodiester and methylphosphonate stretches in a single sequence have been shown to increase the specificity (see Section 4.3.2). Methylphosphonoamidites are commercially available and can be used with a standard coupling procedure. But high yields are more difficult to reach with MP. The deprotection procedure has to be modified to avoid the cleavage of the phosphonate backbone by ammonium hydroxide. A combination of ethylene ammonium hydroxide diamine and ethanol treatment is used instead (21). Reverse-phase HPLC is the preferred method for the purification of MP oligonucleotides due to their non-ionic nature (*Protocol 2*).

3.3 2′-*O*-methyloligoribonucleotides

2′-*O*-methyloligoribonucleotides possess properties that have proved very useful in antisense applications; they are much more chemically stable than oligodeoxyribonucleotides, and are relatively resistant to degradation by

nucleases (22). These analogues are easily synthesized using standard phosphoramidite chemistry. 2'-*O*-methylamidites are commercially available. The
standard synthesis cycle of DNA can be used, the coupling reaction time
being increased up to 15 min, with tetrazole as activator. A protocol is given
below (*Protocol 4*) that we have used in our laboratory in particular for the
synthesis of chimeric sequences (see Section 4.3.2). The use of supplementary
ports available on most synthesizers allows the synthesis of mixed sequences.

Protocol 4.   Synthesis of 2'-*O*-methyloligoribonucleotides

Equipment and reagents
- 2'-*O*-methylphosphoramidites (Glen Research, Millipore)
- Dry acetonitrile
- *n*-butanol
- Oligo synthesizer fitted with standard reagents

Method

1. Dissolve 2'-*O*-methylphosphoramidites (250 mg) in 3.5 ml of dry acetonitrile to give a 0.1 M solution.

2. Transfer the solution to standard vials and screw the four vials onto
   the synthesizer.

3. Purge the solutions twice.

4. Type the sequence to be prepared using port number instead of classical
   base symbols.

5. Purify according to *Protocols 1* and *2* which work for 2'-*O*-methyl
   derivatives too. Precipitation by *n*-butanol allows high yield of purification. It seems to work better than the classical precipitation with
   ethanol/Mg^{2+} for these derivatives.

3.4 Terminal modification of oligonucleotides

Many ligands have been proposed in order to:

- increase the affinity of the antisense sequence for the target
- improve its stability toward nucleases
- achieve highly sensitive detection

The simplest method to introduce new chemical functions on an oligo
involves reaction at the 5' terminus of the chain when still bound to the solid
support. Numerous phosphoramidite derivatives are commercially available
allowing an increase in affinity (acridine), cross-linking properties (psoralen),
better uptake by cells (cholesterol), or for use as hybridization probes
(biotin). The standard protocol of coupling is used for the oligo. An additional
reagent flask is generally connected to the synthesizer to allow synthesis of a

new function. Introduction of a primary amino group at the 5′ end is also a very useful method for further derivatization by a wide range of ligands. Phosphoramidite synthons have been proposed, which after the deprotection step by ammonium hydroxide lead to such 5′ amino groups. The protocol presented below (*Protocol 5*) has allowed us to introduce 5′ chemical groups (cholesterol, aminolink) with good yields.

An alternative procedure for 5′ functionalization involves the chemical activation of the terminal hydroxyl group of the oligo still bound to the solid support. The free OH group is activated with 1,1′ carbonyl diimidazole and immediately treated with a suitable diamine linker such as ethylene diamine or tetraethylene diamine. These steps are conducted on the solid support, further derivatization is then achieved in solution after ammonium hydroxide treatment of the column. We have used this method to introduce a palmitic chain at the 5′ end of a phosphorothioate antisense sequence (see *Protocol 7*).

Protocol 5.   Introduction of 5′ modification using phosphoramidite derivatives[a]

Equipment and reagents
- Amidites
- Dry acetonitrile
- Oligo synthesizer fitted with standard reagents

Method

1. Dissolve 100 μmoles amidites in 1 ml of dry acetonitrile.

2. Transfer the solution to a standard vial and screw it onto the synthesizer (port 5). Among the various available derivatives, we generally choose those which incorporate a trityl group. This allows either an easy purification by reverse-phase chromatography (amino modifiers) or a control of the last coupling step (cholesterol, acridine, fluorescein).

3. Purge the solutions twice.

4. Type the sequence to be prepared using port number 5 for the amidite to introduce at the 5′ end.

5. Follow *Protocol 2* for purification. For modified oligos bearing chromophors (e.g. fluorescein, psoralen) light absorbance of the ligand, at 260 nm, must be taken into account in the calculation of the extinction coefficient of modified oligos. Alternatively, the molar extinction coefficient of the ligand can be used, at a wavelength at which nucleotides do not absorb.

[a] The synthesis cycle is the same as for 2′-*O*-methyloligoribonucleotide (see *Protocol 4*). An additional port is used to introduce the amidite derivative.

---

4. Properties of oligonucleotides

The biological effect produced by an antisense oligo depends on many criteria:

(a) Its affinity for the target sequence and for non-target ligands (proteins, mismatched sequences).

(b) Its lifetime in the biological milieu (growth medium, cell) which contains nucleases.

(c) Its ability to induce RNase H activity.

   Most of these parameters can be quickly and simply evaluated in order to understand the behaviour of oligos and to optimize the choice of the antisense sequence.

4.1 Affinity

The properties of an antisense oligo derive from the binding of the oligo to its complementary (sense) sequence. In other words, the biological effect is driven by the amount of complex formed. This is related to the binding constant and to the concentration of the two partners, according to the law of mass action. It is therefore of interest to evaluate the affinity of the oligo for its target, although *in vitro* measurements do not strictly reflect the situation in more complex biological systems.

4.1.1 UV monitored melting curves

Two major types of interaction contribute to the free energy of binding of complementary sequences: interstrand hydrogen bonding between bases (usually Watson–Crick pairs) and stacking between adjacent pairs along the double helix. Disruption of structured nucleic acids results in unstacking which can be monitored by UV spectrophotometry, as this leads to an absorbance increase (hyperchromism). Monitoring the absorbance at 260 nm of a mixture of sense and antisense sequences versus temperature generates a sigmoidal curve. The temperature of mid-transition, called T_m (melting temperature) corresponds to 50% duplex and 50% single-stranded species and is related to the stability of the duplex. If we assume a two-state transition, analysis of the curve provides thermodynamic parameters (ΔH, ΔG, ΔS). A complete description can be found in ref. 23. In short, the melting transition allows the determination of the fraction of double-strand, i.e. the association constant K_a for each temperature, from which a Van't Hoff plot ln K_a versus $1/T$ can be drawn (T is the temperature in °K). Slope and y intercept of this linear plot allows calculation of the enthalpy ΔH and the entropy ΔS according to the equation:

$$\ln K_a = \frac{\Delta H}{R}\frac{1}{T} + \frac{\Delta S}{R}\qquad\text{[1]}$$

where R is the gas constant.

The free energy of binding ΔG is then obtained from the standard equation:

$$\Delta G = \Delta H - T\Delta S \qquad [2]$$

It is important to note that ΔG and not T_m is the true thermodynamic parameter which describes the stability of the [antisense oligo:sense RNA] duplex. However, trends observed for ΔGs are generally reflected in T_ms. Therefore, useful information can be obtained from T_m measurement which are experimentally easier to get and more accurate. Nevertheless, it should be kept in mind that the difference ΔT_m between the mid-point transitions obtained for two duplexes will depend on the oligo concentration and on the intrinsic ΔH and ΔS values for the particular analogue used, in contrast to the difference $\Delta\Delta G$ between the free energy of the two duplexes, which is exclusively related to $\Delta\Delta H$ and $\Delta\Delta S$ (24).

Experimentally, running a melting curve requires a spectrophotometer, the cell holder of which is thermostated. Several companies offer equipment devoted to thermal denaturation studies that allows multicell analysis at several wavelengths while different temperature ramps can be set-up. Some of them even offer a program to extract thermodynamic parameters. Though very convenient, these devices are expensive and are not necessary unless a large number of samples have to be routinely analysed. For occasional use a thermostated cell holder connected to a water-bath externally driven by a temperature controller is enough, the only requirement being to record, simultaneously, the absorbance and the temperature.

A few points have to be kept in mind in running a thermal denaturation experiment to obtain meaningful T_m values:

(a) One assumes that an equilibrium between double-stranded and single-stranded forms is reached at each temperature. Consequently a slow rate of temperature increase should be selected. Usually 0.5 °C/min gives good results with double-stranded nucleic acids.

(b) The process should be reversible: the same mid-point temperature is expected for dissociation and reannealing processes. If not, this indicates either an artefact (e.g. air bubble, evaporation) or a slow equilibrium. Slow association is not observed with short double-stranded moieties, but this is characteristic of triple-stranded structures, particularly at low ionic strength. In this case, a melting curve analysis should be performed with a lower temperature increment per minute (slower than 0.1 °C/min); consequently, a thermal denaturation and reannealing cycle can take up to 10–20 hours. For such a long time spent at high temperature, evaporation should be prevented. This can be achieved by covering the solution with a film of paraffin. Cuvettes must then be cleaned with a solution of detergent at high temperature, followed by washing with chloroform.

(c) Nucleic acids are polyanions. Therefore, the association of complementary strands will lead to electrostatic repulsion. The addition of salt will reduce these interactions and stabilize multistranded structures. As a consequence, T_m values will be ionic strength-dependent, with a $dT_m/d[\log \text{Na}^+]$ of about 15 °C for phosphodiester DNA. Such a variation is not expected with neutral analogues of nucleic acids such as methylphosphonates.

(d) The thermodynamic parameters ΔH and ΔS vary with the chemical nature of the nucleic acid strand. If RNA is the biological target it is highly recommended to use RNA and not DNA to monitor the binding of antisense oligos. Although synthetic DNA is easier (and cheaper) to get than RNA, no significant conclusion can be drawn from comparison of oligo:DNA complexes if RNA is the physiological target. Substantially different T_m values can be obtained for duplexes formed by oligo derivatives with either DNA or RNA (*Table 1*): methylphosphonate and 2'-O-methyl 12-mers bound to the complementary DNA sequence give a similar T_m value. The ΔT_m is about –8 °C compared to the unmodified duplex. In contrast, with an RNA target, the 2'-O-methyl derivative gives a higher T_m ($\Delta T_m = +12$ °C) whereas methylphosphonate leads to a lower value ($\Delta T_m = -19$ °C) than that of the unmodified DNA:RNA hybrid. Moreover, differences may vary from sequence to sequence.

(e) The parameter of interest is the absorbance variation between the bound and dissociated state of the hybrid. The method will not allow accurate determination of T_m values from complexes between an antisense oligo (say 15–20 nt long) and a long RNA target (more than about three times the length of the antisense sequence) as the relative absorbance of the complex decreases as the length (i.e. the absorbance) of the target

Table 1. T_m and T_c values for antisense oligo:DNA or :RNA hybrids

Antisense[c]	$T_m{}^a$ ($\pm$ 0.5 °C) 12-mer DNA	$T_m{}^a$ ($\pm$ 0.5 °C) 12-mer RNA	$T_c{}^b$ ($\pm$ 1 °C) RNA[b]
PO	47.6	46.1	39
α-PO	42.8	43.1	33
MP	40.3	27.4	ND[d]
PS	37.9	36.1	33
2'-O-Me	39.0	58.4	55
2'-O-Al	35.7	54.0	54

[a] Each strand was 3 μM (in 0.1 M NaCl, 10 mM cacodylate pH 7.0) (from ref. 33).
[b] 5 pmol of *in vitro* transcribed RNA (259 nt) was hybridized with 20 pmoles of antisense oligo in 6 × SSC:10 × Denhardt's at room temperature. Elution was performed with 6 × SSC, while a temperature gradient of 1.2 °C/min was applied.
[c] The antisense sequence 5' ACACCCAATCT was synthesized as unmodified phosphodiester (PO), methylphosphonate (MP), or phosphorothioate (PS) oligodeoxynucleotide or 2'-O-methyl (2'-O-Me), or 2'-O-allyl oligoribonucleotide (2'-O-Al). The alpha phosphodiester α-PO sequence was prepared in the reverse orientation as such analogues give rise to parallel-stranded duplexes.
[d] ND: not determined. Methylphosphonates give rise to high background on nitrocellulose and nylon filters.

increases. In addition, long RNA sequences may adopt structures that will also melt, leading to complicated multistep transitions. Thus, it might also be of interest to check the melting of individual strands to make sure that they do not exhibit melting transitions originating from self-structures.

4.1.2 Thermoelution of filter bound hybrids

We have developed a method, based upon the formation of hybrids with an immobilized polynucleotide target that allows monitoring the binding of oligos to full-length RNA or DNA (25). This is described in *Protocol 6*. The presence of non-canonical bases, or of structured regions are taken into account, in contrast to the situation encountered with short synthetic target oligoribonucleotides used to run UV monitored melting curves.

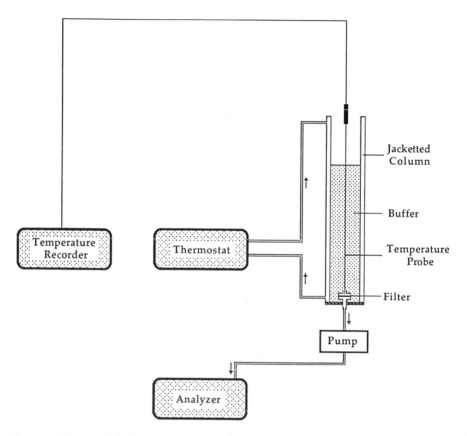

Figure 4. Scheme of the thermal elution device (adapted from ref. 25). The jacketed chromatographic column (2.5 × 40 cm), the peristaltic pump, fraction collector and chart recorder can all be obtained from Pharmacia; the filter holder (13 mm diameter) from Swinnex, Millipore; the cryothermostat (ministat) and the temperature programmer (Model PD410) from Huber; the temperature probe from Bioblock.

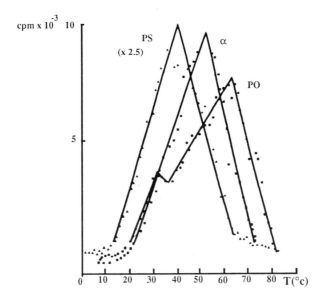

Figure 5. Thermal elution profile of (rabbit globin mRNA:synthetic 17-mers) hybrids. Filters were prepared as described in *Protocol 6* with 1 μg of mRNA. Hybridization was performed with 20 pmoles ^{32}P-labelled oligos 5' CACCAACTTCTTCCACA, targeted to nt 113–129 of the β-globin message, in 6 × SSC:10 × Denhardt's at 4°C for 15 h. Filters were eluted with 6 × SSC buffer: (PO) unmodified, α-phosphodiester, (PS) phosphorothioate. The profile obtained with the unmodified PO 17-mer displays a secondary maximum at about 30°C which corresponds to the binding of the oligo to a partially complementary site (likely a sequence allowing the formation of 13 base pairs out of 17). α-oligomers are nuclease-resistant analogues which do not elicit RNase H activity. They bind in a parallel orientation to the target sequence. They are made of α-isomers of the nucleoside unit.

Briefly, a filter on which a nucleic acid (the target) has been layered and immobilized is incubated in a solution containing the radiolabelled oligo (the antisense sequence). The filter is then mounted in a home-made device (*Figure 4*), immersed in a thermostated buffer. Pumping across the membrane while the temperature of the buffer is increased allows collection of the antisense molecules which dissociate from the target. A plot of the eluted radioactivity versus the temperature yields a bell-shape curve whose maximum corresponds to the critical temperature, T_c, characterizing the stability of the filter bound complex (*Figure 5*).

Protocol 6. Thermoelution of filter bound complexes

All steps require RNase-free conditions. Water is treated with diethyl pyrocarbonate as described in Chapter 7, Section 3. Glassware, tools, and components of the thermoelution device are immersed for 2 h in 0.2 M NaOH and rinsed with water.

Protocol 6. *Continued*

Equipment and reagents

- 6 × SSC: 6 × (0.15 M NaCl, 0.015 M sodium citrate)
- Sephadex G10 spin columns
- Nitrocellulose or nylon filters: 0.45 μm pore size, 13 mm diameter

- Thermal elution device
- 10 × Denhardt's: 10 × [0.2 mg/ml BSA (bovine serum albumin) (RNase-free), 0.2 mg/ml Ficoll, 0.2 mg/ml PVP (polyvinyl-pyrrolidone)]
- Peristaltic pump

Method

1. Boil filters for 3 min in water. Briefly immerse them in 6 × SSC buffer (nitrocellulose filters) or in 150 mM ammonium acetate (nylon filters).

2. Load the RNA of interest (4–8 μl of aqueous solution containing about 1 pmole of target sequence) on the wet filter.

3. Immobilize the RNA by UV irradiation (nylon filters) or by incubation at 80°C for 2–3 h. Store the filters[a] or transfer to the hybridization buffer (6 × SSC, 10 × Denhardt's).

4. ^{32}P end-label 20 pmoles of the antisense oligo.

5. Phenol extract and dilute the solution up to 100 μl with water prior to purification on a 1 ml Sephadex G10 spin column. Elute with water.

6. Check the purity of the antisense probe by gel electrophoresis on urea–polyacrylamide gels.

7. Add 20 pmoles of gel filtered, ^{32}P end-labelled oligomer to the hybridization mix (10^6–10^7 c.p.m./ml) and incubate for 5–15 h.[b]

8. Place the drained filter into the filter holder. Connect it to the top part of the jacketed column (which is the bottom part of the thermal elution device) (*Figure 4*) and assemble the device. Start the circulating water-bath in order to get the desired temperature in the jacketed column.

9. Fill the column with 6 × SSC buffer. Gently shake the filter holder to remove air bubbles. Adjust the temperature probe close to the filter.

10. Start the peristaltic pump (flow rate about 15 ml/h) when the temperature of the buffer is equilibrated. Collect a few fractions,[c] (generally 10–15 fractions of 300 μl each is enough) to wash out the non-bound antisense sequence.

11. Apply and record the temperature gradient (1°C/min). Collect fractions and count these by Cerenkov counting.

12. Plot the eluted radioactivity as a function of temperature.[d] The profile allows determination of the T_c (see *Figure 5*).

[a] More reproducible results are obtained from filters prepared at the same time. It is therefore recommended to load multiple filters at a time. Filters can then be stored dry at –20°C for months.
[b] Better results are obtained when incubation is performed 15–20°C below T_c. This can be estimated from *Equation 3*, but we generally run a first experiment to get a T_c value from which we set the hybridization and the starting temperature for a second trial.

c Alternatively an on-line radioactivity counter can be used.
d The volume of the tubing between the filter holder and the fraction collector should be measured to accurately determine the temperature at which a given fraction was collected. This can be done by monitoring the front of the buffer when the pump is turned on for the washing step.

It is important to note that T_c is not equivalent to T_m. In contrast to melting curves performed in solution, thermal elution of filter bound complexes is not an equilibrium method as the freed oligo is continuously removed and therefore cannot reassociate. Although there is no clear-cut meaning of the T_c value with respect to thermodynamic parameters, intuitively, T_c is related to the off-rate constant. However, the general trends observed for T_m are also reflected in T_c (*Table 1*):

(a) The binding is specific: no signal is detected when the immobilized nucleic acid does not contain the binding site of the antisense oligo and the presence of mismatches reduces the T_c value.

(b) T_c is linearly related to the logarithm of the length (*l*) and to the G + C content (*x*) of the oligo, at least for duplexes up to 30 bp (26).

$$T_c = 111 \log(l + x) - 103 \qquad [3]$$

(c) The T_c value is ionic strength-dependent: increasing the salt content also increases T_cs. However, for an acridine-linked 10-mer we found that the increment $dT_c/d(\log[Na^+])$ is lower (6.5 °C) than the $dT_m/d(\log[Na^+])$ (16 °C) obtained in solution (26).

In addition to these properties which are common to both melting curves (T_m) and thermal elution profiles (T_c), the latter method displays additional features:

(d) The size of the target-bearing nucleic acid can be infinite compared to the few tens nucleotides of the antisense sequence. This means, in particular, that crude mixtures of nucleic acids can be used to investigate the behaviour of antisense oligos. We have routinely analysed the binding of oligos, targeted to a 39 long nucleotide sequence of trypanosome mRNA, on total RNA extracted from cultured parasites (26).

(e) The presence of both perfect and mismatched binding sites can be detected as they will result in a broad peak or even in individually resolved peaks (*Figure 5*) (27).

(f) T_c 'describes' more than the information embodied in the primary sequence. This is exemplified in the aforementioned study with trypanosome RNA. A number of antisense sequences yielded T_cs that fell outside the linear plot derived from *Equation 3*. This is fairly well explained by the presence of modified bases and by a hairpin structure in the target RNA sequence (26).

(g) Thermal elution profiles can be derived from nucleic acids targets immobilized under 'native' or denaturing conditions.

One should also be aware of the limitation of thermal elution of filter bound complexes:

(h) The sensitivity of the method is limited by the amplitude of the signal. On the one hand this depends on the amount of immobilized target. We standardized our experiments with 1–2 pmoles of target per membrane and find that a tenfold reduction reaches the limit of sensitivity. On the other hand sensitivity is also related to the affinity of the oligo for its target. With conventional oligodeoxynucleotides we were not able to detect a signal with antisense sequences shorter than ten bases. This requires handling the membrane at a temperature close to 0°C before the elution starts.

(i) The reproducibility is fairly good. The peak position is identical within ± 1.5°C for totally independent experiments. The results are even more reproducible (± 1°C) for a series of filters prepared at the same time. Filters can be reused up to five times without significant loss of binding capacity. (Filters are kept dry at –20°C for several months.)

(j) This method can be used with chemically modified oligonucleotides α-phosphodiester, phosphorothioate (*Figure 5*), 2'-*O*-alkyl oligoribonucleotides (*Table 1*), or oligos linked to different ligands. A combination of oligos can also be studied: we have monitored the binding of two oligos targeted to two adjacent sequences. The T_c value for oligo 1 (radiolabelled) was increased in the presence of oligo 2 (cold). However, the use of the thermal elution procedure must be restricted to sequences that do not interact with the membrane; we have never obtained good results with methylphosphonate oligonucleosides that give rise to high background.

4.1.3 Binding of chemically modified oligonucleotides

Most antisense oligos used in cultured cells or *in vivo* are chemically modified with the aim of producing more effective antisense activity. However, any modification introduced to improve a given parameter might also modify other properties of the oligo. For example, analogues designed to display nuclease-resistance exhibit altered affinity compared to unmodified oligos. Among the currently available derivatives, 2'-*O*-methyl oligos are the only ones found to yield increased T_m and T_c values with complementary RNA (*Table 1*). In contrast, both phosphorothioate and methylphosphonate:RNA complexes show lower T_m (and T_c), reflecting a lower affinity for the target (these are general trends, as substantial differences can be observed from one sequence to the other). As a consequence, larger amounts of phosphorothioate or methylphosphonate oligonucleosides have to be used to give a similar amount of complex. However the overall biological efficiency results from multiple parameters and not only from the binding constant.

Chemical modifications have been introduced to purposefully increase the affinity of the antisense oligo. An intercalating agent covalently linked to an oligo provides additional binding energy due to stacking interactions between the aromatic ligand and base pairs of the sense:antisense duplex (28). The free energy of binding for such a tethered oligo is roughly (the entropic term is not paid twice) the sum of the free energies for the isolated molecules. Acridine-linked oligos have been reported to induce stronger antisense effects in cell-free extracts and in cultured cells than the parent compound (27, 29). The relative contribution of the intercalating dye vanishes when the length of the oligo increases. No significant improvement is seen for oligos longer than 20 bases.

4.2 Specificity

The association of two complementary strands, via the formation of base pairs, is assumed to be specific if the sequence of the duplex is long enough. At a statistical basis, an AT sequence 19 nt long is unique in the human genome at the DNA level. For a GC sequence the figure is 15 nt (29). The difference between the two numbers originates in the over-representation of AT pairs (60%) in the human genome. If we assume that only 0.5% of DNA is transcribed, uniqueness at the RNA level is reached with 15-mer and 11-mer for AT and GC targets, respectively. The values are significantly different if the frequency of dinucleotides are used for calculation instead of that of mononucleotides. For instance, due to the under-representation of the dinucleotide 5' CpG in the human genome, the statistical occurrence of the decanucleotide GGCATCGTCG is only one on the dinucleotide basis compared to 22 on the mononucleotide basis.

To increase the chances of targeting a unique sequence, one might be tempted to lengthen the antisense oligo. This leads to a result opposite to the expected one, i.e. decreased specificity: longer antisense sequences will generate a larger number of partially complementary sequences. Such long duplexes will tolerate mismatches and mismatched complexes might be a substrate for RNase H (see below).

As stated above, antisense efficiency is determined by the fraction of target sites bound to the antisense sequence. Similarly, specificity is related to the ratio of saturated target sites to non-target sites. Therefore, this will depend on the association constants $K_{a(t)}$ and $K_{a(nt)}$ for target and non-target sequences, respectively, and on the oligo concentration C. Consequently, high values of K_a C product will likely result in non-specific effects. This was clearly demonstrated for an antisense sequence targeted to the point mutated region of the *Ha-ras* gene: a 17-mer allowed discrimination between the mutated and the wild-type sequences whereas a 19-mer did not (24). The thermodynamic cost for any mismatch can be calculated. Unfortunately, most of the time, the number and sequences of non-target sites are unknown. It is difficult therefore to predict the optimal antisense sequence. Nevertheless, the selective tuning of gene expression implies that:

(a) Different oligos have to be designed, depending on the experimental conditions. Shorter sequences should be used in *Xenopus* oocytes grown at 18 °C than in cultured mammalian cells at 37 °C.

(b) The length of the antisense sequence should be shorter for analogues that bind strongly to RNA (2′-*O*-methyl oligoribonucleotide for instance) than for derivatives that have low affinity (phosphorothioate for instance).

(c) The amount of antisense oligo required to saturate the target depends on the target RNA concentration. Therefore, if one can choose between two (or more) different messages, the best choice from this point of view would be the less abundant RNA.

(d) A 'quality' factor should be introduced to complete the picture of specificity. Target sequences are not equivalent for 'physiological' reasons. For instance, inhibition of translation by antisense oligos occurs via two different mechanisms: hybrid-arrest and RNase H-mediated degradation (30, 31). Both mechanisms can take place in the 5′ leader region, upstream of the AUG start site whereas only the second one is efficient downstream. Non-target sites located in the coding region will likely generate limited non-specific effects, in particular with antisense derivatives that do not elicit RNase H (see below). At first, we suggest targeting the 5′ leader region (the cap site or the AUG initiation region are the favourite target sites of many antisense users) although in several instances the most efficient sequence is located outside the 5′ leader. This could be in part related to RNA structures which are known to interfere with antisense oligo binding. Therefore, the 'quality' factor is largely beyond control and will vary from gene to gene. A 'trial and error' method is the best way to identify a good site, but the full scanning of a message is restricted to a few teams working in Biotech companies, but see also Chapter 5.

   Given all these points, how does one make sure that the biological effect induced by an oligo is an antisense one? In other words: what are the appropriate controls? A good control oligo should be as close as possible to the antisense one. A given oligo is characterized by sequence (i.e. length and base composition) and chemistry. This eventually gives rise to internal structure. We recommend using several controls with the same chemistry as that of the antisense sequence, listed below:

(a) The sense sequence. The base composition is not conserved. In addition, this sequence might be a binding site for some factor playing a role in the regulation of the target gene. Therefore the sense sequence should not be used as the only control.

(b) Scrambled sequences. The same base composition is arranged in a different order. In this case structural features are lost.

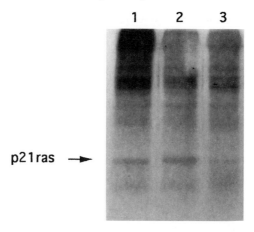

Figure 6. Inhibition of *ras* gene in *Xenopus* oocyte by antisense oligodeoxynucleotides. *Xenopus* oocytes were either non-injected (lane 1) or injected with 20-mers at a final intracellular concentration of about 20 µM (lanes 2, 3). Lane 2: control oligo (5′ ACACC-CAATTCTGAAAATGG) non-complementary to *ras* mRNA. Lane 3: antisense oligo (5′ CGCACTCTTGCCCACACCGA) complementary to the last two nucleotides of the 12th codon and downstream of the activated *Ha-ras* human oncogene. 3 h after injection, oocytes were incubated for 5 h in modified Barth's saline containing [³⁵S]methionine. After homogenization, the p21 *ras* protein was immunoprecipitated with monoclonal antibody Y13–259 and protein A–Sepharose coated with rabbit anti-rat IgG, then electrophoresed and autoradiographed as described in ref. 72. Since this experiment was performed, the sequence of the *Xenopus ras* gene was published (73) revealing that our selected antisense oligo formed a duplex with six interspersed mismatches, with the *Xenopus* mRNA. This illustrates both the efficiency of antisense oligos injected in *Xenopus* oocytes and the limit of specificity.

(c) Mismatched sequences. The introduction of one or several mismatches in the antisense sequence should weaken the binding and, consequently, the amplitude of the antisense effect. The base composition is lost.

(d) Mixed sequences. A mixture of all possible sequences is achieved by the introduction of the four bases at each position. Some sequences are expected to display a strong effect but this will be averaged by the number of different molecules (a mixed 20-mer corresponds to $4^{20} = 10^{12}$ sequences).

(e) The inverted sequence (5′ to 3′). It is the only one which keeps constant the base composition and most of the structural peculiarities. However this cannot be used for palindromic or quasi-palindromic sequences.

4.3 Nuclease sensitivity

4.3.1 DNases

Gene regulation by antisense oligos implies that the inhibitory agent remains intact a long enough time with respect to the time-scale of the process one

61

wants to block, as the amount of [sense:antisense] hybrid depends on the pool of free antisense sequences. Cell growth media containing fetal serum are contaminated with 3′ exonucleases (32). Both endo- and exonucleases are present in cells. Therefore, unmodified oligos are of limited interest because they are rapidly degraded, except in a cell-free extract (33). Numerous analogues have been derived which display resistance to DNases. This includes compounds in which the phosphodiester linkage has been modified: phosphorothioate and methylphosphonate oligonucleosides belong to this class. The use of alpha isomers of the nucleoside units (*Figure 2*) or of analogues in which the 2′ position has been modified also give rise to oligos of increased lifetime. N3′, phosphoramidates and peptide nucleic acids (PNA) are promising nuclease-resistant derivatives (*Figure 2*) which display increased affinity for complementary RNA.

As, at least outside the cell, 3′ exonuclease is the major activity, chemical modification at the 3′ end of an otherwise non-modified oligo substantially increases the stability of the antisense sequence: a 3′ acridine-linked 9-mer was found to have a tenfold increased lifetime in cell growth medium compared to the unmodified parent compound (32). Similarly, a stretch of a few nucleotides resistant to DNases at the 3′ end of the oligo can slow down the degradation of the antisense sequence (34). Such 'capped' oligos allow restriction of the number of modified residues which might be responsible for non-desired effects. In addition, these oligos might also be a means to keep the concentration of modified residues at a low level. One should be aware that, although more resistant to nucleases, chemically modified oligos are degraded. The breakdown products might then be reused by the cell machinery, leading to the incorporation of nucleotide analogues into DNA or RNA. For long-term *in vivo* experiments (and more importantly for therapeutic use), this might be important as, up to now nothing is known about mutagenic or carcinogenic properties of modified derivatives.

An interesting alternative is the use of structures that increase resistance to degradation: short hairpins at the end of the sequence provide unmodified oligos with increased lifetime as double-stranded regions are less prone to nuclease digestion than single-stranded ones. Circular oligos have also been suggested: endless sequences will not be a substrate for exonucleases (35). However, these possibilities require longer oligo sequences which might also be a source of non-specific inhibition, mainly due to RNase H (see below). The major use of such structured oligos might reside in the design of decoys rather than for antisense effects: in the so-called sense approach, a double-stranded sequence is provided to trap a protein such that it can no longer bind to the natural target and therefore exert its physiological function. Dumbell molecules have been suggested to deplete transcription factors (36).

One should also mention here the action of phosphatase, although this activity does not have the same consequence as nuclease on the biological effect of antisense oligos (37). This nevertheless should be taken into account when uptake or compartmentalization studies are carried out. ^{32}P 5′ end-labelling is a convenient method to monitor oligo behaviour in cultured cells. However, one has to check that labelled material actually corresponds to the initial compound and not to the label incorporated into nucleic acids following dephosphorylation. For this purpose the use of internally labelled oligos is recommended.

4.3.2 RNases H

RNases H are ubiquitous enzymes that recognize RNA:DNA hybrids and cleave the RNA moiety (38). The use of unmodified oligos in pioneering work on antisense oligodeoxynucleotide led to the discovery that this activity plays a key role in translation inhibition in cell-free media (wheat germ extract) and in injected *Xenopus* oocytes (31, 39). This is also true for the inhibition of reverse transcription (40). In this process the target RNA is irreversibly eliminated. As long as the oligo remains intact, a single antisense sequence can mediate multiple rounds of destruction of the target. In other words the combination of oligo and RNase H can lead to a pseudo-catalytic control of gene expression. The involvement of RNase H in other eukaryotic cells, suspected for long, on the basis of indirect evidence, was recently demonstrated (41).

Most nuclease-resistant derivatives (methylphosphonates, 2′-*O*-alkyl analogues, alpha isomers) do not allow RNase H activity. Consequently they display a translational inhibitory effect only when targeted to the leader region of mRNA. In contrast, phosphorothioate oligonucleotide:RNA duplexes are substrates for RNase H. At first sight, these latter analogues are good antisense candidates as they combine long lifetime with the ability to elicit RNase H. However, due to the low stringency of RNase H activity this might be a source for non-specific effects. Very short heteroduplexes (four base pairs for the *E. coli* enzyme I) are substrates for RNase H. The human RNase HI is even able to cleave at the level of a single ribo-residue in a DNA context for both strands of the duplex (42). Moreover, mismatches are tolerated by RNase H (31). Therefore, phosphodiester oligos can trigger the cleavage of non-target RNA strands. This is a far more important problem with phosphorothioate analogues which remain intact in the cell for a longer time. This is likely the reason of some non-specific effects of phosphorothioates in *Xenopus* oocytes (43).

Chimeric or 'sandwich' oligos have been designed to preserve the ability to mediate the cleavage of the target site on the one hand and to reduce the activity of RNase H at non-complementary sequences on the other hand. Oligos in which a central window of unmodified phosphodiester nucleotides is placed between two stretches of methylphosphonate residues has been shown to display a higher specificity than the homologous unmodified parent sequence (44, 45). This is due to a reduced ability of RNase H to cleave partially complementary sequences, mainly related to a lower affinity of the sandwich oligo for the perfect and mismatched targets. Therefore, RNase H is one more reason to adapt the oligo concentration and the association constant (i.e. the length and the chemistry of the oligo sequence) in order to saturate a significant fraction of the target RNA and to minimize the binding to non-target sequences.

4.4 Triplex formation

The amount of target sequence determines the amount of antisense oligo required to saturate the target. From this point of view the mini-exon sequence of trypanosomatids is the worst case one can imagine for the antisense strategy as this 39 nt long sequence is present at the 5' end of every message of these parasites (32)! A single sequence complementary to the mini-exon will potentially interfere with the synthesis of all parasite proteins, but this will only be achieved at very high concentration of antisense oligo. The ideal target would be represented only once in the cell; this is never observed at the RNA level even for low abundant transcripts. But this is generally the case if the DNA is targeted. In this case the oligo should be able to bind a double-strand. More than 30 years ago physical studies of homopolynucleotides led to the conclusion that triple-stranded structures can be formed. This gave rise recently to the triplex strategy in which a double-strand formed of purines on one strand and of pyrimidines on the second strand can accommodate a third strand in the major groove (see ref. 46 for a review). The association involves so-called Hoogsteen (or reverse Hoogsteen) hydrogen bonds between purines of the second strand and pyrimidines of the third strand. The two canonical triplets are T•AxT and C•GxC$^+$ (where '•' and 'x' stands for Watson–Crick and Hoogsteen bonding, respectively). Therefore:

(a) Not every DNA sequence is a target as triple-strands require a long enough purine/pyrimidine stretch (typically more than 15 base pairs are required to yield stable triplexes under physiological conditions).

(b) The cytosine should be protonated in the third strand which occurs at acidic pH. This last problem can be circumvented by the use of analogues modified on the C5 position. Alternatively, oligopurines can constitute third strands leading to T•AxA and C•GxG triplets.

A few other possibilities can be exploited, including the use of non-natural bases. The description of these developments is beyond the scope of this chapter, and the use of modified bases is mostly restricted to the team that synthesized them. The reader interested in this strategy should refer to recent reviews of the field (7, 46). Presently, this extremely attractive approach faces the constraints upon the target sequence even though one can now play with sophisticated solutions allowing one to switch from one strand to the second one. Aside from these problems, most of the parameters described for the antisense approach are relevant for the triplex approach: nuclease-resistance, affinity, specificity are grossly similar.

5. *In vitro* use of antisense oligonucleotides

5.1 In cell-free extracts

The use of cell-free extracts to assess the efficiency of oligos aimed at inhibiting translation of a mRNA fulfils several purposes:

(a) Very limited quantities of oligos are needed as these assays are performed in small volumes (typically 30–50 µl). This allows one to screen a number of oligos synthesized on an economical scale (0.1 µmole) among which a subset of the better performing oligos are selected for studies on cultured cells, or for *in vivo* applications, which require large scale synthesis (1 µmole or more). Such screening will allow selection of sites on the mRNA which are sensitive to inhibition, and *prima facie* good targets for *in vivo* experiments, even if some of those will not be sensitive to antisense in real cellular conditions, as extrapolation from *in vitro* conditions is not straightforward.

(b) Oligos directed to the same mRNA sequence, but displaying different chemical modifications can also be assayed. In cell-free extracts, one can assess the inhibitory potential of modified oligos directly on translation without other effect (cellular uptake, subcellular compartmentalization, and metabolic stability, for instance). This does not imply that oligos non-inhibitory to translation are useless, as they may inhibit other processes, such as splicing for example.

(c) Toxicity or non-specificity of oligos can be rapidly evaluated. This can result from poisoning of the translation machinery, or because the target is present on several messenger RNAs, even though a first screening in data bank should retain oligos which are, a priori specific. A general decrease of the translational efficiency can result:

   (i) From the presence of impurities (for example organic compounds which are used during the oligo synthesis). This problem can be easily solved by purifying further, for example by addition of two rounds of ethanol precipitation.

   (ii) From the interaction of the oligo with component(s) of the translation machinery as documented for phosphorothioate oligos (31). A judicious adjustment of the oligo concentration can limit the problem if there is a range of concentration for which the translation of the target mRNA is inhibited without large deleterious effect on general translation. We recommend checking the absence of inhibition of non-target mRNAs.

The complexity of the mRNA population can be varied at will from the pure mRNA (for example produced by *in vitro* transcription of cloned cDNA) to the whole population of mRNAs extracted from the cell under study. A genuine specific antisense oligo will be such that it will decrease the translation of the target mRNA, without effect on the synthesis of other proteins relative to control without oligo. The sensitivity of the analysis can be increased by using two-dimensional electrophoresis instead of the usual SDS–polyacrylamide gel electrophoresis.

Two cell-free protein synthesizing systems are frequently used, the rabbit reticulocyte lysate (RRL) and the wheat germ extract (WGE). The preparation and the use of these systems are described in detail by Clemens (47). Home-made lysates or extracts can prove to be economical if a very large number of assays is planned. However commercial extracts are very convenient and time-saving as they are already optimized for translation efficiency (in practice, only the potassium acetate concentration has to be adjusted by the user as its optimal concentration varies from one message to another). We have routinely used lysates and extracts from Promega, but reduced the final volumes to 25 μl and 30 μl for the WGE and the RRL respectively, in order to economize on mRNA, lysate, and oligos used for these assays.

Although both extracts are convenient, several differences, which might be important for antisense experiments, have to be stressed:

(a) The translational efficiency of the RRL is generally higher than that of the WGE whereas the RRL, less prone to premature termination, is generally more suitable for the translation of long mRNAs.

(b) Bands of low molecular weight proteins are distorted on polyacrylamide gels (in the 25–15 kDa range) due to the large amount of endogenous globins present in the rabbit reticulocyte (the protein concentration is around four times higher in the RRL than in the WGE), so that for short mRNAs the latter extract is preferred.

(c) The recommended temperature is 25°C and 30°C for the WGE and the RRL, respectively. If a low temperature is expected to favour hybridization of the oligo to its target, it will also stabilize secondary structures on both the mRNA and the oligo. A low temperature will also promote the binding of the antisense sequence to partially complementary sequences present on non-target mRNAs (see Section 4.2).

(d) Finally, the participation of RNase H in the inhibition of translation greatly differs between the two systems: although both contain endogenous RNase H, this does not contribute (or only marginally contributes) to the action of antisense oligos in the RRL, whereas it plays a crucial role in the WGE (39, 48). The ultimate reason for these differences of RNase H activity under conditions of translation may possibly be related to the nature of these enzymes (49). Consequently, oligos which allow RNase H action (e.g. phosphodiester, phosphorothioate) will be inhibitory in the WGE irrespective of their target on the mRNA by inducing its cleavage. In the RRL, due to the lack of RNase H action, oligos will generally be inhibitory only if targeted to the initiation codon or upstream sequences (30, 50). However, low levels of RNaseH can be occasionally detected in some RRL batches. Hybridization to the coding sequence is ineffective due to the presence of an unwinding activity associated with the translating ribosome (51, 52). However, cleavage can be induced by adding *E. coli* RNase H to the extract (39). Alternatively, oligos that trigger a permanent modification of the RNA can be used (53).

A comparison of results obtained in the RRL and in the WGE can give information on the mechanism by which antisense oligos interfere with the translation process.

5.2 In injected *Xenopus* oocytes

Microinjected *Xenopus* oocytes are another useful system for assessing the potency of oligos directed to selectively inhibit the translation of a target mRNA. This allows the evaluation of oligos directed inside a genuine cell, the oocyte serving simply as a living test-tube, circumventing problems of uptake, as the oligo as well as the target mRNA can be injected inside the cytoplasm. However, the metabolic stability of oligos is an important factor for the final result. For example, phosphorothioates, which are nuclease-resistant, can be specific inhibitors at a concentration ten times lower than their phosphodiester equivalent (31, 54), the latter being extremely rapidly degraded in the oocyte (48, 55). Moreover, *Xenopus* oocytes allow the translation to be pursued for several hours, or even days, at rates which are several orders of magnitude higher than those of cell-free systems, so that enough protein can accumulate to be detected by any suitable bioassay (56). This allows a more stringent test of the effect of the antisense oligo upon translation inhibition. Several procedures for microinjection, some with a sophisticated set-up, can be found in the literature. We especially recommend those of Colman (57).

Antisense oligos induce RNase H-mediated degradation of both exogenous microinjected mRNAs and endogenous mRNAs either stored or actively translated. This has been used to deplete oocytes of some of its mRNAs and to examine the consequence of this ablation for example on oocyte maturation (58) or on the embryo development (59). Concentrations effective to get a nearly complete ablation appear to be quite variable (internal concentration, calculated with the assumption that the diffusion-free compartment of a fully grown oocyte is about 0.5 μl, ranges from 0.05–20 μM). This is very likely related to target accessibility, due to local secondary structure and to competition with endogenous proteins, and to RNA target abundance. In contrast to cell-free extracts, oligos with modifications which prevent RNase H action (methylphosphonates or alpha oligonucleotides) are not very effective even when targeted to the leader part of the message, although an alpha oligo linked to an acridine derivative was found effective when targeted to the very 5' end of the message (30). Non-specific effects can arise from hybridization of antisense oligos with sequences showing partial complementarity: non-target RNAs can be destroyed by RNase H (a class I enzyme, present in the cytoplasm corresponds to about 5% of the total activity (60). It requires four to six contiguous paired nucleotides for cleavage) (61). So an effect can only be ascribed to a given mRNA:

- if different oligos targeted to different regions of this RNA give a similar result
- if several control oligos (see Section 4.2) do not lead to this phenotype

5.3 In cultured cells

Early attempts on cultured cells were reported in 1978 by Zamecnick and Stephenson (3). In 1987, several investigators described the inhibition of the *myc* oncogene by antisense oligos (62). Since that time, a very large number of attempts have been reported on a variety of cell lines and genes including viral genes. Similar procedures were used for these experiments: the oligo is added to the culture medium bathing the cells and the effect of the oligo is then monitored at the mRNA and protein levels encoded by the target gene, together with relevant parameters of the cell physiology. Here again, several antisense oligos and controls have to be employed before a conclusion can be reached.

First, the problem of nuclease degradation in the extracellular medium has to be solved (see Section 4.3) as:

(a) This will lower the amount of oligo available for antisensing and so will either necessitate the use of very high amounts of oligos or frequent changes of medium with fresh intact oligos.

(b) The degradation products can have adventitious effects on cell growth or lead to artefacts (for example, affecting the thymidine pool which will lead to erroneous determination of cell proliferation based on tritiated thymidine incorporation).

This can be achieved:

(a) By using nuclease-resistant oligos, such as phosphorothioates, 2'-O-methyl oligoribonucleotides, or chimeric oligos containing a stretch of nuclease-resistant analogues at their 3' end.

(b) By the increased inactivation of exonucleases present in the serum. Typically, this can be done by incubating the serum for 30 min at 65°C (instead of 56°C, usually used to inactivate the complement).

The problem of intracellular degradation of oligos appears to be quite variable between different cell lines. For cells having very active nucleolytic activities the use of nuclease-resistant oligos remains the solution.

Uptake of oligos by cells is an active process mediated in part by one (or several?) receptors. The rate of uptake appears to be quite variable between cells but, as a general rule, established cell lines internalize oligos much more efficiently than primary cell cultures. For example uptake is more efficient for spleen B cells than for T cells, and the uptake in pre-B cells and pre-myeloid cells is high compared to propre-B cells and pre-T cells (63). Although uptake is heterogeneous, it can be induced by mitogens. Concerning cell lines, if uptake is generally high for rapidly growing cells, culture at high density or under serum starvation can severely decrease the efficiency of uptake.

Uptake can be divided into several stages:

(a) Adsorption on the cell surface. Studies with fluorescently labelled oligos have indicated that fixation of oligos on cells incubated on ice, to prevent receptor-mediated endocytosis, is maximum for phosphorothioates and minimum for oligos containing methylphosphonate 'caps' at their ends, whereas unmodified oligos as well as oligos capped with phosphorothioates are in between (64).

(b) Internalization by an endocytic process. Oligos accumulate until a plateau is reached, typically after 1–2 h at 37 °C, for unmodified oligos. Phosphorothioates appear to follow the same pathway, as deduced from competition experiments, but with slower kinetics and with a higher final concentration. Several lines of evidence (effects induced by lysosomotropic agents such as chloroquine, observation of punctate patterns in cells when fluorescent oligos are used, etc.) indicate that a great amount of oligo is concentrated in endosomal particles from which some escapes to reach the cytoplasm, and/or the nucleus, where it might play its expected antisense role (65).

Methylphosphonate oligos, which are uncharged, enter cells via an active process too, but along another route, which seems to correspond to non-specific endocytosis, either fluid phase, or adsorptive endocytosis (66).

Several means aimed at increasing cell uptake have been reported: encapsidation into liposomes (67), adsorption onto nanoparticles (68), covalent linkage to a polylysine tail (69), or covalent linkage of the oligo to a lipophilic moiety such as cholesterol (70) or palmitate (71). In these last two cases, the oligo linked to the lipid can be associated with low density lipoproteins (LDL) and the resulting complex can be internalized via LDL receptor-mediated endocytosis. We have recently used this property to enhance the uptake of a phosphorothioate oligo by macrophages infected by *Leishmania*, a protozoan parasite (see detailed procedure in *Protocol 7*). The oligo being complementary to a sequence of mini-exon present on all messenger RNAs of this parasite, the treatment was able to cure macrophages, or at least to significantly decrease the number of parasites per infected cell. An oligo linked to a palmitate chain has been found to be much more efficient than its non-conjugated counterpart. Uptake was greatly increased by using native LDL; it can be further improved by using oxidized LDL which associates with scavenger receptors whose endocytosis is not regulated by internal LDL concentration.

Protocol 7. Preparation of complexes between LDL and palmitate–oligonucleotide conjugates

Equipment and reagents

- Oligo of interest
- Carbodiimidazole
- Dioxane
- Ethylene diamine
- Acetonitrile
- Palmitic anhydride

- Pyridine
- Glutathione
- Caproic acid
- Chloramphenicol
- Phosphate-buffered saline (PBS)
- Oligo synthesizer

A. *Synthesis of a 5' palmityl oligonucleotide*

1. Synthesize the oligo of interest as usual (see for instance *Protocol 1*), including detrytilation at the last step.

2. Remove the cartridge from the synthesizer and dry extensively under vacuum. Complete the synthesis manually on the column with the help of two 1 ml (insulin-type) syringes, one filled with the solution to be injected on the column. Connect both to each extremity of the column. Transfer and mix the reactants by alternative use of the plungers.

3. Add to the column 1 ml of a carbodiimidazole solution (60 mg/ml in dry dioxane) and incubate for 1 h at room temperature with occasional mixing.

4. Withdraw the solution and wash five times with 1 ml dioxane. Dry the column in vacuum.

5. Add ethylene diamine (25 mg/ml dioxane:water, 9:1) and incubate for 2 h at room temperature, with occasional mixing.

6. Wash two times with 1 ml dioxane, then twice with 1 ml methanol, and finally twice with 1 ml acetonitrile. Dry in vacuum overnight.

7. Add a saturated solution of palmitic anhydride in pyridine (100 mg/ml); incubate for 3 h at 40°C (to avoid precipitation of the anhydride) with occasional mixing.

8. Withdraw the solution; wash five times with 1 ml pyridine, and twice with 1 ml acetonitrile.

9. Deprotect and remove the oligo from support as described in *Protocol 1*.

10. Analyse and purify the oligo dissolved in water as described in *Protocol 2*. This protocol can be used as palmitate exhibits hydrophobic properties similar to the DMT group. Palmityl–oligonucleotide conjugate is eluted with a much longer retention time than the parent compound (typically 41 min compared to 22 min, respectively).

B. *Preparation of low density lipoprotein (LDL)*

Commercially available LDL (for example from Sigma) are typically partially oxidized. If native LDL are needed, they can be prepared from
human blood.

1. Collect 35 ml of blood in EDTA tubes (tubes containing EDTA as anti-
   coagulant), after lunch if possible. Centrifuge the tubes at 2000 *g* for
   15 min at room temperature. Collect the plasma under sterile
   conditions with a pipette and pool it in a sterile 100 ml graduated
   cylinder.

2. Add 1% (v/v) 'Allopovic' preservative solution[a] and then 3.43 g KBr for
   every 100 ml of plasma. This brings the density up to *d* = 1.030.

3. Transfer the plasma to ultracentrifuge tubes, cleaned with 95%
   ethanol, and dried in a sterile cabinet on a layer of blotting paper
   prior to use. Centrifuge at 4°C for 20 h at 100 000 *g*.

4. Carefully remove, with a sterile pipette, the upper part (containing
   chylomicrons and KBr) by a slow circular movement along the tube
   walls. Then collect the lower part (which contains HDL and LDL)
   taking care not to disturb the pellet of fibrinogen, to a sterile 50 ml
   graduated cylinder.

5. Add 1% of the 'Allopovic' preservative solution,[a] then 3.43 g of KBr
   for every 100 ml. The density is now *d* = 1.053. Proceed as in step 3.

6. Collect carefully LDL at the upper part of the tubes and transfer to a
   sterile 25 ml graduated cylinder. Add 1% 'Allopovic' solution.[a]

7. Dialyse against sterile PBS containing 20 μM butylated hydroxy-
   toluene,[b] with four changes for 24 h at 4°C.

8. Collect LDL and transfer to sterile ampules (example: vacule®
   Wheaton of 5 ml). Apply a brief stream of sterile nitrogen, and seal
   the ampules. (They can be stored for up to two weeks at 4°C.)

9. Determine the LDL concentration by protein measurement.

C. *Preparation of palmityl modified oligonucleotide–LDL complexes*

1. Mix LDL with the oligo conjugate in PBS (pH 7.4) containing 2 mM
   EDTA at final concentration of 5 mg/ml (that is 10 μM if we assume a
   M_r of ≈ 500 kDa for the apolipoprotein) and of 100 μM respectively.
   Incubate for 2 h at 37°C.

2. Add LDL loaded with oligo to the culture medium at the desired final
   concentration, usually in the range of 0–1 μM LDL. If a small part of
   oligo has remained free, it will become associated to LDL contained
   in the serum present in the culture medium.

[a] Allopovic solution. Solution A: dissolve 0.5 g glutathione, 1.3 g caproic acid, and 0.25 g EDTA
in 9 ml sterile water. Solution B: 0.02 g chloramphenicol in 1 ml ethanol. Mix A and B. This

6

###.

###########

Jean-Jacques Toulmé et al.

Protocol 7. *Continued*

solution cannot be stored for long, and is prepared and used just at the time of LDL preparation. Keep at 4°C.
[b] This antioxidant should not be added if LDL are intended to be oxidized.

Acknowledgements

We are grateful to Dr E. Saison (INSERM U 201, Paris) for sharing unpublished results. Our work was supported by the Conseil Régional d'Aquitaine, by the Association pour la Recherche contre le Cancer, and by the Direction des Recherches, Etudes et Techniques.

References

1. Belikova, A. M., Zarytova, V. F., and Grineva, N. I. (1967). *Tetrahedron Lett.*, **37**, 3557.
2. Miller, P. S., Braiterman, L. T., and Ts'o, P. O. P. (1977). *Biochemistry*, **16**, 1988.
3. Zamecnik, P. C. and Stephenson, M. L. (1978). *Proc. Natl. Acad. Sci. USA*, **75**, 280.
4. Offensperger, W. B., Offensperger, S., Walter, E., Teubner, K., Igloi, G., Blum, H. E., *et al.* (1993). *EMBO J.*, **12**, 1257.
5. Skorski, T., Perrotti, D., Nieborowska-Skorska, M., Gryaznov, S., and Calabretta, B. (1997). Antileukemia effect of c-myc N3'->P5' phosphoramidate antisense oligonucleotides *in vivo. Proc. Natl. Acad. Sci. USA*, **94**, 3966.
6. Zon, G. (1995). Brief overview of control of genetic expression by antisense oligonucleotides and *in vivo* applications—prospects for neurobiology. *Mol. Neurobiol.*, **10**, 219.
7. Hélène, C. and Toulmé, J. J. (1990). *Biochim. Biophys. Acta*, **1049**, 99.
8. Demesmaeker, A., Haner, R., Martin, P., and Moser, H. E. (1995). Antisense oligonucleotides. *Account. Chem. Res.*, **28**, 366.
9. Cohen, J. S. (1989). *Oligodeoxynucleotides: antisense inhibitors of gene expression*, p. 255. Macmillan Press, London.
10. Murray, J. A. H. (1992). *Antisense RNA and DNA*, p. 401. Wiley-Liss, New York.
11. Crooke, S. T. and Lebleu, B. (1993). *Antisense research and applications*, p. 579. CRC, Boca Raton, FL.
12. Goodchild, J. (1990). *Bioconjugate Chem.*, **1**, 165.
13. Eckstein, F. (ed.) (1991). *Oligonucleotides and analogues: a practical approach*. IRL Press, Oxford.
14. Beaucage, S. L. and Iyer, R. P. (1993). *Tetrahedron*, **49**, 6123.
15. Beaucage, S. L. and Caruthers, M. H. (1981). *Tetrahedron Lett.*, **22**, 1859.
16. Puglisi, J. D. and Tinoco Jr., I. (1989). Absorbance melting curves of RNA. In *Methods in enzymology* (ed. J. N. Abelson and M. I. Simon), Vol. 180, pp. 304–25. Academic Press, New York.
17. Stein, C. A. and Cheng, Y. C. (1993). Antisense oligonucleotides as therapeutic agents—is the bullet really magical. *Science*, **261**, 1004.
18. Vu, H. and Hirschbein, B. L. (1991). *Tetrahedron Lett.*, **32**, 3005.

19. Iyer, R. P., Egan, W., Regan, J. B., and Beaucage, S. L. (1990). *J. Am. Chem. Soc.*, **112**, 1253.
20. Miller, P. S., Yano, J., Yano, E., Carroll, C., Jayaraman, K., and Ts'o, P. O. P. (1979). *Biochemistry*, **18**, 5134.
21. Hogrefe, R. I., Vaghefi, M. M., Reynolds, M. A., Young, K. M., and Arnold, L. J. (1993). *Nucleic Acids Res.*, **21**, 2031.
22. Inoue, H., Hayase, Y., Asaka, M., Imura, A., Iwai, S., Miura, K., *et al.* (1985). *Nucleic Acids Res. Symposium Series*, 165–8.
23. Breslauer, K. J. (1994). In *Protocols for oligonucleotide conjugates* (ed. S. Agrawal), pp. 347–72. Humana Press, Totowa.
24. Freier, S. (1993). In *Antisense research and applications* (ed. B. Lebleu and S. T. Crooke), pp. 67–82. CRC, Boca Raton.
25. Porumb, H., Verspieren, P., Rayner, B., Imbach, J. L., Malvy, C., and Toulmé, J. J. (1992). *Methods Mol. Cell. Biol.*, **1**, 10.
26. Verspieren, P., Loreau, N., Thuong, N. T., Shire, D., and Toulmé, J. J. (1990). *Nucleic Acids Res.*, **18**, 4711.
27. Toulmé, J. J., Krisch, H. M., Loreau, N., Thuong, N. T., and Hélène, C. (1986). *Proc. Natl. Acad. Sci. USA*, **83**, 1227.
28. Asseline, U., Thuong, N. T., and Hélène, C. (1983). *C R. Acad. Sci. Paris*, **297 (III)**, 369.
29. Hélène, C. and Toulmé, J. J. (1989). In *Oligodeoxynucleotides: antisense inhibitors of gene expression* (ed. J. S. Cohen), pp. 137–72. Macmillan Press, London.
30. Boiziau, C., Kurfurst, R., Cazenave, C., Roig, V., Thuong, N. T., and Toulmé, J. J. (1991). *Nucleic Acids Res.*, **19**, 1113.
31. Cazenave, C., Stein, C. A., Loreau, N., Thuong, N. T., Neckers, L. M., Subasinghe, C., *et al.* (1989). *Nucleic Acids Res.*, **17**, 4255.
32. Verspieren, P., Cornelissen, A. W. C. A., Thuong, N. T., Hélène, C., and Toulmé, J. J. (1987). *Gene*, **61**, 307.
33. Morvan, F., Porumb, H., Degols, G., Lefebvre, I., Pompon, A., Sproat, B. S., *et al.* (1993). *J. Med. Chem.*, **36**, 280.
34. Tidd, D. M. and Warenius, H. M. (1989). *Br. J. Cancer*, **60**, 343.
35. Dolinnaya, N. G., Blumenfeld, M., Merenkova, I. N., Oretskaya, T. S., Krynetskaya, N. F., Ivanovskaya, M. G., *et al.* (1993). *Nucleic Acids Res.*, **21**, 5403.
36. Clusel, C., Ugarte, E., Enjolras, N., Vasseur, M., and Blumenfeld, M. (1993). *Nucleic Acids Res.*, **21**, 3405.
37. Boiziau, C. and Toulmé, J. J. (1991). *Biochimie*, **73**, 1403.
38. Crouch, R. J. and Toulmé, J.-J. (1997). *Ribonucleases H*. INSERM, John Libbey, Paris, in press.
39. Minshull, J. and Hunt, T. (1986). *Nucleic Acids Res.*, **14**, 6433.
40. Boiziau, C., Thuong, N. T., and Toulmé, J. J. (1992). *Proc. Natl. Acad. Sci. USA*, **89**, 768.
41. Giles, R. V., Spiller, D. G., and Tidd, D. M. (1995). Detection of ribonuclease H-generated mRNA fragments in human leukemia cells following reversible membrane permeabilization in the presence of antisense oligodeoxynucleotides. *Antisense Research Development*, **5**, 23.
42. Eder, P. S. and Walder, J. A. (1991). *J. Biol. Chem.*, **266**, 6472.
43. Woolf, T. M., Melton, D. A., and Jennings, C. G. B. (1992). *Proc. Natl. Acad. Sci. USA*, **89**, 7305.
44. Giles, R. V. and Tidd, D. M. (1992). *Nucleic Acids Res.*, **20**, 763.

45. Larrouy, B., Blonski, C., Boiziau, C., Stuer, M., Moreau, S., Shire, D., *et al.* (1992). *Gene*, **121**, 189.
46. Thuong, N. T. and Hélène, C. (1993). *Angew. Chem. Int. Ed. Engl.*, **32**, 666.
47. Clemens, M. J. (1984). In *Transcription and translation: a practical approach* (ed. B. D. Hames and S. J. Higgins), pp. 231–70. IRL Press, Oxford.
48. Cazenave, C., Loreau, N., Thuong, N. T., Toulmé, J. J., and Hélène, C. (1987). *Nucleic Acids Res.*, **15**, 4717.
49. Cazenave, C., Frank, P., and Büsen, W. (1993). *Biochimie*, **75**, 113.
50. Bertrand, J. R., Imbach, J. L., Paoletti, C., and Malvy, C. (1989). *Biochem. Biophys. Res. Commun.*, **164**, 311.
51. Liebhaber, S. A., Cash, F. E., and Shakin, S. H. (1984). *J. Biol. Chem.*, **259**, 15597.
52. Shakin, S. H. and Liebhaber, S. A. (1986). *J. Biol. Chem.*, **261**, 16018.
53. Kean, J. M., Murakami, A., Blake, K. R., Cushman, C. D., and Miller, P. S. (1988). *Biochemistry*, **27**, 9113.
54. Fakler, B., Herlitze, S., Amthor, B., Zenner, H. P., and Ruppersberg, J. P. (1994). *J. Biol. Chem.*, **269**, 16187.
55. Woolf, T. M., Jennings, G. B., Rebagliati, M., and Melton, D. A. (1990). *Nucleic Acids Res.*, **18**, 1763.
56. Soreq, H. (1985). *CRC Crit. Rev. Biochem.*, **18**, 199.
57. Colman, A. (1984). In *Transcription and translation: a practical approach* (ed. B. D. Hames and S. J. Higgins), pp. 271–302. IRL Press, Oxford.
58. Sagata, N., Oskarsson, M., Copeland, T., Brumbaugh, G., and Vande Woude, G. F. (1988). *Nature*, **335**, 519.
59. El-Baradi, T., Bouwmeester, T., Giltay, R., and Pieler, T, (1992). *EMBO J.*, **10**, 1407.
60. Cazenave, C., Frank, P., Toulmé, J. J., and Büsen, W. (1994). *J. Biol. Chem.*, **269**, 25185.
61. Dagle, J. M., Walder, J. A., and Weeks, D. L. (1990). *Nucleic Acids Res.*, **18**, 4751.
62. Heikkila, R., Schwab, G., Wickstrom, E., Loke, S. L., Pluznik, D. H., Watt, R., *et al.* (1987). *Nature*, **328**, 445.
63. Krieg, A. M., Gmelig-Meyling, F., Gourley, M. F., Kisch, W. J., Chrisey, L. A., and Steinberg, A. D. (1991). *Antisense Research Development*, **1**, 161.
64. Krieg, A. M. (1993). *Clin. Chem.*, **39**, 710.
65. Léonetti, J. P., Mechti, N., Degols, G., Gagnor, C., and Lebleu, B. (1991). Intracellular distribution of microinjected antisense oligonucleotides. *Proc. Natl. Acad. Sci. USA*, **88**, 2702.
66. Shoji, Y., Akhtar, S., Periasamy, A., Herman, B., and Juliano, R. L. (1991). *Nucleic Acids Res.*, **19**, 5543.
67. Thierry, A. R. and Dritschilo, A. (1992). *Nucleic Acids Res.*, **20**, 5691.
68. Chavany, C., Le Doan, T., Couvreur, P., Puisieux, F., and Helene, C. (1992). *Pharm. Res.*, **9**, 441.
69. Lemaître, M., Bayard, B., and Lebleu, B. (1987). *Proc. Natl. Acad. Sci. USA*, **84**, 648.
70. Boujrad, N., Hudson, J. R., and Papadopoulos, V. (1993). *Proc. Natl. Acad. Sci. USA*, **90**, 5728.
71. Ramazeilles, C., Mishra, K. R., Moreau, S., Pascolo, E., and Toulmé, J. J. (1994). *Proc. Natl. Acad. Sci. USA*, **91**, 7859.
72. Saison-Behmoaras, T., Tocque, B., Rey, I., Chassignol, M., Thuong, N. T., and Hélène, C. (1991). *EMBO J.*, **10**, 1111.
73. Andeol, Y., Gusse, M., and Mechali, M. (1990). *Dev. Biol.*, **139**, 24.

4

In vitro RNAs

UTE WEBER and HANS J. GROSS

1. Introduction

RNA can be synthesized *in vitro* in two different ways which complement each other. Chemical synthesis yields oligoribonucleotides of a maximal length of 40 nucleotides, in contrast to the 100 to 150 nucleotides possible for oligodeoxyribonucleotides (Chapter 3). Nevertheless, the chemical approach does not impose any restrictions on the nucleotide sequence, including the use of modified building blocks.

Here we provide methods for the second approach, enzymatic RNA synthesis, which allows the production of RNA ranging in length from 12 to thousands of nucleotides with almost no sequence restrictions. Different *in vitro* transcription systems are available, using phage DNA-dependent RNA polymerases. Most common are the SP6 and T7 RNA polymerase-based systems. The T7 RNA polymerase has some advantages over the SP6 enzyme: it accepts a single-stranded DNA template with only the 18 bp promoter region double-stranded and it can be prepared in a very rapid and efficient way.

Large amounts of RNA (milligram yields) can easily be synthesized by the standard procedure of *in vitro* transcription. The *in vitro* synthesized RNA can be introduced into cell cultures and by microinjection into cells analogous to the procedures used for oligodeoxyribonucleotides.

In this chapter we present a simple procedure for the purification of high yields of T7 RNA polymerase, an *in vitro* transcription protocol including cloning strategies for the template DNA, and a survey of modified transcripts and their applications.

2. Rapid purification of T7 RNA polymerase

T7 RNA polymerase is available from several commercial sources, however, it can also be purified from *E. coli* BL21/pAR1219, the expression system of Studier (1). The original procedure of Grodberg and Dunn (2) involved growth of the cells, induction of T7 RNA polymerase synthesis by addition of isopropyl-β-D-thiogalactopyranoside (IPTG), collection and lysis of the cells, followed by polymine P precipitation, ammonium sulfate precipitation, dialysis, and chromatography on three different ion exchange columns.

The purification scheme described in *Protocols 1–4* (3) allows the purification of at least 10 mg (about 4 000 000 U) of highly active, pure T7 RNA polymerase per 1 g of *E. coli* cells (net weight) within three days. The purified enzyme is much more concentrated than commercially available enzyme. For some applications this is very important.

Protocols 1–3 are essentially as described by Grodberg and Dunn (2), with only minor modifications, and *Protocols 4* and *5* are according to Zawadzki and Gross (3).

Protocol 1. Growth medium, cell culture, and induction of expression

Equipment and reagents

- *E. coli* BL21/pAR1219
- Refrigerated centrifuge with large bucket rotor (e.g. Sorvall GS-3)
- 37 °C shaking incubator
- Bactotryptone
- IPTG

Method

1. Sterilize 1 litre of double distilled water containing 10 g Bactotryptone and 5 g NaCl.

2. Sterilize 50 ml of double distilled water containing 1.5 g KH_2PO_4, 3 g Na_2HPO_4, and 0.5 g NH_4Cl (M9 salts).

3. Sterilize 20 ml of 20% glucose.

4. Mix the components. Add 1 ml of sterile 1 M $MgSO_4$ and 40 mg ampicillin.

Two 500 ml cultures of *E. coli* BL21/pAR1219 are grown in 1 litre flasks, shaking at 37 °C in tryptone broth supplemented with M9 salts, 0.4% glucose, and 40 μg/ml ampicillin.

When the cultures have reached an A_{600} of 0.5, IPTG is added to a final concentration of 0.5 mM in order to induce the synthesis and accumulation of T7 RNA polymerase in the cells. After 4 h of induction, the cells are collected by centrifugation (Sorvall GS-3 rotor) at 10 000 g for 10 min, and washed twice with 100 ml cold 20 mM Tris–HCl pH 8.1, 20 mM NaCl, 2 mM Na_2 EDTA. The packed cell pellet, usually about 8 g per litre of culture, can be stored for many months at –80 °C.

Protocol 2. Cell lysis

Equipment and reagents
- Buffers as listed in *Table 1*
- Lysozyme

- Deoxycholate
- Sonicator

Method

All steps are performed at 0 to 4°C.

1. Resuspend the fresh or frozen cells (8 g) in 24 ml cold buffer B.

2. Initiate lysis by addition of a freshly prepared solution of 9 mg egg white lysozyme in 6 ml of the same buffer. Shake gently for 20 min.

3. Add 2.5 ml of 0.8% sodium deoxycholate. Allow the mixture to stand for another 20 min to complete lysis.

4. Reduce the viscosity of the solution by controlled sonication in an ice-bath (50 W, three times, 5 sec).

5. Add 5 ml 2 M ammonium sulfate and bring the lysate to 50 ml with buffer B.

In the next step, the lysate is fractionated by polymine P precipitation. T7 polymerase is precipitated from the supernatant with ammonium sulfate, and desalted by dialysis.

Table 1. Buffers for the purification of T7 RNA polymerase

Buffer A	20 mM	Tris–HCl pH 8.1
	20 mM	NaCl
	2 mM	Na_2 EDTA
	1 mM	dithiothreitol
Buffer B	Buffer A with:	
	0.19 mM	phenylmethylsulfonyl fluoride (PMSF)
	0.1 mM	benzamidine
	10 μg/ml	bacitracin
Buffer C	20 mM	sodium phosphate pH 7.7
	1 mM	Na_2 EDTA
	1 mM	dithiothreitol
	5%	glycerol
	20 μg/ml	PMSF
Dilution buffer	10 mM	$K_nH_mPO_4$ pH 7.9 (prepare 0.5 M stock solutions of KH_2PO_4 and K_2HPO_4, mix, and adjust the pH to 7.9 by the ratio of one solution to the other)
	0.1 mM	Na_2 EDTA
	200 mM	KCl
	5 mM	dithiothreitol
	50%	glycerol

Protocol 3. Precipitation and dialysis of crude polymerase

Equipment and reagents

- Buffers as listed in *Table 1*
- Refrigerated centrifuge with fixed angle rotor (e.g. Sorvall SS-34)
- Polymine P (polyethylenimine; Sigma)
- Dialysis tubes

Method

1. Add 5 ml of a 10% solution of polymine P (adjusted to pH 8.0 with 32% HCl) slowly while stirring on ice.

2. Remove the precipitate after 20 min by centrifugation for 15 min at 39 000 *g* (Sorvall SS-34 rotor) and discard it.

3. Mix the supernatant slowly on ice with 0.82 vol. of a saturated solution of ammonium sulfate (saturated at room temperature, adjusted to pH 7.0 with Tris base, passed through a 0.45 μm filter, chilled to 4°C).

4. Stir gently on ice for 15 min. Collect the precipitate by centrifugation (10 min, 12 000 *g*), and dissolve in 15 ml of buffer C containing 100 mM NaCl.

5. Dialyse overnight against two changes of 1 litre 100 mM NaCl in buffer C. Remove the precipitate by centrifugation (see step 4) and discard it.

The supernatant from the dialysis procedure is prepared for column chromatography and further purified by passage over S-Sepharose FF.

Protocol 4. S-Sepharose chromatography and specific precipitation of polymerase

Equipment and reagents

- Buffers as listed in *Table 1*
- S-Sepharose FF (Pharmacia)
- Conductometer
- Equipment for SDS–polyacrylamide gel electrophoresis
- Centricon concentrator (Amicon)
- Dialysis tubes
- Refrigerated centrifuge with fixed angle rotor (e.g. Sorvall SS-34)

Method

1. Dilute supernatant from dialysis with buffer C (without NaCl, about 16 ml) until it has the conductivity of 50 mM NaCl in buffer C.

2. Apply the solution (about 35 ml) onto a 1.5 cm (i.d.) × 10 cm (height) column of S-Sepharose FF (Pharmacia) equilibrated with 50 mM NaCl in buffer C.

3. Wash the column extensively with at least ten column volumes of the

---

same buffer to remove even traces of non-binding proteins. The flow rate should be around 4–5 drops/min.

4. Elute T7 RNA polymerase with 200 mM NaCl in buffer C and collect fractions of about 1 ml.

5. Analyse fractions by SDS–10% polyacrylamide gel electrophoresis and staining with Coomassie Brilliant Blue R-250 (Serva Blue R).

6. Pool only the three fractions with the highest protein, i.e. T7 RNA polymerase concentration (fractions 11–13 in *Figure 1*). This is important! If other fractions with lower T7 RNA polymerase content are added (e.g. fraction 14 in *Figure 1*), the precipitation of the enzyme will not occur or will be incomplete and thus reduce the overall yield. If the enzyme does not precipitate or if a higher enzyme yield is needed, more fractions may be pooled but have to be concentrated with a Centricon concentrator (Amicon) before dialysis (below).

7. Dialyse the pooled fractions against 2 × 500 ml buffer C containing 10 mM NaCl for at least 15 h at 4°C.

8. Collect the precipitated T7 RNA polymerase by centrifugation at 10 000 g for 5 min.

9. Dissolve the enzyme precipitate in 10 ml 100 mM NaCl in buffer C.

10. Dialyse overnight against 900 ml 100 mM NaCl in buffer C containing 50% glycerol.

11. Analyse the resulting dialysate (about 3 ml) by SDS–polyacrylamide electrophoresis (fractions a and b, *Figure 1*).

The enzyme has a specific activity of about 400 000 U/mg and is stable for years if stored in aliquots at –20°C. We recommend preparing appropriate dilutions (e.g. 1 mg/ml) of the enzyme using T7 RNA polymerase dilution buffer (see *Table 1*). The purified enzyme and the diluted enzyme aliquots should be supplemented every 6–12 months with dithiothreitol to a final concentration of 5 mM to avoid loss of enzymatic activity by oxidation (4).

3. Template DNA

3.1 Essential features

The template DNA sequence has to be placed directly behind the T7 promoter consensus sequence, which comprises the nucleotides in position –17 to –1 TAATACGACTCACTATA and the +1 to +6 nucleotides of the transcript GGGAGA. The first three guanosines are most important for the efficiency of transcription, whereas a change of the nucleotides +4 to +6 has only minor effects. At least the very first guanosine is essential, an exchange of G at the second or third position is less detrimental for transcription efficiency. T7 RNA polymerase prefers the nucleotides G > A > C>>> U for initiation

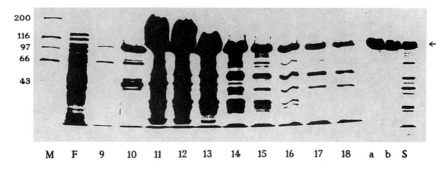

Figure 1. Analysis of T7 RNA polymerase fractions by SDS–10% polyacrylamide gel electrophoresis. M, molecular weight markers (kDa); F to 18, S-Sepharose FF chromatography fractions; F, flow-through; 9 to 18, fraction numbers. 20 μl aliquots were analysed. a, 140 μg; b, 90 μg purified T7 RNA polymerase as obtained upon specific precipitation (see *Protocol 4*). S, supernatant after dialysis against 10 mM NaCl in buffer C. Staining was with Coomassie Brilliant Blue R-250. The *arrow* indicates the position of T7 RNA polymerase.

of transcription (4). Furthermore it has to be considered that A:U base pairs in the initiation sequence lead to low yields of transcripts, no matter whether there are guanosines in the first positions. But even with the optimal initiation sequence the transcripts have to be purified because of the presence of abortive initiation products.

The enzymatic reaction will be terminated at the physical end of the DNA template. A suitable restriction endonuclease recognition site has therefore to be placed at the 3' terminus to ensure run-off transcription with a high yield of the desired RNA after digestion of the template DNA.

3.2 Synthetic templates

If the transcript has to be very short (12 to about 50 nucleotides), we recommend using the template system established by Uhlenbeck and co-workers (5). An oligodeoxynucleotide comprising the coding strand of the template DNA and the T7 promoter at its 3' end and a second DNA oligonucleotide with the T7 promoter sequence –17 to +1 are annealed prior to the transcription reaction (6).

Protocol 5. Preparation of DNA template oligonucleotides

Equipment and reagents
- Electrophoresis equipment as in *Protocols 9* and *10*
- Heating block

Method

1. Purify the DNA oligos by preparative 20% polyacrylamide–8 M urea gel

---

electrophoresis (see *Protocols 9* and *10*) or column chromatography (e.g. Qiagen or Nucleobond from Macherey-Nagel, tip-5).

2. Mix template and promoter DNA oligos in 10 mM Tris–HCl pH 8.0, 0.1 mM MgCl$_2$.

3. Heat for 5 min at 65°C and allow to cool slowly to room temperature. Place on ice.

4. Proceed with *Protocol 8*.

3.3 Cloned templates

If the transcript is too long for the system described above, it is necessary to clone the template sequence into a transcription vector. Most common is the use of 'transcription plasmids', which possess a T7 promoter sequence in front of the multiple cloning site (7). Some vectors (e.g. Bluescript) contain a second, different bacteriophage promoter at the other end of the polylinker so that the DNA template can be transcribed alternatively in sense or antisense direction. The template DNA can be cloned by standard procedures into the multiple cloning site using, either a subcloned fragment, or chemically synthesized oligodeoxynucleotides. Transcripts will have additional 5' sequence consisting of plasmid sequence between the promoter and the actual cloning site. It has to be considered whether this will interfere with the application of the RNA.

The situation at the 3' end of the template DNA is quite different. If this sequence is of no importance for the application, any convenient recognition site of the multilinker may be chosen to restrict the DNA for run-off transcription. Enzymes which leave 5' overhanging ends should be preferred over those leaving blunt-ends, 3' protruding termini should not be used because they lead to extraneous transcription products (8). To avoid additional, undesired sequences at the 3' end, we recommend using a class IIS restriction enzyme. *Fok*I (New England Biolabs), for example, cleaves at a distance of 9/13 nucleotides from its recognition site and produces four nucleotides 5' cohesive ends (9). Hence the recognition site can be placed downstream of the template sequence, thus avoiding any polylinker sequence in the transcript.

3.4 Preparation of the plasmid DNA template for run-off transcription

The DNA template should be free of contaminants; especially ribonucleases and RNA should be excluded. It is sufficient to purify the plasmid DNA after large scale preparation by an anion exchange column step (e.g. Nucleobond from Macherey-Nagel or Qiagen). The 3' terminus of the template DNA for run-off transcription is created by digesting the DNA with the appropriate restriction endonuclease (see Section 3.3). Multiple cuts in the plasmid do not affect transcription yields. It is not necessary to isolate the fragment containing the promoter and the template, but it is essential to digest the DNA to completion. Uncut DNA will serve as template for run-around transcripts, thus using up the NTPs in the assay (10).

Protocol 6. Preparation of the plasmid DNA template

Equipment and reagents
- Equipment for agarose gel electrophoresis
- Microcentrifuge
- Spectrophotometer
- Chloroform:isoamyl alcohol (24:1, v/v)
- Phenol:chloroform:isoamyl alcohol (25:24:1, by vol.)
- Ethanol
- 2 M NaOAc pH 5.5

Method
1. Cleave the plasmid DNA with a suitable endonuclease.
2. Check the result of the reaction by electrophoresis of an aliquot of the restricted DNA (max. 1 µg) on a 0.8% agarose gel.
3. If the reaction is complete, extract the DNA with 1 vol. of buffer saturated phenol:chloroform:isoamyl alcohol (25:24:1, by vol.). Separate the phases by centrifugation in a microcentrifuge for 5 min.
4. Recover the upper (aqueous) phase and extract it once with chloroform:isoamyl alcohol (24:1, v/v).
5. Precipitate the DNA by addition of 0.1 vol. of 2 M NaOAc pH 5.5 and 2.5 vol. of ethanol. Mix thoroughly and keep on dry ice for 15 min or overnight at –20°C.
6. Collect the DNA pellet by centrifugation at 13 000 g for 15 min at 4°C. Wash once with 70% ethanol, dry briefly, and redissolve the DNA in a suitable volume (estimated 1 µg/µl or less) of TE buffer.
7. Measure the A_{260} of the DNA solution and calculate the DNA concentration. Store at –20°C.

---

4. *In vitro* transcription with T7 RNA polymerase

4.1 Optimizing the transcription assay

It is recommended that the system be tested on an analytical scale prior to large scale preparations. Thus the following questions may be addressed:

(a) Is the enzyme preparation active?
(b) Is the template DNA efficiently transcribed, i.e. is the initiation sequence sufficiently close to the consensus?
(c) Is there only one transcript or are there several bands? Does the main product have the desired length?
(d) Can the reaction conditions be optimized?

Two or more bands of the size of the expected product may result from 3′ terminal heterogeneity since T7 RNA polymerase sometimes adds one or more nucleotides not encoded by the template. This cannot be avoided completely by optimization of the assay and may depend on the overall structure of the template DNA (5). Very long transcription products may be due to incomplete restriction digestion of the plasmid DNA template. Transcripts which are longer than expected may also result from RNA-directed RNA synthesis. It has been reported that T7 RNA polymerase can efficiently replicate a self-primed RNA template with sufficient self-complementarity at its 3′ end (11, 12).

Suggestions for the improvement of the transcription assay are as follows:

(a) Try different template concentrations (5–500 nM DNA).
(b) Try different polymerase concentrations (4000–80 000 U of T7 RNA polymerase/ml, excess of enzyme may cause a lower yield).
(c) Try different NTP concentrations (0.5–4 mM NTPs each from stock solutions of pH 8.1).
(d) Try different $MgCl_2$ concentrations (6–40 mM $MgCl_2$, usually 6 mM more than the NTP concentration).

There are a number of optional components and parameters for the transcription assay, which are recommended for very long transcripts or when the standard procedure yields no satisfying results. The use of DEPC (diethyl pyrocarbonate) treated water, RNasin (human placental RNase inhibitor, 100–1000 U/ml), and inorganic pyrophosphatase (0.15–5 U/ml) are highly recommended, if the transcripts are longer than 100 nucleotides and/or the reaction time is extended. The reaction is complete after 4 h of incubation. Additional incubation only increases the risk of RNA degradation even by traces of contaminating RNases.

Further optional ingredients of the transcription assay are DMSO (5–10%), PEG 8000 (5–8%), Triton X-100 (0.01%), and/or bovine serum albumin (50–100 μg/ml). Moreover, the use of synthetic polyamines instead of spermidine may stimulate transcription up to 12-fold (13). The incubation temperature can be optimized between 37°C and 42°C, nevertheless 37°C is, in general, sufficient and yields fewer incomplete transcripts (14). If there are too many incomplete transcripts, it is further recommended to increase the concentration of nucleotides as they could be limiting for the reaction.

4.2 Small scale *in vitro* transcription

All enzymatic RNA syntheses described here have been performed with T7 RNA polymerase purified as described in Section 2. The conditions for analytical transcription assays are in general the same as for the large scale procedure (see Section 4.3). The components of the reaction mixture are scaled down to a final volume of 20 μl (concentrations are maintained). The transcription products are analysed by electrophoresis on a denaturing (8 M urea) 20 × 20 × 0.04 cm polyacrylamide gel. Product bands are visualized by toluidine blue staining as described in *Protocol 7*. This is a sensitive method, which is faster and easier than silver staining. It is also less hazardous than ethidium bromide staining and the results can easily be documented.

Protocol 7. Toluidine blue staining of polyacrylamide gels

Equipment and reagents

- Staining solution: 40% (v/v) methanol, 10% (v/v) acetic acid, 0.4% (w/v) toluidine blue O (Serva)
- Washing solution: 35% (v/v) methanol, 10% (v/v) acetic acid
- Slab gel vacuum dryer

Method

1. After electrophoresis remove one glass plate and transfer the gel on the remaining glass plate into a plastic tray.

2. Soak with 250 ml 40% (v/v) methanol, 10% (v/v) acetic acid, 0.4% (w/v) toluidine blue O (Serva). Shake gently for 15 min. Remove the staining solution (can be used several times).

3. Wash with 250 ml 35% (v/v) methanol, 10% (v/v) acetic acid for 5 min. Remove the solution.

4. Repeat step 3 once or twice, until the bands are set-off sufficiently from the background.

5. Dry the stained gel on a slab gel vacuum dryer between Whatman 3MM paper and Saran Wrap, or between two sheets of cellophane for overhead projection.

4.3 Large scale *in vitro* transcription

The conditions of the reaction are valid for transcripts of < 40 to 1000 nucleotides length. They can easily be optimized for each template (see suggestions for optimization in Section 4.1). *Protocol 8* describes a 500 μl transcription assay for a transcript of 100 nucleotides length. For ethanol precipitation the assay volume may be divided between two microcentrifuge tubes.

Protocol 8. Large scale transcription

Equipment and reagents

- Heating block
- 10 × T7 polymerase reaction buffer: 400 mM Tris–HCl pH 8.1, 120 mM MgCl$_2$, 50 mM DTT, 10 mM spermidine
- 0.5 M Na$_2$ EDTA
- NTP mix: 10 mM ATP, CTP, GTP, and UTP pH 8.1
- Gel loading buffer: 90% (v/v) deionized formamide, 0.04% bromophenol blue, 0.04% xylene cyanol, 10 mM Na$_2$ EDTA

Method

1. Mix 50 μl 10 × T7 RNA polymerase reaction buffer and 100 μl NTP mix with the appropriate volume of ddH$_2$O, and pre-warm at 37°C.

2. Add 25 μg template DNA as prepared in *Protocol 6* and start the re-action with 5 μg of T7 RNA polymerase (specific activity about 400 000 U/mg prepared as described in Section 2). Mix and incubate at 37°C for 1 h in a heating block or water-bath.

3. Stop the reaction by adding 20 μl 0.5 M Na$_2$ EDTA (excess of Na$_2$ EDTA relative to MgCl$_2$ dissolves a potential magnesium pyrophosphate precipitate which may form during the reaction).

4. Purify the RNA by phenol:chloroform extraction and subsequent ethanol precipitation (see *Protocol 6*).

5. Resuspend the dried RNA in 20 μl gel loading buffer.

4.4 Purification of the transcripts

We recommend purifying the transcripts from the DNA template, from un-incorporated nucleotides, abortive initiation products, and any other RNAs with undesired length. A variety of methods for purification is available: poly-acrylamide or agarose (for transcripts of > 1000 nucleotides length) gel electro-phoresis, column chromatography, HPLC, and others. Since it is a common method with standard laboratory equipment, we will describe the isolation of the transcript by preparative gel electrophoresis in a denaturing (8 M urea) polyacrylamide gel. In this case a DNase digest after the transcription re-action is not necessary. Appropriate gel concentrations for short transcripts (40 nucleotides) to transcripts of intermediate length (400 nucleotides) are in the range of 20% to 6% polyacrylamide, respectively. The gels are prepared as described elsewhere (15). The use of long gels (20 × 40 × 0.1 cm) is recom-mended for optimal band separation with single nucleotide resolution.

Protocol 9. Electrophoresis of RNA transcripts

Equipment and reagents
• Standard vertical polyacrylamide gel electrophoresis equipment

Method

1. Heat the RNA at 95°C for 2 min, chill on ice immediately, centrifuge briefly, and load the sample onto a 10% (w/v) polyacrylamide gel.

2. Start electrophoresis at 20 W until the dyes have migrated into the gel. Increase to 40 W (~ 1000 V, 40 mA). In order to reduce 'smiling' of the gel bands by uncontrolled heating of the gel, an aluminium plate is

---

Protocol 9. *Continued*

tightly fixed to one of the glass plates. Make sure that the aluminium plate is not in contact with the electrophoresis buffer for safety reasons.

3. Do not terminate the electrophoresis before the bromophenol blue front has reached the bottom of the gel.

The RNA can be recovered from the preparative gel by various methods, e.g. diffusion or electroelution. Transcripts are then precipitated with ethanol. In *Protocol 10* we describe the most simple diffusion method which can be performed with a minimum of equipment.

Protocol 10. Elution of RNA from polyacrylamide gels

Equipment and reagents

- Silicagel thin-layer plate (TLC) with fluorescence indicator (F254) (Schleicher & Schüll)
- UV illuminator
- Sterile scalpel
- Dry ice
- Elution buffer: 0.5 M ammonium acetate, 0.1 M Na_2 EDTA, 1 mM $MgCl_2$, 0.1% SDS
- Microcentrifuge
- Spectrophotometer

Method

1. After electrophoresis remove one glass plate and cover the gel with Saran Wrap. Remove the other glass plate from the gel and cover the other side of the gel with Saran Wrap.

2. Detect the RNA bands by UV shadowing: place the gel on a silicagel thin-layer plate with fluorescence indicator (F254). Locate the RNA bands under UV light (254 nm) and cut them out with a sterile scalpel. Work fast to avoid UV damage to the RNA.

3. Collect the gel pieces in 1.5 ml microcentrifuge tubes and freeze in dry ice for 10 min in order to destabilize the gel matrix.

4. Add 300 µl elution buffer, shake 4 h or overnight at room temperature.

5. Remove the supernatant and add another 150 µl elution buffer to the gel pieces. Shake for 2 h at room temperature.

6. Combine the supernatants. Purify the RNA by phenol:chloroform extraction (optional step). Precipitate the RNA with 800 µl cold ethanol for at least 20 min on dry ice.

7. Centrifuge at 13 000 *g* for 20 min (4°C) in a microcentrifuge and dry the pellet.

8. Dissolve the RNA pellet in 50–100 µl TE buffer or ddH_2O. Determine the concentration of the transcripts by UV spectrophotometry at 260 nm. Store at –20°C.

Typical yields of the transcription assay described above are in the range of 50 μg. The assay can easily be scaled up to millilitre volumes for the production of milligram quantities of RNA. The *in vitro* transcripts are stable at –20°C and can be stored for years, if repeated freeze–thaw cycles are avoided. The RNA lacks modifications like, for instance, the 5′ cap of mRNA (which can be added *in vitro*, see Section 5.2) or the modified nucleosides of ribosomal or transfer RNA. Nevertheless, in most cases the T7 transcripts are biologically active in the same way as native RNA. This has been shown for a wide range of applications, including the characterization of structure and function of *in vitro* transcribed tRNA by Sampson and Uhlenbeck (16) and Hall *et al.* (17).

5. Modified transcripts

Modified RNAs can be obtained by two basically different methods. Oligoribonucleotides may be synthesized chemically with phosphoramidites protected by alkylsilyl and other groups (18) as the most common procedure. A variety of modified building blocks is available for the introduction of a particular modification at any given site (19). The other approach is the enzymatic synthesis of RNA with:

- end-modification of standard transcripts
- co-transcriptional modification by substituting one of the nucleotides in the transcription assay
- use of initiator oligoribonucleotides containing modified residues

Here we summarize the different types of modifications, which can often be introduced by various methods (see also *Table 2*). Whenever a nucleotide

Table 2. Modified nucleotides for T7 transcription

	Concentration of modifier	Concentration of NTPs	Application
GMP	4 mM	1 mM	5′pG instead of 5′pppG
ApG (initiator)	1 mM	0.5 mM	A+1 instead of essential G+1
m⁷G(5′)ppp(5′)G	0.5 mM	0.5 mM (GTP at 0.05 mM)	5′ cap structure
[α-³²P]NTPs	10–20 μCi/0.5 ml assay	Standard	Radiolabelled transcripts
Biotin-16-UTP, digoxigenin-11-UTP, fluorescein-12-UTP	Same or higher as NTPs	Standard	Non-radioactive labelling
4-thio-UTP, 5-bromo-UTP	As above	Standard	Cross-linking
NTPαS (Sp isomers), 2′-deoxy NTPs	As above	Standard	RNA stability

analogue is substituted in a transcription assay, it is always advisable to optimize the reaction as described in Section 4.1. In particular, the ratio of modified to unmodified nucleotides must be tested for each application, see our suggestions below. Substitution in general means that there is no unmodified nucleotide added.

5.1 The 5' terminus of the RNA

The enzymatic synthesis of RNA results in transcripts with a 5' triphosphate terminus. If this 5' pppG is not desirable it can be dephosphorylated with alkaline phosphatase as described elsewhere (15). Alternatively, either GDP or GMP can serve to initiate the transcription reaction. In this case the initiator nucleotide is added to the transcription assay at 4 mM with the four nucleotides at 1 mM each, resulting in a very efficient incorporation of GMP (16).

If the desired initiation sequence of the transcript is GAA(GAG) rather than GGG(AGA), the SP6 RNA polymerase-based transcription system should be applied instead of the T7 system, as the former sequence is the +1 to +6 consensus sequence of the SP6 promoter.

If the +1 guanosine is completely undesirable, a chemically synthesized oligoribonucleotide can serve as a kind of 'primer' for the initiation of transcription. Pitulle and co-workers (20) have shown that initiator oligonucleotides of two to six nucleotides length with one template encoded guanosine at the 3' end are efficiently incorporated. By this combination of chemical and enzymatic RNA synthesis any 5' terminal sequence with the first guanosine in position +2 to +6 can be obtained. These oligoribonucleotides are only used in the initiation, not in the elongation reaction. The use of ApG for initiating the T7 transcription (see *Protocol 11*) has been shown to be more efficient than any other oligoribonucleotide.

Protocol 11. T7 transcription with the ApG initiator

Equipment and reagents

- ApG (e.g. Sigma), it is not necessary to phosphorylate the initiator
- Reagents for transcription as specified in *Protocol 8*

Method

1. Use a standard transcription assay as described in *Protocol 8* with 0.5 mM NTPs each (final concentration), add ApG to a final concentration of 1 mM.
2. Purify the transcript as described in Section 4.4.
3. The yield should be at least as high as without the initiator.

5.2 5' cap structure

The 5' cap structure is not only necessary for the translation of eukaryotic mRNAs, it is also essential for the stability of RNA injected into oocytes. Two different methods for 5' capping of RNA can be applied. The 5' cap may be added enzymatically to *in vitro* synthesized RNA by guanylyl transferase, or the transcript can be synthesized in the presence of a cap analogue, which serves as initiator for the transcription. The latter method is especially recommended for large amounts of RNA. The 'cap nucleotide' $m^7G(5')ppp(5')G$ (e.g. Boehringer Mannheim) is added to the transcription assay at 500 μM, the same concentration as for all NTPs except for GTP, which should be added at a very low concentration of 50 μM (10).

5.3 Radiolabelled RNA

Transcripts can be ^{32}P-labelled in two different ways. They may be end-labelled, e.g. at the 5' terminus using $[\gamma\text{-}^{32}P]ATP$ and T4 polynucleotide kinase (15). The other possibility is to substitute one or more nucleotides in the transcription assay by the appropriate $[\alpha\text{-}^{32}P]NTP$. It might be useful to add the unlabelled NTP(s) as well, at 1/10 of the standard NTP concentration. For a large scale transcription as described above (see Section 4.3) it is sufficient to apply 10–20 μCi (specific activity 410 Ci/mmol or about 15 TBq/mmol, Amersham) of a $[\alpha\text{-}^{32}P]NTP$. Important: It is recommended that GMP be added as initiator nucleotide when radiolabelled GTP is used, otherwise yields could be low because the radiolabel may be used up in abortive initiation reactions (4).

5.4 Non-radioactive labels for RNA

RNA is labelled non-radioactively in the following way: nucleotide analogues with non-radioactive labels attached by an 11–16 carbon chain (Boehringer Mannheim) can be introduced during enzymatic synthesis. UTP is substituted by either:

• biotin-16-UTP

• digoxigenin-11-UTP

• fluorescein-12-UTP

at the same or higher concentration as the other nucleotides. These nucleotide analogues are good substrates for all phage-derived RNA polymerases. The labelled RNA can be detected by specific antibody/enzyme systems with subsequent staining, or directly by its fluorescence in case of fluorescein. Even polylabelled oligoribonucleotides can be obtained by chemical synthesis of RNA using modified nucleoside phosphoramidites, as shown for the biotin label (21).

The initiator oligonucleotide system described in Section 5.1 (20) can be applied for short synthetic oligoribonucleotides with non-radioactive labels as well. Logsdon and co-workers (22) have developed a system for selective 5' modification of T7 RNA polymerase transcripts: GTP is replaced by GTPγS (same concentration). Hence every transcript contains a 5'-γ-thiophosphate group which can be modified with a thiol-specific reagent, e.g. a fluorescent group.

5.5 Modified RNA for cross-linking experiments

RNA containing 4-thio-UTP or 5-bromo-UTP can form specific covalent bonds with associated proteins when irridated with long-wavelength ultraviolet light (> 300 nm). These nucleotide analogues are substrates for T7 RNA polymerase, so they can be substituted for UTP in the transcription assay (23).

5.6 Modifications stabilizing the RNA

RNA is not as stable as would be convenient for some *in vivo* applications. It might be advisable to incorporate nucleotide analogues which make the polymer more resistant, e.g. phosphorothioates (24). NTPαS (Sp isomers) are substrates for T7 RNA polymerase and make the transcript resistant to hydrolysis. At least it has been shown that phosphorothioates at terminal positions can protect RNA against 3' exonuclease activity (25). Especially for hammerhead ribozymes, a number of ribose 2' hydroxyl modifications have been tested, e.g.

- 2'-deoxy NTPs
- 2'-amino-2'-deoxy NTPs
- 2'-fluoro-2'-deoxy NTPs
- 2'-O-alkyl NTPs

Only 2'-deoxy NTPs are substrates of T7 RNA polymerase, yet not very efficient ones. The other modified NTPs can be used for the chemical synthesis of oligoribonucleotides or for combined chemical and enzymatic synthesis (see Section 5.1).

In the case of ribozymes, the RNA has to be stabilized without interference with its catalytic function (25). This is not a problem for many antisense applications. 2'-O-alkyl RNA as an antisense reagent has proven to be resistant to alkali and nuclease-induced cleavages. Furthermore it hybridizes more stably to complementary sequences (26), making it the tool of choice for efficient antisense hybridization.

Acknowledgements

We thank Dr M. Famulok for establishing the Centricon concentration step before dialysis (*Protocol 4*) and Professor H. Beier for critically reading the manuscript. This work was supported by Fonds der Chemischen Industrie, SFB 165, and DFG-Schwerpunkt RNA-Biochemie.

References

1. Davanloo, P., Rosenberg, A. H., Dunn, J. J., and Studier, F. W. (1984). *Proc. Natl. Acad. Sci. USA*, **81**, 2035.
2. Grodberg, J. and Dunn, J. J. (1988). *J. Bacteriol.*, **170**, 1245.
3. Zawadzki, V. and Gross, H. J. (1991). *Nucleic Acids Res.*, **19**, 1948.
4. Milligan, J. F. and Uhlenbeck, O. C. (1989). In *Methods in enzymology* (ed. J. E. Dahlberg and J. N. Abelson), Vol. 180, pp. 51–62. Academic Press, London.
5. Milligan, J. F., Groebe, D. R., Witherell, G. W., and Uhlenbeck, O. C. (1987). *Nucleic Acids Res.*, **15**, 8783.
6. Stump, W. T. and Hall, K. B. (1993). *Nucleic Acids Res.*, **21**, 5480.
7. Krupp, G. (1988). *Gene*, **72**, 75.
8. Schenborn, E. T. and Mierendorf, R. C. Jr. (1985). *Nucleic Acids Res.*, **17**, 6223.
9. Sugisaki, H. and Kanazawa, S. (1981). *Gene*, **16**, 73.
10. Yisraeli, J. K. and Melton, D. A. (1989). In *Methods in enzymology* (ed. J. E. Dahlberg and J. N. Abelson), Vol. 180, pp. 42–50. Academic Press, London.
11. Konarska, M. M. and Sharp, P. A. (1990). *Cell*, **63**, 609.
12. Cazenave, C. and Uhlenbeck, O. C. (1994). *Proc. Natl. Acad. Sci. USA*, **91**, 6972.
13. Frugier, M., Florentz, C., Hosseini, M. W., Lehn, J.-M., and Giegé, R. (1994). *Nucleic Acids Res.*, **22**, 2784.
14. Krieg, P. A. and Melton, D. A. (1987). In *Methods in enzymology* (ed. R. Wu), Vol. 155, pp. 397–415. Academic Press, London.
15. Beier, H. and Gross, H. J. (1991). In *Essential molecular biology: a practical approach.* Vol. II (ed. T. A. Brown), pp. 221–36. IRL Press, Oxford.
16. Sampson, J. R. and Uhlenbeck, O. C. (1988). *Proc. Natl. Acad. Sci. USA*, **85**, 1033.
17. Hall, K. B., Sampson, J. R., Uhlenbeck, O. C., and Redfield, A. G. (1989). *Biochemistry*, **28**, 5794.
18. Scaringe, S. A., Francklyn, C., and Usman, N. (1990). *Nucleic Acids Res.*, **18**, 5433.
19. Usman, N. and Cedergren, R. (1992). *Trends Biochem. Sci.*, **17**, 334.
20. Pitulle, C., Kleineidam, R. G., Sproat, B., and Krupp, G. (1992). *Gene*, **112**, 101.
21. Teigelkamp, S., Ebel, S., Will, D. W., Brown, T., and Beggs, J. D. (1993). *Nucleic Acids Res.*, **21**, 4651.
22. Logsdon, N., Lee, C. G. L., and Harper, J. W. (1992). *Anal. Biochem.*, **205**, 36.
23. Tanner, N. K., Hanna, M. M., and Abelson, J. (1988). *Biochemistry*, **27**, 8852.
24. Eckstein, F. and Gish, G. (1989). *Trends Biochem. Sci.*, **14**, 97.
25. Heidenreich, O., Benseler, F., Fahrenholz, A., and Eckstein, F. (1994). *J. Biol. Chem.*, **269**, 2131.
26. Lamond, A. I. and Sproat, B. S. (1993). *FEBS Lett.*, **325**, 123.

Catalytic antisense RNA based on hammerhead ribozymes

MARTIN TABLER and GEORG SCZAKIEL

1. Introduction

1.1 Origin of hammerhead ribozymes

Some plant viruses are accompanied by so-called satellite RNAs which cannot replicate autonomously but require a helper virus for replication (1). Some of these satellite RNAs, especially the circular satellite RNAs, which are also called virusoids, undergo Mg^{2+}-catalysed autocatalytic self-cleavage at a defined position of their RNA genome. A specific phosphodiester bond is chemically cleaved in an intramolecular *trans*-esterification reaction resulting in a $2',3'$ cyclic phosphate located $5'$ to the cleavage site, whereas the $3'$ cleavage product has a $5'$ hydroxyl group. A similar type of reaction can be found in two plant viroids which, unlike the satellite RNAs, do not depend on the presence of a helper virus for replication, and in the RNA transcript of a satellite DNA 2 of newt. A feature common to all these RNAs is the presence of a so-called 'hammerhead structure' (2).[a] The hammerhead structure consists of several conserved nucleotides and three helices that can be drawn such that the resulting secondary structure resembles the head of a carpenter's hammer. A generally accepted numbering system has been defined (4) which can be seen in *Figure 1*. Despite the generality of the hammerhead structure and the conservation of the nucleotides that are involved in catalysis, there is considerable flexibility with regard to loop size and length of the helices.

It should, however, be kept in mind that the representation of this class of catalytic RNAs as a two-dimensional 'hammerhead' is a simplification. Recently, the three-dimensional structure of the hammerhead has been de-

[a] It is often stated that hammerhead RNAs are 'derived from plant viroids'. However, to date hammerhead structures and self-cleavage reactions have only been found in two exceptional viroids, the avocado sunblotch viroid (ASBVd) (2) and the peach latent mosaic viroid (PLMVd) (3). The majority of the viroids neither contain a hammerhead structure, nor is there any substantiated evidence for a self-cleavage reaction.

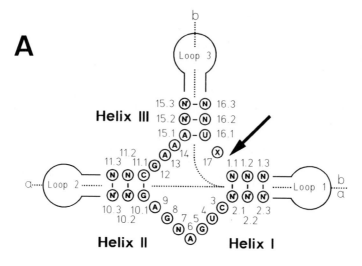

A

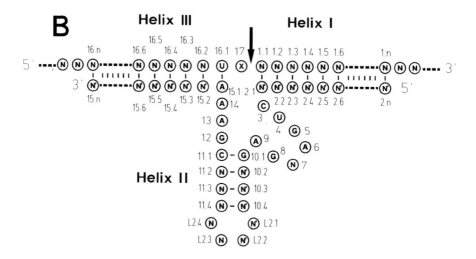

B

lineated by two different approaches. McKay and co-workers have described the X-ray structure of a hammerhead ribozyme, co-crystallized with a DNA substrate (5), and Eckstein and co-workers have used fluorescence energy transfer to determine a three-dimensional model for the hammerhead RNA in solution (6). Both groups showed that there is a sharp turn between helices I and II identical to the uridine turn of transfer RNA (for details see also ref. 7). Helix III points away, so that the tertiary structure can be described as a 'wishbone' (5) or as a 'Y' (6).

Figure 1. (A) The hammerhead structure with the numbering system introduced by Hertel *et al.* (4). The conserved nucleotides that are essential for catalysis are indicated, the residual nucleotides are given as N, and nucleotides that are complementary as N'. The cleavage occurs between X_{17} and $N_{1.1}$. None of the loops is essential since the hammerhead RNA can be open at any of the loop positions. The size of any of the three helices is variable. The dotted line (a) shows the subdivision of the hammerhead into two individual RNA molecules which cleave in *trans* as described by Uhlenbeck (8). The dotted line (b) shows the division of Haseloff and Gerlach (9) which also results in two *trans*-cleaving RNAs. Besides the UX ↓ motif all the conserved nucleotides that are essential for catalysis of the hammerhead are found in the ribozyme half. (B) Another representation of the subdivision according to Haseloff and Gerlach. The upper part of it corresponds to the target or substrate RNA and the lower part to the hammerhead ribozyme, which in that case contains a helix II of four base pairs and a four-base loop 2, representing the most common design. The region from nucleotide 3–14 is also called 'catalytic domain', but it has to be kept in mind that nucleotides 15.1 and 16.1 also contribute to catalysis.

1.2 Exploitation of the hammerhead structure for generation of ribozymes that act as specific endoribonucleases

The naturally occurring hammerheads initially described cleave in *cis*, i.e. in an intramolecular self-cleavage reaction. Uhlenbeck (8) was the first to try to generate a hammerhead consisting of two different RNA molecules and demonstrated that cleavage can also occur in *trans*, i.e. in an intermolecular cleavage reaction (delineated by axis 'a' in *Figure 1A*). Here, the RNA molecule that is cleaved is the substrate or target RNA and the RNA molecule that is required to cause the cleavage is the RNA enzyme or ribozyme. Haseloff and Gerlach (9) described another way to subdivide the hammerhead structure into a substrate and a ribozyme molecule (delineated by axis 'b' in *Figure 1A*). This subdivision of the catalytically active hammerhead has the advantage that practically all the nucleotides essential for the cleavage reaction are part of the ribozyme. The substrate RNA just contains a three-nucleotide 'cleavable motif' which is usually described as NUX ↓, wherein N can be any nucleotide and X any but G. The target motif may also be given just as UX ↓. However, the efficiency of cleavage depends on the nature of N and thus the usage of a three-nucleotide sequence is preferred.

These limited sequence requirements allow design of hammerhead ribozymes directed against practically any target RNA and provide a great degree of flexibility regarding the selection of a target site. First a NUX ↓ motif must be selected in the target RNA. Then, a ribozyme for that motif and its specific sequence context is designed. The resulting ribozymes consist of three domains (see *Figure 1B*):

(a) The helix I-forming region (nucleotides 5' 2.n–2.1) which is complementary to the target RNA downstream of the NUX ↓ motif (nucleotides 5' 1.1–1.n).

(b) The catalytic domain (nucleotides 5′ 3–14, including the helix II and loop 2) which does not base pair with the target RNA.

(c) The helix III-forming region (nucleotides 5′ 15.1–15.n that are complementary to the target RNA upstream of the NUX ↓ motif (nucleotides 5′ 16.n–16.1, wherein 16.1 is the U of the NUX ↓).

The specificity of the ribozyme originates from the antisense arms (the regions that form helices I and III). These domains establish the association of the ribozyme with its target and enable the correct positioning of the catalytic domain such that the hammerhead conformation is formed and the target RNA can be cleaved. The antisense arms, as such, are not catalytically active but are required for association with the target RNA and thus confer the specificity of the ribozyme. Gene suppression by hammerhead ribozymes according to this design has been successfully tested in various biological systems (reviewed in ref. 10).

2. The ribozyme reaction

2.1 The catalytic cycle

Hammerhead ribozymes can be considered as specialized antisense RNAs which carry the catalytic domain that can function as a 'warhead'. Like antisense RNAs they associate with their target via Watson–Crick base pairing, but, unlike antisense RNA, they have the ability to cleave the target, most likely irreversibly.

There are several ways to design and apply hammerhead ribozymes which need to be considered in more detail. One of the most fascinating features of hammerhead ribozymes is their ability to act catalytically like ordinary enzymes. Since the ribozyme molecule itself remains unchanged during the cleavage reaction it acts as a highly specific endoribonuclease and, in line with this, one ribozyme molecule can cleave, as a true catalyst, multiple substrate RNA molecules. Kinetic studies have indeed shown that the cleavage reactions of a hammerhead ribozyme can be described by Michaelis–Menten kinetics, similar to an ordinary enzymatic reaction (11, 12).

The ribozyme reaction can be subdivided into at least three steps as shown in *Figure 2*. First is the association during which the two RNA partners form the duplex region of helix I and III and eventually the active hammerhead conformation. Secondly, the actual chemical cleavage occurs, resulting in the two cleavage products that remain bound to the ribozyme RNA. For multiple turnover, it is thirdly necessary that the ribozyme effectively dissociates from the cleaved RNA in order to enter a new round of catalysis with an uncleaved substrate RNA molecule. *In vitro* experiments with short model substrate RNAs have shown that such a reaction cycle indeed proceeds, resulting in K_m values of about 0.5 mM and k_{cat} values in the order of about 1 min^{-1} (11). In contrast to these *in vitro* studies, *in vivo* applications of such ribozymes in

Substate RNA

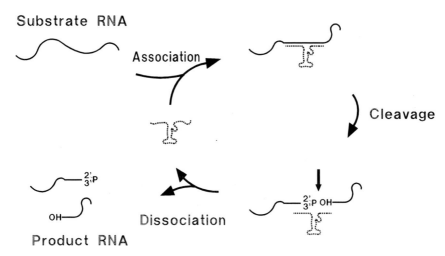

Association

Cleavage

Dissociation

Product RNA

Figure 2. The catalytic cycle for a hammerhead reaction can be divided in three major steps: the association step where the two RNA partners form helices I and III via their complementary regions. The next step is the actual cleavage reaction where a 2′,5′ cyclic phosphate is formed at nucleotide 17 (see *Figure 1*), which is the 3′ terminal nucleotide of the 5′ cleavage product. The final dissociation step releases the two cleavage products and the ribozyme can then enter a new round of catalysis.

several cases have been less successful since a high molar excess has been found to be required to observe gene suppression.

As in any reaction cycle, each individual step can be rate-limiting. For a potent ribozyme reaction it would be desirable to have effective association of the two RNAs, followed by rapid cleavage and fast dissociation of the cleaved products.

Since it is of great importance to the design a hammerhead ribozyme, and to use its full potential, it is worth considering the individual steps of the reaction scheme in greater detail.

2.1.1 Association
Association of the ribozyme with the substrate RNA is a key step. Obviously, if there is little or no efficient association there will be no ribozyme effect. If the antisense flanks are sufficiently long, the formation of the duplex RNA with the unpaired catalytic domain of the hammerhead is the most stable conformation that the two RNAs can assume. However, in the case of long stretches of RNA, duplex formation is not controlled thermodynamically but kinetically. Both the substrate and the ribozyme will assume one or several intramolecular secondary structures, the destruction of which requires activation energy before duplex formation can proceed. If both the RNAs are small, the degree of secondary structure will be relatively unstable, so that the activation energy is low and duplex formation is favoured. That may change if

at least one of the two RNA partners (usually the target RNA) is longer and/or highly structured.

For any *in vivo* application, the ribozyme must be designed such that it is able to resolve the secondary structure of the endogenous target RNA which clearly cannot itself be modified. If the cleavable motif is embedded in a stable secondary structure the hammerhead ribozyme cannot efficiently associate, even at high molar excess (*Figure 3A*).

For example, Cotten *et al.* (13) have reported that an U7 RNA-directed ribozyme required a 1000-fold molar excess versus the target RNA for complete inhibition of U7-dependent processing *in vivo* while a longer antisense

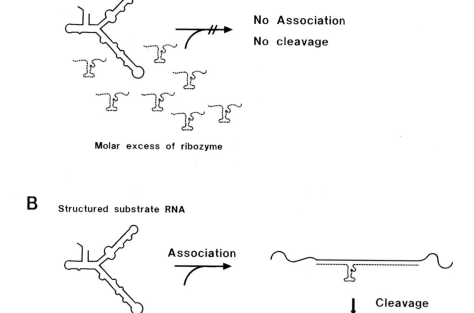

Figure 3. Potential ways of hammerhead ribozymes to associate with a structured substrate RNA depending on the length of the antisense arms. Long target RNAs as found *in vivo* are more or less structured. (A) Hammerhead ribozymes with short antisense arms may not be able to resolve the secondary structure, even if present at molar excess. (B) Catalytic antisense RNA with long antisense arms (at least one long antisense arm) are able to unravel the secondary structure and induce cleavage. Due to lack of dissociation, each catalytic antisense RNA can cleave only one substrate molecule and thus follows a stoichiometric cleavage mode.

RNA was effective at only a sixfold molar excess, suggesting that binding and not cleavage is the rate-limiting step in living cells. Similarly, Cameron and Jennings (14) found that a 1000-fold molar excess of hammerhead ribozyme was required to inhibit chloramphenicol acetyltransferase in monkey cells and L'Hullier *et al.* (15) similarly used a 1000-fold molar excess to inhibit α-lactalbumin in mouse cells. Recently, there have also been other examples of more efficient gene suppression with short hammerhead ribozymes (reviewed in ref. 10).

The association rates can be determined experimentally (Section 5.2 and *Protocol 4*), and—as will be outlined later (Section 5.3 and *Protocol 5*)—optimized.

2.1.2 The cleavage reaction

The efficiency of the cleavage reaction is dictated by the catalytic domain of the hammerhead. Many attempts have been undertaken to study the role and influence of individual nucleotides of the catalytic domain using mutagenesis or chemical modification. These studies have helped to clarify which nucleotides are absolutely required for the cleavage reaction, which have a supportive role, and which have little or no influence upon the reaction (for review, see ref. 11). However, to date no significant increase of the catalytic activity of the hammerhead has been obtained by these studies, indicating that nature has already 'optimized' the cleavage step of the hammerhead.

Another factor that influences the cleavage efficiency is the type of the cleavable motif. Most of the naturally occurring hammerhead RNAs have a GUC ↓ motif (position 16.2, 16.1, 17) (16) and, in accordance with this, the majority of the hammerhead ribozymes made for gene suppression have been designed for GUC ↓ motifs. Other motifs are also cleaved but there are differences in the efficiency. For example GUA ↓, GUU ↓, CUC ↓, and UUC ↓ have also been described as effective motifs, whereas AUC ↓ has been described as uncleavable in one case (17) whereas it was cleavable in another sequence context (18). Two systematic studies have been carried out in which all twelve cleavable motifs were tested in a similar or the same sequence context (34, 35). Disregarding the influence of different motifs on binding rates, both studies showed that GUC ↓ and AUC ↓ were the strongest motifs. Similarly, GUA ↓ and AUA ↓ cleaved relatively well according to both studies. The motif CUC ↓ was found well cleaving in one case (34) and UUC ↓ in the other (35). It should be stressed, however, that even a GUC ↓ does not guarantee efficient cleavage. GUC ↓ motifs that vary greatly in their cleavability have been observed (19), strongly indicating that the neighbouring nucleotides also influence the cleavage reaction in a manner as yet not understood.

In order to find out whether there is reasonable cleavage, it is necessary to determine the cleavage rate of a catalytic antisense RNA (see Section 5.4 and *Protocol 6*).

2.1.3 Dissociation

In order to be available for cleaving a second substrate RNA, the ribozyme has to dissociate from the two cleavage products. If dissociation is to occur, binding in the duplex regions must be sufficiently weak. Recently, Hertel *et al.* (12) have shown, for a ribozyme RNA with eight complementary nucleotides on either side, that dissociation is rate-limiting for the catalytic cycle, at least at higher concentrations of the product RNA. This implies that the flanking anti-sense sequences must not be chosen too long to allow a multiple catalytic cycle.

2.2 Stoichiometric cleavage

As has been outlined for the multiple-turnover reaction (Section 2.1), there is a conflict of interest regarding the length of the antisense regions: they should be sufficiently long to assure effective association and they need to be short enough to allow effective dissociation from the cleaved RNAs.

One way out of the dilemma is to optimize the hammerhead ribozyme just for association, taking into account only a single-turnover mode of cleavage due to lack of dissociation. The reaction scheme for a stoichiometric cleavage reaction is given in *Figure 3B*. An RNA that cannot dissociate from its cleaved substrate RNA would resemble a conventional antisense RNA with the additional feature of being able to cleave the target RNA. We propose to call this type of RNA 'catalytic antisense RNA'. The most important question is whether the presence of the catalytic domain is any improvement over the conventional antisense RNA. This is not immediately evident since the mechanism of gene suppression by antisense RNA itself is not absolutely clear; but there are indications that at least a part of the suppressive effect is mediated by RNases that are specific for double-strands (20). Since antisense RNA may finally result in the RNase-mediated destruction of the target RNA, one might argue that the additional feature of catalytic activity will not add any value to the antisense RNA.

However, comparisons of antisense RNA and catalytic antisense RNA for inhibiting the replication of human immunodeficiency virus type 1 (HIV-1) have shown that there is indeed a marked difference between whether the catalytic domain is included or not (21). So far we do not know whether these catalytic antisense RNAs indeed just have single-turnover when tested in living cells. It could be that proteins assist in dissociation under *in vivo* conditions, so that again a catalytic cycle could be established which might be responsible for the more efficient suppression of the target RNA.

Irrespective of whether or not catalytic antisense RNA has a multiple-turnover mechanism of cleavage in living cells, it should be added that the increased inhibition potential could also be explained under the assumption of a solely stoichiometric cleavage reaction (one ribozyme binds and cleaves one substrate RNA). At low RNA concentrations near to the value of the dissociation constant K_D for the complex formed between antisense and target

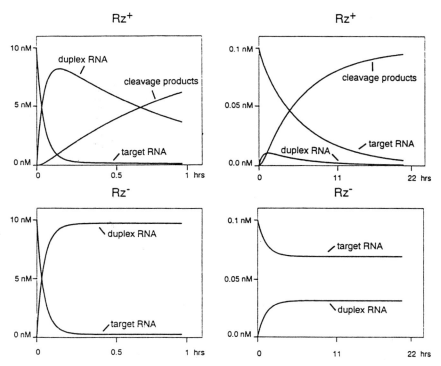

Figure 4. Kinetic models for antisense- and ribozyme-mediated inactivation of target RNA. Using the rate constants for association and cleavage as determined by Homann *et al.* (21) for an *in vitro* reaction, the decline in the concentration of free target RNA was simulated using the program *KSIM* by Neil Millar. The two panels on the *left* use a start concentration of 10 nM for the target RNA (and a fourfold molar excess of ribozyme). The *top* panel shows the reaction with a catalytic antisense RNA, whereas the *bottom* reaction is with an ordinary antisense RNA. In the case of the catalytic antisense RNA the duplex RNA eventually become cleaved. The decline in the concentration of free target RNA is similar. The *right* two panels show the situation for a starting concentration of 0.1 nM of target RNA (and a fourfold molar excess of ribozyme). In this case the target RNA is much more rapidly removed if a catalytic antisense RNA is used (*top*) compared with the non-catalytic RNA (*bottom*). (From ref. 22, with permission.)

RNA there should be a difference depending on the presence of a catalytic domain. Computer simulation shows (22) that the quasi irreversible site-specific hydrolysis of the substrate strand results in more effective inactivation of the target RNA than via conventional duplex formation which is followed by degradation in living cells (see *Figure 4*). It must be added that at such low RNA concentrations, in combination with the rate constants determined for the *in vitro* reaction, RNA duplex formation is extremely slow (*Figure 4*, right panel), so that these data are not compatible with the effects observed in living cells. This changes if one assumes the existence of factors, such as certain nuclear proteins, which enhance RNA association.

Regardless of what the mechanism may be, we have observed in many different cases that the incorporation of the catalytic domain into an antisense RNA increases its inhibitory effect. Moreover, we have shown recently that catalytic antisense RNAs are particularly effective when applied in the nucleus. Conventional ribozymes with shorter antisense sequences were found to be almost ineffective under those conditions. Conversely, short-chain ribozymes were found to be superior to long-chain ribozymes when applied in the cytoplasm. The high effectiveness of catalytic antisense RNA in the nucleus recommends its application where endogenous expression is desired.

2.3 Conclusion of theoretical considerations

In summary we believe that catalytic antisense RNAs, i.e. hammerhead ribozymes with long antisense arms are advantageous, especially for endogenous expression. They are highly effective in the nucleus (36) and they can associate efficiently with their target RNA which they can irreversibly cleave in the next step. However, this stoichiometric mode requires a molar excess of the ribozyme to the target RNA.

3. Design of the ribozyme and target site selection

3.1 Cleavable motifs

Depending on the length of the target RNA there should be plenty of cleavable motifs available. Since the general sequence has been described as NUX ↓, wherein X is anything but G, statistically any RNA sequence should contain a suitable motif every 12 bases. In practice however, it is recommended to select a 'strong' cleavable motif, such as GUC ↓, AUC ↓, GUA ↓, GUU ↓ and possibly also CUC ↓ and UUC ↓ (34, 35). It must be added that cleavability certainly depends also on the sequence context surrounding the cleavable motif. For example, we have encountered GUC ↓ motifs that vary about 50-fold in their cleavability depending on their sequence context (19).

Therefore, we recommend testing the cleavage efficiency of the catalytic antisense RNA *in vitro* before expressing it *in vivo* (see Section 5.4, *Protocol 6*).

3.2 Prediction of accessible target regions by computer analysis

As outlined in Section 2.1.1, association of catalytic antisense RNA with target RNA is one of the critical steps which depends on local target accessibility for complementary sequences. Sczakiel *et al.* (23) have described a way to analyse for the local folding potential of an RNA molecule by using the algorithm of Zuker and Stiegler (24) for calculating the lowest folding energies. The idea is to dissect a long RNA molecule into smaller (artificial) RNA fragments for which the lowest possible folding energies are calculated separately. For example, an RNA molecule of 3000 bases is far too long to perform any reasonable structural calculations. For the analysis of the local folding poten-tial it is, however, not necessary to use the entire sequence for a single struc-tural calculation. Instead one can calculate the lowest possible free energy of the secondary structure of a hypothetical RNA consisting, for example, of the first 100 nucleotides. Then, the window of 100 nucleotides is moved and the energy of the next hypothetical RNA molecule ranging from nucleotide 2–101 is calculated and so forth, until finally the theoretical RNA ranging from nucleotide 2901–3000 is calculated. The corresponding values for the folding energies of certain RNA windows are plotted versus the position within the RNA sequence as shown in *Figure 5*. This type of analysis may be repeated with window sizes ranging from 50 to about 400 nucleotides. Maxima of these plots correspond with low folding potential and indicate accessibility of this domain for antisense RNA and/or catalytic antisense RNA.

The value of this technique should not be overestimated but it might pro-vide useful information as to where to direct the catalytic antisense RNA. In the case of HIV-1, it was found that antisense-mediated inhibition rates were high, when regions of the target RNA with low folding potential were tar-geted. Also other examples, such as the mRNA of tap1 or p53 confirm these results (23), which underline the potential of this type of analysis.

The program for the prediction of the local folding potential of RNAs may be obtained from either of the authors. However, the program most likely needs adaptation if run on a different computer. It can be used directly within the Heidelberg Unix Sequence Analysis Resources (HUSAR).

3.3 The length of the antisense arms

The design of a catalytic antisense RNA requires decisions regarding which domain and to which cleavable motif within the target RNA it should be directed. Based on this decision, the length of the antisense arms also have to be determined. A similar question arises in the case of ordinary antisense RNA and there is no general rule as to how long the antisense RNA has to be or should be in order to enable effective association and gene suppression. There are examples of short complementary antisense RNAs that are effect-ive in gene suppression, so that it is not necessary to invert the entire cDNA for expression as an antisense gene. At first view, the decision concerning the length of the antisense arms within a catalytic antisense RNA depends on the cloning strategy and practical considerations such as availability of restriction sites. This does not imply that the length of the antisense arms is of minor importance.

An indication may be that the naturally occurring antisense RNAs in pro-karyotes are about 70 nucleotides in size. This would be in line with the obser-vation of Crisell *et al.* (25) who have described the optimal length of the antisense arms in the order of about 30 nucleotides. But it remains to be seen whether that rule can be generalized or applies for this special sequence only.

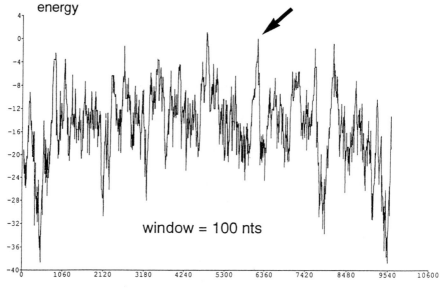

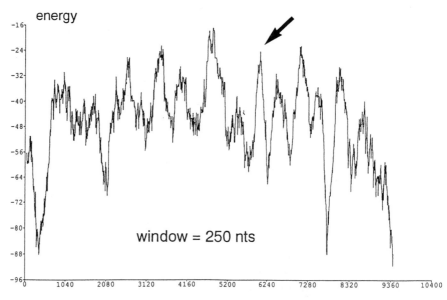

Figure 5. Folding potential indicated by free energy values for the genomic HIV-1 RNA strand. Δ*G* values for the step width of one nucleotide and a window size of 100 (*top*) or 250 (*bottom*) are indicated in kcal/mol. The HIV-1 subregion involved in strong experimental antisense RNA-mediated inhibition (position 5800–6000) is indicated (modified from ref. 23, with permission).

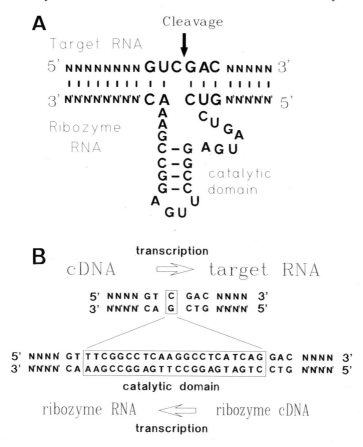

Figure 6. Hammerhead ribozyme directed against the GUC motif within a *Sal*I recognition sequence. (A) Hammerhead conformation between a target RNA containing a *Sal*I recognition sequence (GUCGAC) and its corresponding ribozyme RNA with the 'catalytic domain' 5' CUGAUGAGGCCUUGAGGCCGAA 3'. Compared to the satellite RNA of tobacco ring spot virus (sTobRV) (28, 29), the sequence of the catalytic domain has been modified at positions 10.2 and 11.2. This modification introduced the recognition sequence for *Stu*I (AGG/CCT) into the domain. The sequence GUC of the substrate RNA is the 'cleavable motif' and the site of cleavage is indicated. (B) Sequence of the double-stranded DNA templates from which target and ribozyme RNA are synthesized upon transcription in the directions are indicated by arrows. The two DNAs differ only in the boxed parts. For conversion of a cDNA into an antisense:ribozyme construct, the boxed C:G pair, corresponding to the third position within the GUC cleavable motif, or nucleotide 17 (see *Figure 1*) has to be replaced by the boxed 22 base pairs representing the catalytic domain (modified from ref. 27).

The length of the catalytic antisense RNA certainly influences the association kinetics, but it is hard to predict a fast hybridizing species and we recommend analysis of association kinetics experimentally (*Protocol 4*) because, for antisense RNAs, it has been shown that fast association *in vitro* correlated with biological effectiveness in living cells (26) (Section 5.2, *Protocol 4*). Selection of a long antisense arm, especially in the form of an asymmetric hammerhead ribozyme (see below) has the advantage for optimization of association through an *in vitro* selection procedure (Section 5.3, *Protocol 5*).

4. Construction strategies for catalytic antisense RNAs

4.1 Incorporation of DNA cassettes

If it is desired to generate a catalytic antisense RNA regardless of any structural considerations (see Section 3.2), it is possible to convert an existing cDNA fragment into a ribozyme construct by incorporating some DNA

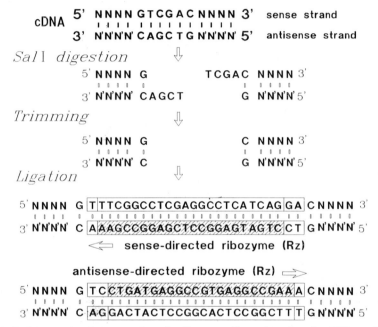

Figure 7. Strategy for the generation of a ribozyme directed against the GUC motif within a *Sal*I recognition sequence. The four internal nucleotides TCGA of the *Sal*I site correspond to nucleotides 16.1, 17, 1.1, and 1.2 according to the numbering systems for hammerhead ribozymes (see *Figure 1*). These nucleotides are removed by *Sal*I digestion and subsequent trimming of the protruding ends. Then, a *Sal*I-specific DNA Rz-cassette is inserted. The cassette consists of the catalytic domain (hatched part), as outlined in *Figure 6*, but nucleotide L2.2 has been changed to C in order to ceate an *Xho*I (C/TCGAG) site. In addition to the sequence of the catalytic domain, the cassette contains additional terminal nucleotides, which replace three of the four substrate-specific nucleotides (16.1, 1.1, and 1.2), previously removed from the protruding *Sal*I ends by trimming. Depending on the orientation of insertion, an RNA can be transcribed from the recombinant DNA, which is able to cleave the sense strand of the target RNA (sense-directed) or its antisense strand (antisense-directed). This insertion strategy can be used for the restriction enzymes listed in *Table 1* (from ref. 27, with permission).

cassettes as described by Tabler and Tsagris (27). This strategy exploits the fact that the recognition sequence of some restriction enzymes contains a part of the cleavable motif. *Figure 6A* shows a substrate RNA that contains the recognition sequence (at the DNA level) for *Sal*I, which itself contains the GUC ↓ target motif, base paired with a catalytic antisense RNA. *Figure 6B* shows that the DNAs that encode these two RNAs differ by the two boxes and the direction of transcription. Using the strategy outlined in *Figure 7* it is possible to incorporate a synthetic DNA cassette into the cDNA that has been cleaved with *Sal*I, followed by trimming the protruding ends. The DNA cassette contains the sequence of the catalytic domain plus three of the four nucleotides that have been removed by trimming. Depending on the orientation of insertion a sense or antisense-directed ribozyme is obtained. This strategy can be modified so that the insertable cassette contains a selectable gene, for example the tetracycline-resistance gene, to allow easy selection for the desired clone following ligation and transformation (27).

This kind of strategy can be applied with about 20 different restriction enzymes as listed in *Table 1*.

Protocol 1.   Incorporation of synthetic DNA ribozyme cassettes into cDNA

Equipment and reagents

- Standard equipment for molecular biology work
- A cDNA fragment of the target RNA, cloned into a plasmid vector allowing *in vitro* transcription of both strands of the cDNA by T3, T7, or SP6 RNA polymerase
- Two synthetic DNA oligonucleotides that can form a double-stranded DNA cassette according to *Table 1*, dissolved at a concentration of about 1 OD$_{260}$ U/ml
- Nuclease S1 buffer: 225 mM NaCl, 30 mM potassium acetate pH 4.5, 200 μM ZnSO$_4$, 5% (v/v) glycerol
- Nuclease S1 stop buffer: 300 mM Tris–HCl, 50 mM EDTA pH 8.0
- Nuclease S1 (Boehringer Mannheim)
- TE buffer: 10 mM Tris–HCl, 1 mM EDTA pH 8.0
- Blunt-end ligation buffer: 50 mM Tris–HCl pH 7.5, 10 mM MgCl$_2$, 5% (v/v) PEG 8000, 1 mM DTT, 100 μM ATP (store as 5 × buffer without ATP)
- T4 DNA ligase
- T4 DNA polynucleotide kinase buffer: 500 mM Tris–HCl pH 7.6, 100 mM MgCl$_2$, 50 mM dithiothreitol, 1 mM spermidine, 0.25 mg/ml BSA (Molecular Biology grade), 1 mM ATP
- T4 DNA polynucleotide kinase

A. Selection of target site

1. Select from *Table 1* a suitable restriction site that is present once in your cDNA fragment. Make sure that this restriction enzyme will not cut the vector DNA. If your cDNA is cleaved twice by the enzyme in question, consider subcloning a shorter fragment.

2. Synthesize the DNA oligos required for the DNA cassette.

B. Annealing of the two DNA oligonucleotides

1. Take 2 μl of the synthetic DNA oligos dissolved at a concentration of

---

Protocol 1. *Continued*

    1 OD_{260} U/ml (this corresponds to about 8 pmole) and dissolve in 20 μl 1 $\times$ T4 DNA polynucleotide kinase buffer.

2. Add 5 U of T4 DNA polynucleotide kinase and incubate at 37°C for 30 min.

3. Add another 5 U T4 DNA polynucleotide kinase and incubate for another 30 min.

4. Mix equal volumes (for example 10 μl) of the two kinase reactions.

5. Heat mixture to 100°C for 1 min and let cool down slowly to room temperature within about 1 h.

6. Use annealed mixture directly for ligation.

C. *Insertion of the DNA cassette*

1. Digest at least 2 μg of the recombinant plasmid with the restriction enzyme selected from *Table 1*.

2. Check an aliquot for complete digestion on an agarose gel.

3. Dissolve 2 μg of the cleaved plasmid in 2 μl 1 $\times$ S1 buffer and pre-incubate on ice for about 10 min.

4. Dilute nuclease S1 in 1 $\times$ S1 buffer to a final concentration of 1 U/μl and also pre-incubate on ice.

5. Add 2 U of the nuclease S1 to the DNA, mix quickly, but do not centrifuge down to avoid any heating of the sample (unless a centrifuge in the cold room is used).

6. Incubate for 20 min on ice.

7. Add 5 μl of S1 stop buffer, incubate for 10 min at 65°C, phenol extract the DNA, extract with chloroform:isoamyl alcohol 24:1 (v/v), and recover by precipitation with ethanol or isopropanol.

8. Dissolve the DNA in TE buffer and load an aliquot on an agarose gel for estimation of recovered material.

9. Ligate the annealed synthetic DNA cassette into the cDNA. Take about 200 ng of trimmed plasmid DNA dissolved in 1 $\times$ ligation buffer and make two ligations, one with a threefold (approx.) molar excess of DNA cassette, and a second one with a tenfold (approx.) molar excess. Keep the volume as small as possible (5–10 μl), and incubate at 12°C for 3–12 h with about 2 U of T4 DNA ligase.

10. Transform any suitable *E. coli* strain.

11. Screen colonies for the presence of recombinant plasmids by digesting minipreps with *Stul* and/or *Xhol* which will be introduced by the DNA cassette.

12. Make a quick cleavage test (*Protocol 3*) to test for functionality and the orientation of the DNA cassette.

13. Confirm correct insertion of the DNA cassette by sequencing.

Table 1. Summary of different synthetic DNA cassettes suitable for insertion into restriction sites for generation of ribozyme-encoding DNAs

Number	Sequence of DNA cassette[a]	Restriction enzyme(s)[b]	Recognition sequence with cleavable motif[c]
1.	5′ ttcggcctcgaggcctcatcag G 3′ aagccggagctccggagtagtc C	*Acc*I[d] *Acc*I[d] *Cla*I *Bst*BI	GT/AGAC GT/CGAC AT/CGAT TT/CGAA
2.	5′ ttcggcctcgaggcctcatcag T 3′ aagccggagctccggagtagtc A	*Acc*I[d] *Acc*I[d]	GT/ATAC GT/CTAC
3.	5′ T ttcggcctcgaggcctcatcag A 3′ A aagccggagctccggagtagtc T	*Bsu*36I[e] *Esp*I[e]	CC/TNAGG GC/TNAGC
4.	5′ GT ttcggcctcgaggcctcatcag 3′ CA aagccggagctccggagtagtc	*Ava*II[f] *Eco*O109I[g] *Ppmu*I[f] *Rsr*II[f]	G/GWCC RG/GNCCY RG/GWCCY CG/GWCCG
5.	5′ GAT ttcggcctcgaggcctcatcag 3′ CTA aagccggagctccggagtagtc	*Bam*HI *Bcl*I *Bgl*II *Bst*YI	G/GATCC T/GATCA A/GATCT R/GATCY
6.	5′ CT ttcggcctcgaggcctcatcag G 3′ GA aagccggagctccggagtagtc C	*Avr*II *Nhe*I *Spe*I *Xba*I[h]	C/CTAGG G/CTAGC A/CTAGT T/CTAGA
7.	5′ GT ttcggcctcgaggcctcatcag C 3′ CA aagccggagctccggagtagtc G	*Acc*65I *Kpn*I *Bsi*WI *Bsr*GI	G/GTACC GGTAC/C C/GTACG T/GTACA
8.	5′ T ttcggcctcgaggcctcatcag GA 3′ A aagccggagctccggagtagtc CT	*Ava*I[i] *Sal*I *Xho*I	C/YCGRG G/TCGAC C/TCGAG
9.	5′ T ttcggcctcgaggcctcatcag GG 3′ A aagccggagctccggagtagtc CC	*Ava*I[i]	C/YCGRG
10.	5′ ttcggcctcgaggcctcatcag ATG 3′ aagccggagctccggagtagtc TAC	*Bsp*HI[k]	(N)T/CATGA
11.	5′ ttcggcctcgaggcctcatcag CGG 3′ aagccggagctccggagtagtc GCC	*Bsp*EI[k]	(N)T/CCGGA
12.	5′ ttcggcctcgaggcctcatcag TAG 3′ aagccggagctccggagtagtc AGC	*Xba*I[k]	(N)T/CTAGA
13.	5′ GT ttcggcctcgaggcctcatcag AC 3′ CA aagccggagctccggagtagtc TG	*Bst*EII[e]	G/GTNACC
14.	5′ T ttcggcctcgaggcctcatcag 3′ A aagccggagctccggagtagtc	*Ase*I	AT/TAATC

[a] The synthetic DNA cassette consists of two DNA oligonucleotides; the double-stranded sequence comprises the catalytic domain (given in lower case letters) plus some additional nucleotide(s) at the 5′ and/or 3′ end(s) which replace nucleotides of the restriction site that will be removed by trimming. The catalytic domain has been modified so that it contains a *Stu*I and an *Xho*I site.
[b] The restriction enzymes which can be used in combination with this DNA cassette.
[c] The recognition sequence is given and the position of the cleavage site is indicated. The cleavable motif corresponds to nucleotides 16.2, 16.1, 17 according to the numbering system for hammerhead RNAs (compare *Figure 1*) and is underlined.
[d] *Acc*I is able to cleave four different sequences; in line with this, cassettes 1 and 2 can be used for this enzyme.
[e] The nucleotide N of the recognition sequence can by any nucleotide but should not be G.
[f] W can be A or T and should be T for using the insertion strategy.
[g] N should be T.
[h] *Xba*I contains a second motif, compare cassette 12.
[i] *Ava*I can be used for the insertion strategy if Y = T and the cassettes 8 or 9 are needed depending whether R = A or R = G.
[k] The nucleotide preceding the actual recognition sequence forms nucleotide 16.2 of the resulting hammerhead.

4.2 Asymmetric hammerhead ribozymes

Recently Tabler *et al.* (30) reported that catalytic antisense RNAs can be designed in a slightly different way. The helix I of the hammerhead was systematically truncated and it was found that only three nucleotides are required for efficient cleavage of the substrate RNA. Therefore, it is possible to use antisense arms which greatly differ in length: a long helix III-forming region which is responsible for association of the ribozyme with the target and a short helix I which is just needed to form the hammerhead. In view of the asymmetry we have called these types of hammerhead ribozymes 'asymmetric hammerhead ribozymes'.

They combine high performance *in vivo* with simple and flexible generation strategies and further potential for *in vitro* optimization with respect to association (see Section 5.3). Techniques for their generation are described below.

4.3 PCR-based generation strategies of asymmetric hammerhead ribozymes utilizing pre-made helix I box vectors

A helix I of only three base pairs allows simplified PCR-based strategies for ribozyme constructions, since helix I and the major part of a catalytic domain can be provided by a pre-made vector. Such a vector consists (see *Figure 8*) of a promoter followed by one or several restriction sites for cloning purposes

---

Figure 8. Strategy for the generation of an asymmetric hammerhead ribozyme. Part (A) shows a cDNA encoding the target RNA which contains a cleavable motif NTX ↓ (NUX ↓) which has been selected for ribozyme construction and which is followed by the tri-nucleotide sequence NNN. As shown in (B) these nucleotides will form nucleotides 16.2, 16.1, 17, 1.1, 1.2, and 1.3 of the hammerhead. Two DNA oligonucleotides are used for PCR amplification. Oligo 1 contains a terminal recognition sequence for *Eco*RI, and oligo 2 contains at the 5′ terminus a C and a G residues followed by an *Xho*I recognition sequence (T/CTGAG), and the sequence that forms nucleotides 11.3–14 of the hammerhead ribozyme (see *Figure 1*), plus sequences that are complementary to the target RNA. The resulting PCR fragment is cleaved with *Eco*RI and *Xho*I and inserted into one out of 64 representative of the pBS-series that is schematically outlined in (C). The pBS-series of plasmids consists of 64 plasmids that are identical but differ in the 64 conceivable helix I boxes. They are further characterized by the presence of a T3 promoter upstream of the helix I box, which is directly followed by nucleotides 3–10.3 of the hammerhead, and an *Xho*I site, corresponding to nucleotides 10.4–11.4 (see *Figure 1*), plus an *Eco*RI site. The final ribozyme RNA that can be obtained by transcription is given in (D), where the helix I-forming region and the 5′ terminal part of the catalytic domain originates from the pBS plasmid, and the residual part of the catalytic domain, as well as the entire helix III-forming region, is derived from the PCR product. Part (E) details the sequence of the plasmid pBS–GAC/CTG, a particular representative of the pBS-series, which could be used for generating an asymmetric hammerhead ribozyme in which the sequence GAC is found in the target RNA downstream of the cleavage site. Sequences corresponding to the catalytic domain are given in lower case letters. *Figure 9* gives a specific example for the construction of an asymmetric hammerhead ribozyme.

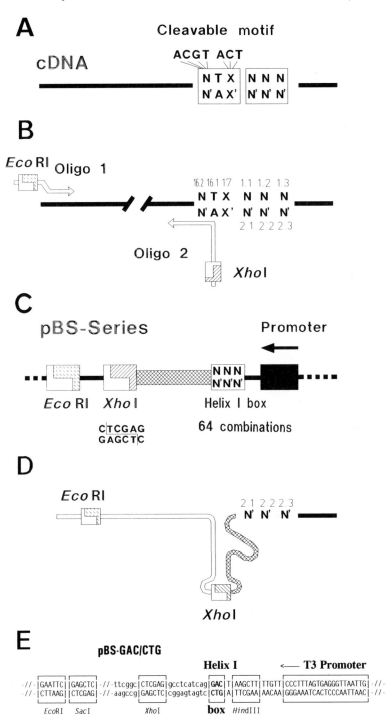

and a three-nucleotide sequence that delivers ultimately the helix I-forming part (helix I box) followed by the 5′ part of the catalytic domain into which a site for a restriction enzymes has been introduced. 64 different vectors are needed representing the 64 conceivable combinations for a helix I.

We are currently in the process of constructing these 64 vectors. Construction of an asymmetric hammerhead ribozyme, as outlined in *Figure 8*, just requires PCR amplification of helix III and subsequent insertion into one of the preformed vectors. Such a strategy provides considerable flexibility to target any conceivable cleavable motif.

Protocol 2. Generation of asymmetric hammerhead ribozymes by PCR amplification and subcloning into a helix I box vector

Equipment and reagents

- Standard equipment for molecular biology work
- PCR machine
- A cDNA fragment of the target RNA
- Two synthetic DNA oligos

- A plasmid vector of the pBS-series. Contact the first author for a suitable plasmid vector.
- *Taq* DNA polymerase
- T4 DNA ligase

A. *Selection of target site and design of the DNA oligos required for PCR*

1. Select a target site $N_{16.2}U_{16.1}X_{17\downarrow}N_{1.1}N_{1.2}N_{1.3}$ taking into account the considerations addressed in Section 3.

2. Check whether a plasmid of the pBS-series that matches to the sequence $N_{1.1}N_{1.2}N_{1.3}$ (helix I box) is available.

3. Plan the synthesis of two DNA oligos required for PCR. The 'downstream' DNA oligo which contains a part of the catalytic domain, and at the 3′ terminus a sequence that is complementary to the region in the target RNA just upstream of the cleavable motif; and the 'upstream' DNA oligo which has the same polarity as the target RNA. The downstream DNA oligo should contain the sequence: 5′ GGCTC-GAGGCCGAA and should be followed by 14–20 nucleotides specific for a certain target RNA (see *Figure 9*). These are the nucleotides 15.1–15.14 or 15.20, respectively according to the nomenclature for the hammerhead ribozymes, which are complementary to nucleotides 16.1–16.20 in the target RNA. Note: since nucleotide 16.1 corresponds to the U of the NUX $\downarrow$ motif, the downstream oligo must continue with an 'A' residue. The length of the specific sequence is influenced on the sequence itself. Pay attention that no self-complementarity occurs and that the 3′ terminus of the DNA oligo is not base paired when forming a dimer with another molecule of the DNA oligo.

B. *PCR amplification and cloning of the asymmetric hammerhead ribozyme*

1. Make a PCR amplification using the two DNA oligos and the cDNA fragment of the target RNA as template. Temperature conditions largely depend on the melting temperatures of the selected DNA oligos; for details consult PCR literature.

2. Check PCR products on an appropriate gel for the expected size.

3. Digest the plasmid of the pBS-series and the PCR fragment with *Eco*RI and *Xho*I.

4. Ligate the PCR fragment into the plasmid of the pBS-series and transform any suitable *E. coli* strain.

5. Screen colonies for the presence of recombinant plasmids.

6. Make a quick cleavage test (*Protocol 3*) to test for functionality and the orientation of the DNA cassette.

7. Confirm correct insertion of the DNA cassette by sequencing.

4.4 Direct PCR amplification

The short length required for helix I allows generation of the desired asymmetric hammerhead ribozyme in a single PCR amplification step using a DNA oligo providing at the 5′ terminus a cloning site, a three-base helix I, the entire catalytic domain, and some nucleotides complementary to the cDNA. The entire length of the required DNA oligo is thus at least 45 nucleotides. Due to the presence of the complete sequence of helix II in this DNA oligo there is internal base pairing which may lead to artefacts during PCR amplification. Meanwhile, we have used the strategy of direct PCR amplification several times without any problems (Liu and Tabler, unpublished data).

5. Testing and optimizing the catalytic antisense RNA *in vitro*

5.1 Quick cleavage assay for testing recombinant DNA encoding catalytic antisense RNA

Regardless of how the DNA that encodes the catalytic antisense RNA has been constructed, the correct manipulations have to be confirmed by sequencing the recombinant DNA. As a preliminary step, a quick ribozyme assay can be performed using RNA transcripts from miniprep DNA. This avoids the sequencing of catalytically inactive constructs and also gives an indication for the overall success of the construction work. This procedure is described in *Protocol 3*.

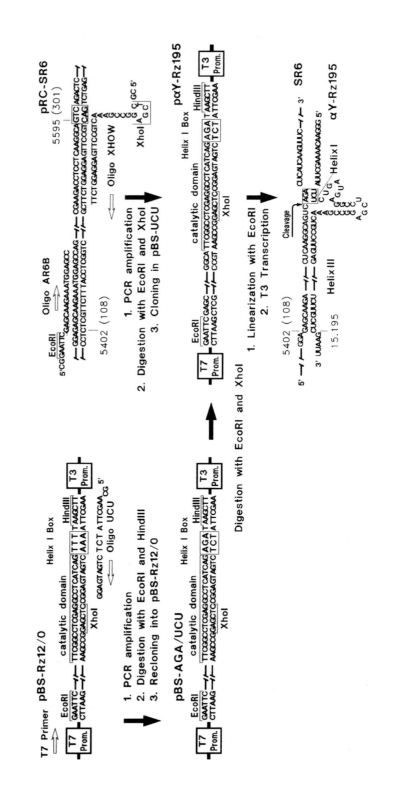

Figure 9. Strategy for simplified generation of asymmetric hammerhead ribozymes. The *left* panel shows the conversion of the plasmid pBS–Rz12/0 that encodes—if transcribed from the T3 promoter—the catalytic domain of a hammerhead ribozyme preceded by an AAA sequence that could form a helix I in an asymmetric hammerhead ribozyme. By using the DNA oligonucleotide UCU and the T7 sequencing primer d(TAATACGACTCAC-TATAGGG), a DNA fragment was PCR amplified and recloned, resulting in plasmid pBS–AGA/UCU with a modified helix I box. The resulting T3 RNA transcript could pair to a target RNA having a cleavable motif that is followed by an AGA sequence. Similarly, all conceivable three-nucleotide helix I boxes can be introduced. The *right* panel shows the conversion of the plasmid pRC–SR6, that contains HIV-1 cDNA sequences (nucleotide numbers refer to sequence of Ratner *et al.* (33); those in parenthesis to the SR6 RNA transcript), into a ribozyme construct. The GTC sequence corresponding to the selected target site is boxed. Two nested DNA oligos AR6B and XHOW were used for PCR amplification. The latter one is given in the conformation of a hammerhead structure for clarity only. The amplified DNA was cloned via the newly introduced *Eco*RI and *Xho*I sites into the plasmid pBS–AGA/UCU that provides the matching helix I box, yielding plasmid pαY195. After transcription with T3 RNA polymerase, the asymmetric ribozyme RNA αY195 is generated which is shown below in its hammerhead conformation together with the target RNA SR6. This hammerhead complex forms a three-nucleotide helix I and a helix III consisting of 195 nucleotides. Thus construction of asymmetric hammerhead ribozymes just requires PCR amplification, followed by subcloning into the matching vector that delivers the catalytic domain plus the target-specific helix I-forming region (from ref. 30, with permission).

Protocol 3. Quick cleavage assay

Equipment and reagents

- Standard equipment for molecular biology work
- A ^{32}P-labelled RNA transcript of the target RNA
- RNase A (DNase-free)
- TE buffer: 10 mM Tris–HCl, 1 mM EDTA pH 8.0
- RNasin (RNase inhibitor)
- DNase I (RNase-free)

Method

1. Prepare a radioactively labelled target RNA by *in vitro* transcription and dissolve it in 10 mM TE buffer.

2. Prepare minipreps of the colonies that might contain DNA that encodes for a catalytic antisense RNA.

3. Check by restriction analysis whether certain restriction sites are present and/or for the expected size of DNA fragments.

4. Cut the miniprep DNA (about 0.5–1 μg) with a restriction enzyme that cleaves downstream of the catalytic antisense RNA. Add about 10 ng of DNase-free RNase to the digestion.

5. Phenol extract twice and precipitate DNA.

6. Use the cut miniprep DNA for synthesis of catalytic antisense RNA by *in vitro* transcription with the appropriate RNA polymerase (SP6, T3,

Protocol 3. *Continued*

T7) in a reaction volume of 20 μl (without any radionucleotides, but in the presence of RNasin in concentration as recommended by supplier) using all the cut miniprep DNA as template.

7. Add 1 U of DNase I (RNase-free) to transcription mixture, incubate for 5 min at 37°C.

8. Inactivate RNA polymerase and DNase I for 10 min at 60°C.

9. Adjust an aliquot of the radioactively labelled substrate RNA to a final concentration of 50 mM Tris–HCl pH 8.0 and 30 mM MgCl$_2$. Take 10 μl of this mixture and add 10 μl of the transcription mixture (obtained with cut miniprep DNA) and incubate for 1 h at 37°C. Use another 10 μl of labelled substrate RNA for control and incubate under the same conditions with 10 μl of 1 × transcription buffer, without any DNA or enzymes added.

10. Add EDTA to a final concentration of 25 mM, phenol extract, add sodium acetate to a final concentration of 0.2 M, and precipitate with ethanol.

11. Analyse reaction products, together with the control RNA, on a denaturating polyacrylamide gel.

12. Check for two defined cleavage products that are not detectable for the control reaction.

5.2 Determination of association rates

The association rate of the sense and the catalytic antisense RNA in a living cell is influenced by a number of parameters, such as intracellular concentration and localization of both RNAs, as well as proteins that interact with the RNA, but also by intrinsic structural features of each of the complementary RNA strands themselves. Measuring the association kinetics of the two RNAs *in vitro* will provide an indication how well a particular catalytic antisense RNA will inhibit the target RNA *in vivo*.

The association of two complementary RNA strands is a bimolecular reaction and follows second order kinetics. The rate of association is determined by the association rate constant k_{ass} and the concentrations of the two RNAs. The reaction proceeds according to *Equation 1*:

$$v = d[\text{T·Cas}]/dt = k_{ass} \times [\text{T}] [\text{Cas}] \tag{1}$$

wherein v is the association rate, and [T] and [Cas] are the concentrations of target RNA and of catalytic antisense RNA, respectively, and [T·Cas] is the association product of both RNAs.

One way to determine the association rate constant is to use one RNA in large molar excess. As a consequence, the concentration of the unpaired

RNA that is present in excess remains unchanged during the course of the association reaction, so that it follows pseudo first order kinetics.

Assuming that the concentration of the target RNA is used in molar excess ([T] >> [Cas]), a constant k_1 can be defined:

$$k_1 = k_{ass} \times [T] \qquad [2]$$

and the association rate v is:

$$v = k_1 \times [Cas] \qquad [3]$$

This is the equation describing (pseudo) first order kinetics and k_1 can be simply determined by measuring the half-life $t_{1/2}$ of the free catalytic antisense RNA [Cas]:

$$k_1 = \ln 2/t_{1/2} \sim 0.69/t_{1/2} \qquad [4]$$

In combination with the known concentration of [T] and *Equation 2*, k_{ass} can be determined according to the final *Equation 5*:

$$k_{ass} = \ln 2/t_{1/2} \times 1/[T] \sim 0.69/t_{1/2} \times 1/[T] \qquad [5]$$

In line with *Equation 5*, the association rate constant k_{ass} has the dimension $M^{-1}sec^{-1}$ or $M^{-1}min^{-1}$.

The determination of the association rate of catalytic antisense RNA and the target RNA requires a gel system in which the double-stranded complex migrates differently from the free RNA that is radioactively labelled. Depending on the size of the two partners each of the RNAs can be radio-actively labelled, but in most cases the catalytic antisense RNA is smaller than the target RNA. By association it will be found in a rather large complex, so that free and complexed RNA may be easier to separate if the catalytic antisense RNA is labelled (see ref. 31 and *Figure 10*).

The association rate can be determined in the presence of $MgCl_2$, even though cleavage occurs under these conditions. The helices of catalytic anti-sense RNA are sufficiently long to prevent dissociation in gel systems under not fully denaturing conditions. When asymmetric hammerhead ribozymes are used, it has to be taken into account that the 3' terminal cleavage product dissociates. In case this might cause any trouble, a Mg^{2+}-free buffer can be used which should contain a higher concentration of NaCl.

Protocol 4. Determination of association rates

Equipment and reagents

- A ^{32}P-labelled RNA transcript of the catalytic antisense RNA and an unlabelled target RNA transcript of known concentration; also the other combination, labelled target and unlabelled catalytic antisense RNA may be used
- Standard equipment for molecular biology work
- Association buffer: 20 mM Tris–HCl pH 7.5, 10 mM $MgCl_2$, 100 mM NaCl
- Polyacrylamide slab gel (4–6%) containing 8 M urea and 22.5 mM TBE buffer

Protocol 4. *Continued*

- Stop buffer: 80% formamide, 50 mM Tris–HCl pH 8.0, 50 mM EDTA pH 8.0, 0.5% SDS, 0.025% xylene cyanol, 0.025% bromophenol blue
- TBE buffer (1 M): 1 M Tris, 830 mM boric acid, 10 mM EDTA, pH adjusts to 8.3

Method

1. Mix in about 30 µl of association buffer the unlabelled RNA at a defined concentration and ranging between 10–100 nM. Select concentration such that the unlabelled RNA is present in an at least five-fold (preferably > tenfold) molar excess over the radiolabelled complementary RNA.

2. Remove aliquots (for example 3 µl) at certain reaction times and add to 10 µl pre-cooled (ice) stop buffer.

3. Apply samples to a polyacrylamide gel without prior boiling of the samples.The percentage of acrylamide depends on the size of the RNAs, and should range between 4% and 6%. The acrylamide gel should contain urea, the concentration may be up to 8 M; the running buffer is 22.5 mM TBE.

4. Determine the amounts of radioactivity found in the complexed and in the free RNA, either by direct image plate analysis or by excising gel slices, and counting in a liquid scintillation counter.

5. Plot percentage of free RNA in a logarithmic scale against time. This representation should give a straight line.

6. Deduce the half-life $t_{1/2}$ from the semi-logarithmic plot and calculate the association rate from the known concentration of the unlabelled RNA according to *Equation 5* in Section 5.2.

5.3 Optimization of the length of helix III

As has been outlined earlier, association of the ribozyme with its target is a key step, especially under conditions of stoichiometric cleavage. The speed of the association is strongly influenced by the overall secondary structures, both of the target RNA as well of the antisense RNA, regardless of whether or not it is catalytic. When engineering an antisense RNA there is a choice to select either a sequence that is complementary to the complete target region or just a subdomain of it. Also this selection will influence the secondary structure of the antisense RNA and thus the association with the target RNA. In line with these considerations, Rittner *et al.* (27) have found that the association rates strongly depend on the length of the antisense RNA, however, in an irregular and unpredictable manner. Small changes in the size of the antisense RNA may result in more than 100-fold increased or decreased association rates. Moreover, the association rates alternate, i.e. they increase and decrease with

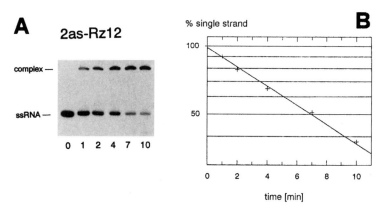

Figure 10. Example of an association kinetics to monitor the formation of duplex RNA. (A) In this case, single-stranded ^{32}P-labelled RNA 2s (ssRNA) was hybridized for the time (minutes) as indicated with a molar excess of ribozyme RNA 2as–Rz12 to form double-strands (complex). Reaction products were separated on a 6% polyacrylamide gel, containing 5.6 M urea. (B) Percentage of remaining single-stranded RNA (logarithmic scale) versus time, for determination of the half-life of free single-stranded RNA (from ref. 21, with permission).

growing chain length of the antisense RNA. Rittner *et al.* (27) have described a technique how to select *in vitro* for fast-hybridizing RNA species from a pool of RNAs that have the identical 5' terminus which is described in *Figure 11*.

This technique is also applicable to asymmetric hammerhead ribozymes. Here, the association is almost completely controlled by the antisense flank that forms helix III, because a three-base pair helix I is too short to initiate hybridization. Variation in the length of helix III will therefore result in asymmetric hammerhead ribozymes with differences in their capabilities for association with the target RNA. In order to identify the fast-hybridizing species it is recommended to first generate an asymmetric hammerhead ribozyme able to form a helix III of about 150–200 nucleotides. Then a pool of asymmetric hammerhead ribozymes must be generated, wherein all RNAs have an identical and labelled 5' end, but differ in the length of the helix III-forming region (*Figure 12*). The pool of asymmetric hammerhead ribozymes is subjected to the *in vitro* selection procedure as outlined for antisense RNA in *Protocol 5*. This will allow one to distinguish one or several fast-associating ribozymes, now with a shorter helix III compared with the parental RNA. For the specific synthesis of a particular fast-hybridizing asymmetric hammerhead ribozyme, it will be necessary to adjust the corresponding DNA construct accordingly. The required truncation of the domain that encodes for helix III is most conveniently done by PCR amplification of the interior, fast-hybridizing subdomain contained in the parental DNA construct. The PCR product may then be recloned or used directly as template for *in vitro* transcription of the fast-hybridizing ribozyme.

119

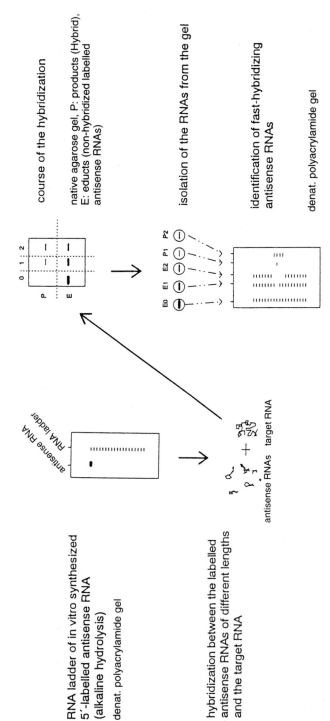

RNA ladder of in vitro synthesized
5′-labelled antisense RNA
(alkaline hydrolysis)

denat. polyacrylamide gel

hybridization between the labelled
antisense RNAs of different lengths
and the target RNA

course of the hybridization

native agarose gel, P: products (Hybrid),
E: educts (non-hybridized labelled
antisense RNAs)

isolation of the RNAs from the gel

identification of fast-hybridizing
antisense RNAs

denat. polyacrylamide gel

Figure 11. Schematic depiction of *in vitro* selection and identification of fast-hybridizing antisense/catalytic antisense RNAs as described in *Protocol 6* (from ref. 26, with permission).

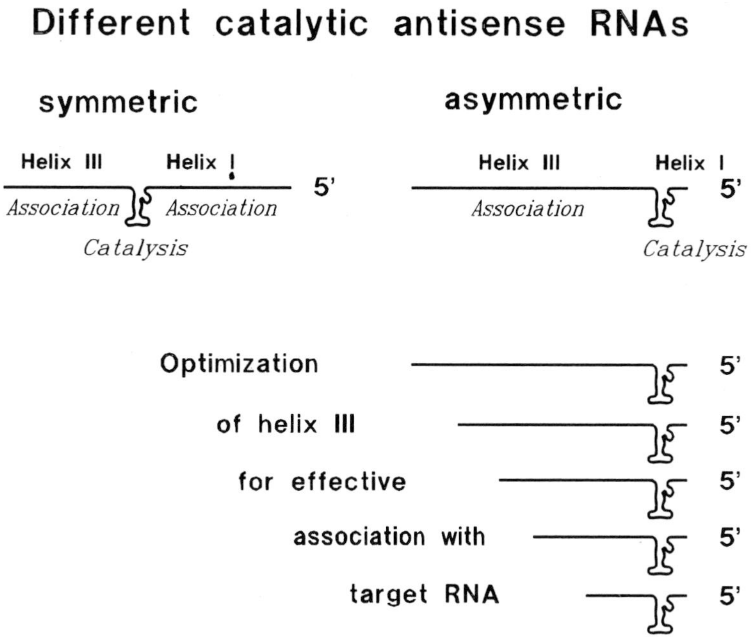

Different catalytic antisense RNAs

symmetric   asymmetric

Figure 12. Different design of symmetric and asymmetric hammerhead ribozymes. By shortening helix III the asymmetric hammerhead ribozyme can be optimized for efficient association with the target RNA.

Protocol 5. *In vitro* selection for fast association

Equipment and reagents

- Standard equipment for molecular biology work
- Phosphoimager (recommended)
- Sephadex G50
- 5′ ^{32}P-labelled asymmetric catalytic antisense RNA
- Unlabelled target RNA, which should be longer than the catalytic antisense RNA

- Hydrolysis buffer: 500 mM NaHCO$_3$
- Standard equipment for running sequencing gels
- Association buffer: 20 mM Tris–HCl pH 7.4, 10 mM MgCl$_2$, 100 mM NaCl.
- Stop buffer: 20 mM Tris–HCl pH 8.0, 10 mM EDTA, 0.5% SDS, 7 M urea, 0.04% bromophenol blue, 0.04% xylene cyanol

Method

1. Subject 5′ ^{32}P-labelled asymmetric catalytic antisense RNA (e.g. 10 ng of a 150-mer) to alkaline hydrolysis by incubation in hydrolysis buffer at 96°C for 10 min.

2. Check extent of hydrolysis by analysis of the products on denaturing polyacrylamide gels (sequencing gels). In case of unequal distribution of hydrolysis products, extend or reduce, respectively, incubation time for an appropriate time interval.

121

Protocol 5. *Continued*

3. Purify hydrolysis products by gel filtration with Sephadex G50 and an elution buffer containing 10 mM Tris–HCl pH 8.0 and 1 mM EDTA. Discard the small hydrolysis products which elute last.

4. Refold RNA species by incubating at 75°C for 10 min and subsequent slow cooling to room temperature.

5. Mix hydrolysis products (*c.* 1 nM final conc.) with an approximately 10- to 100-fold excess of unlabelled target RNA in the association buffer in a final volume of 20 μl at 37°C.

6. Withdraw aliquots (3–5 μl) at appropriate time points, mix with pre-cooled stop buffer (0°C, 40 μl), and analyse the course of the hybridization reaction (association) by analysis of the samples on non-denaturing 1.2% agarose gels in 89 mM Tris–borate buffer pH 8.3.

7. Excise gel slices containing the single-strand fraction (fast mobility) and double-strand fraction (slow mobility due to association with larger sense RNA), respectively, of the hybridization reaction, and elute the RNAs (e.g. by three times freeze–thawing with dry ice and at 37°C, and subsequent centrifugation).

8. Precipitate RNAs with ethanol and redissolve pellets in stop buffer.

9. Analyse the time-dependent distribution of eluted antisense RNA species by polyacrylamide gel electrophoresis under denaturing conditions (sequencing gels).

10. Expose dried gels to X-ray films, or perform image plate analysis for quantification of the time-dependent decrease of antisense RNA species in the single-strand fraction, or the time-dependent increase of antisense RNAs in the double-strand fraction, respectively.

11. Calculate the association rate constants for each individual antisense species according to Section 5.2.

5.4 Determination of cleavage rates

The cleavage reaction is the process that follows the association step, i.e. after the two double-stranded helices I and III of the hammerhead conformation have formed. It is possible that a conformational change is required before the actual chemical step, i.e. the intramolecular *trans*-esterification reaction, can proceed. It might be that this conformational change is actually rate-limiting for the cleavage process (for a more detailed discussion, see ref. 32). The determination of the rate constant describes the overall cleavage process including a possible conformational change. In view of the lack of dissociation in the case of a catalytic antisense RNA, it is necessary to determine the cleavage

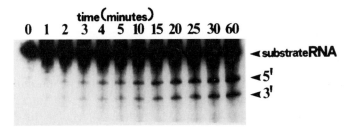

Figure 13. Cleavage analysis of a catalytic antisense RNA. An *in vitro* synthesized and radioactively labelled target RNA derived from Sendai virus of about 220 nucleotides was pre-annealed with a catalytic antisense RNA with deletions in nucleotides 2.2 and 2.3, and incubated in the presence of Mg^{2+} for the times given. The position of the 5' and 3' cleavage products is indicated. Due to the deletion the ribozyme activity was low. Quantitative analysis of the cleavage product in relation to the uncleaved substrate RNA as described in *Protocol 6* determined a half-life of the substrate–ribozyme complex of $t_{1/2} =$ 195 min, which corresponds to a cleavage rate of about 3.6×10^{-3} min^{-1} (from ref. 19, with permission).

activity under conditions of single-turnover. For that purpose, the two RNAs are pre-annealed using a molar excess of the catalytic antisense RNA. After all substrate RNAs are complexed via helices I and III, the cleavage reaction is started by addition of Mg^{2+} and the time course is monitored. Association of Mg^{2+} at concentrations of 10 mM or higher is quick and not rate-limiting. If a preformed hammerhead complex is used, the cleavage reaction is concentration-independent. For a more detailed analysis, it is necessary to confirm that complete cleavage of the target RNA can be achieved. From the proportion of cleaved and uncleaved substrate RNA at different reaction times the half-life for the uncleaved target RNA can be determined. This allows the calculation of the cleavage rate by analogy to *Equation 4* described for the association rate. The cleavage rate constant has the dimension sec^{-1} or min^{-1}. A specific example is outlined in *Figure 13*.

The cleavage rates determined for efficient catalytic antisense RNA are in the order of 10^{-2} min^{-1} at 37°C. It should be added that short hammerhead ribozymes can also be tested under conditions for multiple-turnover reaction where they follow Michaelis–Menten kinetics. In that case the k_{cat} values are in the order of about 1 min^{-1}.

Protocol 6.   Determination of cleavage rate

Equipment and reagents

- Standard equipment for molecular biology work
- ^{32}P-labelled RNA transcript of the target RNA and unlabelled catalytic antisense RNA of known concentration

- Pre-association buffer: 20 mM Tris–HCl pH 7.5, 500 mM NaCl
- Stop buffer: 0.2 M sodium acetate, 4 mM EDTA pH 5.0, 30 μg/ml carrier RNA

Protocol 6. *Continued*

Method

1. Mix in 30 μl of pre-association buffer radioactively labelled target RNA and a molar excess of catalytic antisense RNA (about 50 nM), and incubate for 15 min at 37°C or any other reaction temperature. If association rates are known, the half-life of the free substrate RNA may be calculated, and the time of pre-annealing and/or the NaCl concentration reduced.

2. Remove an aliquot (for example 3 μl) of the mixture as the zero time value.

3. Start ribozyme reaction by adding $MgCl_2$ to a final concentration of 20 mM (for example 3 μl of a 200 mM solution).

4. Remove aliquots of the reaction (for example 3 μl) at certain times (5 min to 2 h), and add the aliquots to 30 μl of stop solution.

5. Add 90 μl of ethanol and collect samples by precipitation and centrifugation.

6. Dissolve samples in 10 μl gel application buffer, put in a boiling water-bath for 1 min, and chill in ice, followed by electrophoretic separation on a polyacrylamide gel (4–6%).

7. Determine for each lane, which represents an individual reaction time, the radioactivity found in the target RNA (T) and in the 5′ and 3′ terminally cleavage products (P5) and (P3).

8. Determine the portion of uncleaved RNA according to the formula:

   % of uncleaved RNA $= 100 - 100(P5 + P3)/(P5 + P3 + T)$.

9. Plot the percentage of uncleaved RNA in a logarithmic scale versus the time axis and deduce the half-life $t_{1/2}$.

10. Calculate cleavage rate according to the formula: $k_{cleav} = \ln 2/t_{1/2}$.

6. Concluding remarks

Most researchers will be interested just in applying the ribozyme technology for suppression of a particular gene, rather than in the technology *per se*. Straightforward strategies have been described in Section 4 for the relatively simple construction of hammerhead ribozymes. Therefore, it is recommended to construct several catalytic antisense RNAs directed against a particular target RNA in parallel, taking the parameters into account as described above in detail. Extensive *in vitro* analysis, applying the various protocols described in Section 5, will allow selection and optimization of a catalytic antisense RNA which performs well with regard to *in vitro* association with target RNA and its cleavage. We believe that this is a prerequisite for the reliable and successful application of the technology. The carefully selected RNA will be most promising when eventually used for target suppression in living cells.

References

1. Matthews, R. E. (1991). *Plant virology*, 3rd edn. Academic Press, San Diego, USA.
2. Forster, A. C. and Symons, R. H. (1987). *Cell*, **49**, 211.
3. Hernández, C. and Flores, R. (1992). *Proc. Natl. Acad. Sci. USA*, **89**, 3711.
4. Hertel, K. J., Pardi, A., Uhlenbeck, O. C., Koizumi, M., Ohtsuka, E., Uesugi, S., *et al.* (1992). *Nucleic Acids Res.*, **20**, 3252.
5. Pley, H. W., Flaherty, K. M., and McKay, D. B. (1994). *Nature*, **372**, 68.
6. Tuschl, T., Gohlke, C., Jovin, T. M., Westhof, E., and Eckstein, F. (1994). *Science*, **266**, 785.
7. Cech, T. and Uhlenbeck, O. (1994). *Nature*, **372**, 39.
8. Uhlenbeck, O. C. (1987). *Nature*, **328**, 596.
9. Haseloff, J. and Gerlach, W. L. (1988). *Nature*, **334**, 585.
10. Marschall, P., Thomson, J. B., and Eckstein, F. (1994). *Cell. Mol. Neurobiol.*, **14**, 523.
11. Symons, R. H. (1992). *Annu. Rev. Biochem.*, **61**, 641.
12. Hertel, K. J., Herschlag, D., and Uhlenbeck, O. C. (1994). *Biochemistry*, **33**, 3374.
13. Cotten, M., Schaffner, G., and Birnstiel, M. L. (1989). *Mol. Cell. Biol.*, **9**, 4479.
14. Cameron, F. H. and Jennings, P. A. (1989). *Proc. Natl. Acad. Sci. USA*, **86**, 9139.
15. L'Hullier, P. J., Davis, S., and Bellamy, A. R. (1992). *EMBO J.*, **11**, 4411.
16. Bruening, G. (1989). In *Methods in enzymology*, Vol. 180, p. 546.
17. Perriman, R., Delves, A., and Gerlach, W. L. (1992). *Gene*, **113**, 157.
18. Koizumi, M., Iwai, S., and Ohtsuka, E. (1988). *FEBS Lett.*, **228**, 228.
19. Zoumadakis, M., Neubert, W. J., and Tabler, M. (1994). *Nucleic Acids Res.*, **22**, 5271.
20. Nellen, W. and Lichtenstein, C. (1993). *Trends Biochem. Sci.*, **18**, 419.
21. Homann, M., Tzortzakaki, S., Rittner, K., Sczakiel, G., and Tabler, M. (1993). *Nucleic Acids Res.*, **21**, 2809.
22. Sczakiel, G. and Goody, R. S. (1994). *Biol. Chem. Hoppe-Seyler*, **375**, 745.
23. Sczakiel, G., Homann, M., and Rittner, K. (1993). *Antisense Res. Dev.*, **3**, 453.
24. Zuker, M. and Stiegler, P. (1981). *Nucleic Acids Res.*, **9**, 133.
25. Crisell, P., Thompson, S., and James, W. (1993). *Nucleic Acids Res.*, **21**, 5251.
26. Rittner, K., Burmester, C., and Sczakiel, G. (1993). *Nucleic Acids Res.*, **21**, 1381.
27. Tabler, M. and Tsagris, M. (1991). *Gene*, **108**, 175.
28. Prody, G. A., Bakos, J. T., Buzayan, J. M., Schneider, I. R., and Bruening, G. (1986). *Science*, **231**, 1577.
29. Buzayan, J. M., Gerlach, W. L., Bruening, G., Keese, P., and Gould, A. R. (1986). *Virology*, **151**, 186.
30. Tabler, M., Tzortzakis, S., Homann, M., and Sczakiel, G. (1994). *Nucleic Acids Res.*, **22**, 3958.

31. Persson, C., Wagner, G. E., and Nordström, K. (1988). *EMBO J.*, **7**, 3279.
32. Homann, M., Tabler, M., Tzortzakaki, S., and Sczakiel, G. (1994). *Nucleic Acids Res.*, **22**, 3951.
33. Ratner, L., Haseltine, W., Patarca, R., Livak, K. J., Starcich, B., Josephs, S. J., *et al.* (1985). *Nature*, **313**, 277.
34. Shimayama, T., Nishikawa, S., and Taira, K. (1995). *Biochemistry*, **34**, 3649.
35. Zoumadakis, M. and Tabler, M. (1995). *Nucleic Acids Res.*, **23**, 1192.
36. Hormes, R., Homann, M., Oelze, I., Marshall, P., Tabler, M., Eckstein, F., and Sczakiel, G. (1997). *Nucleic Acids Res.*, **25**, 769.

6

2–5A-antisense chimeras for
targeted degradation of RNA

R. H. SILVERMAN, A. MARAN, R. K. MAITRA, C. F. WALLER,
K. LESIAK, S. KHAMNEI, G. LI, W. XIAO, and P. F. TORRENCE

1. Introduction

The 2–5A-antisense approach under development in our laboratories seeks to substantially enhance the efficiency of antisense oligodeoxyribonucleotides (oligos) *in vivo* by harnessing the activity of a ubiquitous intracellular endoribonuclease called 2–5A-dependent RNase (RNase L) (1–3). In this strategy, an activator of RNase L, the 2',5'-linked tetraadenylate $pA(2'p5'A)_3$, is conjugated through linkers to antisense oligos. As a result, the cell's ability to control RNA decay is reprogrammed to degrade RNA molecules of choice. We have used this strategy to demonstrate selective and catalytic destruction of target mRNA molecules in assays containing homogeneous RNase L, in crude cell-free systems and in intact human cells. Here we present our protocols to synthesize and analyse the activity of 2–5A-antisense species for the ablation of target RNA molecules. Current work, to be described elsewhere, aims to further enhance the stability, activity, and mode of synthesis of 2–5A-antisense.

2. Background

2.1 The 2–5A system: an RNA degradation pathway that functions in interferon action

The 2–5A system is a controllable RNA degradation pathway that is induced by the interferons (*Figure 1*) (4, 5). Interferon treatment of cells activates genes encoding a group of double-stranded RNA (dsRNA)-dependent synthetases that produce short 5' triphosphorylated, 2',5' phosphodiester-linked oligoadenylates (2–5A) from ATP (6, 7). 2–5A is unstable in cells due to the combined activities of phosphatase and phosphodiesterase and, as a result, effects of 2–5A in cells are transient (8, 9). 2–5A functions by activating the endoribonuclease, RNase L (10, 11), resulting in the degradation of single-stranded RNA with modest specificity after UpN dimers (12, 13). Human RNase L is an 83.5 kDa polypeptide with no measurable RNase activity unless

R. H. *Silverman* et al.

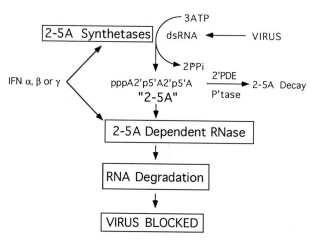

Figure 1. The 2–5A system.

2–5A is present (14). 2–5A binding to RNase L switches the enzyme from its off-state to its on-state. The activators require at least three 2′,5′-linked oligoadenylates and a single 5′ phosphoryl group for maximal activation of the human RNase (14, and references therein). For example, the 3′,5′-linked molecule, $p_3A(3'p5'A)_2$, and the 2′,5′-linked dimer, $p_3A2'p5'A$, are both completely lacking in the ability to activate purified, recombinant RNase L (14). In contrast, $pA(2'p5'A)_3$ (the 2–5A species used in this study), $p_3A(2'p5'A)_2$, or $p_3A(2'p5'A)_3$ each produce optimal activation of RNase L (fully active at nanomolar amounts). The presence of RNase L is observed in a wide range of mammalian cell types. The methods described here involve adapting the 2–5A system for the purpose of developing novel research and therapeutic agents.

2.2 The 2–5A-antisense strategy

We have applied (1–3) the unique and potent 2–5A system (4, 5) to the specific degradation of mRNA through an examination of the biological activity of synthetic adducts between 2–5A and antisense oligos. Such composite nucleic acids, through the antisense domain, could provide a high degree of specificity normally missing from RNase L cleavages to target the chimera to a particular mRNA sequence. The 2–5A component provides a localized activation of the latent RNase L allowing potent and catalytic destruction of the targeted mRNA (*Figure 2*). In the presence of 2–5A-antisense, RNase L was shown to cause the catalytic cleavage of the RNA target (k_{cat} of about 7 sec^{-1}) (15). Suppression of RNase L activity prior to pairing of the antisense to the mRNA target is caused by the linker–DNA parts of the chimeras (15). This effect prevents the non-specific degradation of RNA unless much higher amounts (> tenfold) of the chimeras are used (15). Thus this approach could substantially increase antisense potency without decreasing selectivity. Pre-

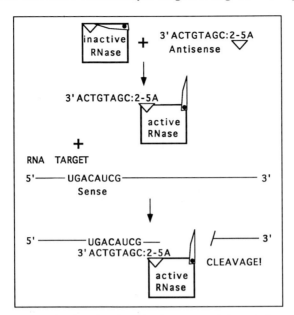

Figure 2. Proposed mechanism for the selective targeted degradation of RNA by 2–5A-antisense chimeras.

sumably, antisense activity by such chimeric molecules could also involve classical mechanisms, such as passive blocking of translation and possibly RNase H degradation (16, 17). The antisense region of the composite molecule might also facilitate uptake of 2–5A since antisense oligos seem to be taken up by intact cells, perhaps by a specific mechanism (discussed in Chapter 3).

3. 2–5A-antisense chimeras: construction, composition, and synthetic approach

There are a number of different possible ways in which 2–5A might be joined to an antisense oligo. Modifications to the 5′ terminus or to most of the functional groups of the sugars or bases may well be expected to adversely affect binding or activation of the RNase L (5, 18, 19). On the other hand, 2′ alterations of the 2–5A molecule usually lead to retention of, or even enhancement of, biological activity (5, 19).

The design of the 2–5A-antisense molecule relies upon information gleaned from studies on the relationship between oligo structure and RNase L activation (5, 18–30). The prototypical 2–5A-antisense chimera consists of an antisense domain made up of oligo(dT)$_{18}$ connected to 2–5A through a linker (1, 2). The 2–5A and antisense moieties are joined by two 1,4-butanediol molecules joined to each other and to 2–5A and antisense by phosphodiester bonds (*Figures 3* and *4*). Linkage to the 2–5A tetramer is at the 2′ terminal hydroxyl

One additional cycle with DMT butanediol phosphoramidite, capping, and oxidation, followed by TCA detritylation.

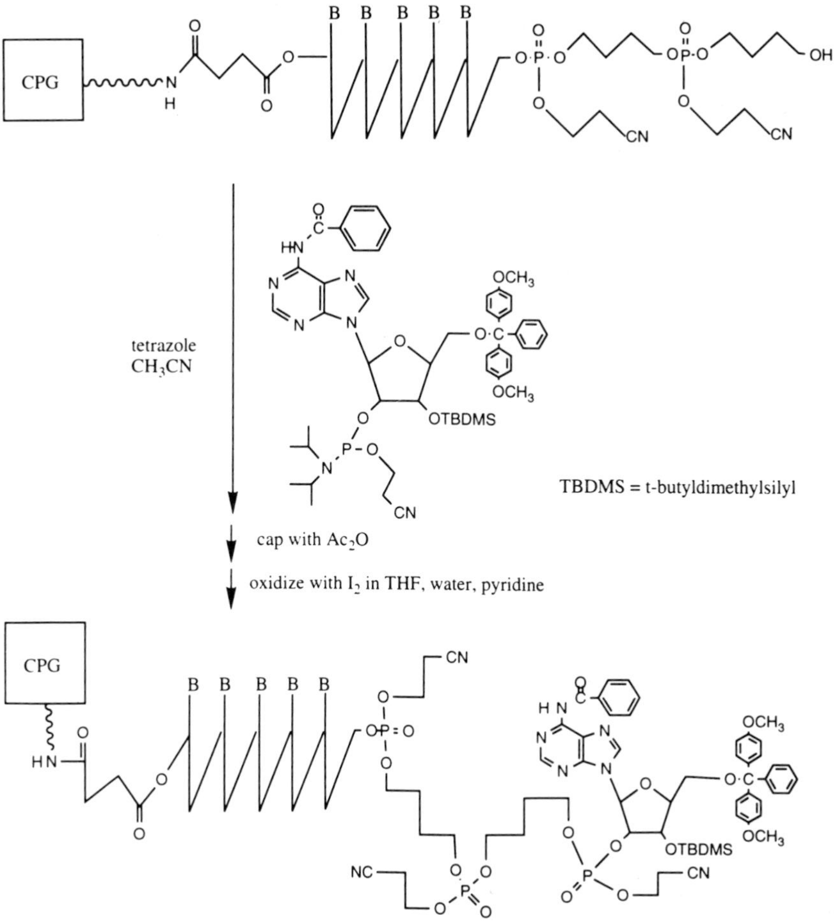

Figure 3. Synthesis of 2–5A-antisense core. The antisense oligodeoxynucleotide, obtained in the course of standard DNA synthesis using the phosphoramidite approach, was 5′ detritylated to allow addition of two units of butanediol linker. Then the antisense deoxyoligonucleotide with two added butanediols served as a substrate for RNA chain elongation using the protected adenosine 2′-phosphoramidite.

and linkage to the antisense oligo is at the 5′ terminal hydroxyl. Thus, the prototypical 2–5A chimera has the following formulation (see *Figure 4* for generic structure):

p5′A2′p5′A2′p5′A2′p5′A2′bupbup5′dT3′(p5′dT3′)$_{16}$p5′dT.

Choice of the illustrated linkage mode permits synthesis of all three domains of the 2–5A-antisense on a solid support using phosphoramidite oligo synthesis technology.

2-5A-ANTISENSE COMPOSITE OLIGONUCLEOTIDES: FINAL DEPROTECTED PRODUCT

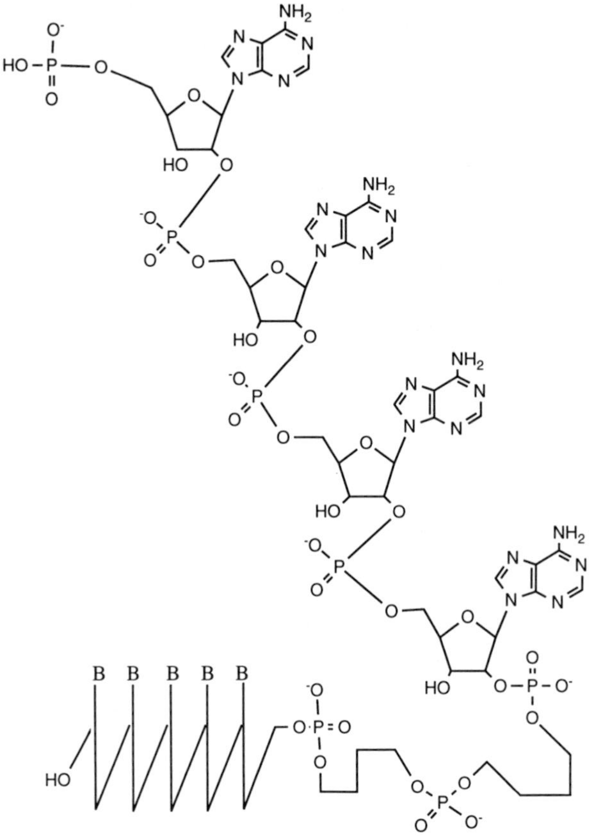

Figure 4. Generic structure for 2–5A-antisense. This product is obtained after 5′ detritylation of the product of *Figure 3*, 5′ phosphorylation and complete deprotection, and HPLC purification (*Figure 5*).

The chemical synthesis (*Figures 3* and *4*) can be performed manually on DNA synthesis columns (1.5 cm American Bionetics, Inc.) loaded with approximately 1.5 mmole of CPG-bound 5′-dimethoxytritylthymidine using adaptors and gas-tight fittings (31). The chemistry is based on the phosphoramidite approach to DNA and RNA synthesis (32–38). Thus, the antisense domain of the chimeric oligo is built up exactly as any DNA oligos would be synthesized with the above technology (39) (see *Figure 3*); however, after the last or 5′ terminal antisense deoxyribonucleotide are added, the intermediate oligo is detritylated and the chain extension continued while the oligo is still attached to the CPG support. Two synthetic cycles (*Figure 3*) are carried out with the phosphoramidite derivative of 4-*O*-(4,4'-dimethoxytrityl)-1,4-butanediol (*Protocol 1*) in order to attach the two butanediol linker moieties. After detritylation of the second added butanediol linker, the 2',5'-oligoadenylate tetramer is added through the use of the phosphoramidite of 5'-*O* - (4,4' - dimethoxytrityl) - 3' - *O* - (*t* - butyldimethylsilyl) - N^6- benzoyladenosine (*Protocols 2–4*).

After the 2',5'-oligoadenylate chain is completed, the oligo is 5' detritylated and then phosphorylated with *bis*(2-cyanoethoxy)(diisopropylamino) phosphine (*Protocol 5*). Final deprotection follows standard procedures for RNA and DNA synthesis. Purification is by HPLC (*Figure 5*), and characterization includes digestion with snake venom phosphodiesterase to determine nucleotide ratios (*Protocol 6* and *Figure 6*) and capillary gel electrophoresis to assay homogeneity (*Figure 7*).

Important caution: the chemical procedures described herein are meant to be carried out by trained chemists following appropriate safety precautions including use of protective clothing, gloves, safety glasses, and chemical fume-hoods where warranted. No chemical operation should be regarded as without risk.

Protocol 1. Preparation of 2-cyanoethyl-4-*O*-(4,4'-dimethoxytrityl)butyl *N,N*-diisopropylphosphoramidite, the bridging linker intermediate for connection of 2',5'-oligoadenylate to 3',5'-oligo

Equipment and reagents

- 1,4-Butanediol
- Anhydrous pyridine
- Ethyl acetate
- Anhydrous methylene chloride
- Benzene
- Petroleum ether
- Triethylamine
- Anhydrous magnesium sulfate
- Silica gel
- Ethyldiisopropylamine
- Methanol
- 4,4'-Dimethoxytrityl chloride
- Vacuum pump
- 2-Cyanoethyl-*N,N*-diisopropyl-phospho-ramidic chloride
- Rotary evaporator
- Dry argon or nitrogen
- Magnetic stirrer

Method

1. Weigh out 9.0 g (100 mmol) of 1,4-butanediol (Aldrich, Milwaukee, WI), dissolve it in anhydrous pyridine (approx. 50 ml), and evaporate the pyridine at a temperature below 40°C *in vacuo* on a rotary evaporator attached to a vacuum pump. Admit dry argon or nitrogen to the flask containing the residue, add an additional quantity of dry pyridine, and repeat the procedure. A total of three additions and evaporations of anhydrous pyridine will suffice to dry the butanediol sufficiently for the following step.

Protocol 1. *Continued*

2. Dissolve the dried 1,4-butanediol in 50 ml of anhydrous pyridine. Then add 4,4'-dimethoxytrityl chloride (3.39 g, 10 mmol) and allow the resulting homogeneous mixture to react for 2 h at room temperature in a stoppered flask. After 2 h reaction time, check the formation of product by TLC (which takes about 10 min) using dichloromethane: methanol (99:1) as solvent. The product will appear on silica gel plates with an R_f of 0.48.

3. Terminate the reaction by pouring the entire mixture onto 100 g of ice contained in a beaker. Stir the mixure with the aid of a magnetic stirring bar until the ice is completely melted.

4. Extract the organic product by adding 100 ml of ethyl acetate, shaking the biphasic mixture in a separatory funnel, and separating the organic (top) and aqueous (bottom) layers. Re-extract the aqueous layer with an additional 50 ml of ethyl acetate. Separate the organic layer once more and combine it with the ethyl acetate layer from the first extraction.

5. Dry the organic layer (top layer) by addition of 10 g of anhydrous magnesium sulfate.

6. Filter off the magnesium sulfate and concentrate the ethyl acetate solution to about 10 ml on a rotary evaporator at a temperature less than 40 °C. For this step, the rotary evaporator can be attached to a water aspirator.

7. Add the above concentrated ethyl acetate solution to the top of a silica gel column containing approximately 250 g of silica gel. Elute the column with methylene chloride containing 1% methanol. Monitor 10 ml fractions for product by spotting aliquots on silica gel plates, and checking for the presence of UV-absorbing material using a hand-held UV lamp. Check any fractions containing UV-absorbing material by running a TLC on silica gel plate using a solvent of chloroform:methanol (99:1). The yield from reaction under the above conditions should be in the range of 2.1 g (about 54% overall yield) of 4-O-4,4'-dimethoxytrityl)-1,4-butanediol. Proton NMRs are reported in the following form: proton NMR (solvent used) δ (p.p.m.): signal p.p.m. (multiplicity, J or coupling constant, integrated numbers of hydrogens, assignment). The abbreviations used for multiplicity include, s, singlet, d, doublet, t, triplet, and m, multiplet. Results obtained here resulted in the following: proton NMR (CDCl$_3$, 1% deuteriopyridine) δ (p.p.m.): 1.68 (m, 4H, CH$_2$); 3.10 (t, J = 5.7 Hz, 2H, CH$_2$O); 3.62 (t, J = 5.8 Hz, 2H, CH$_2$OH); 3.76 (s, 6H, CH$_3$O); 6.79–7.46 (m, benzene ring protons).

8. To convert the 4-O-(4,4'-dimethoxytrityl)-1,4-butanediol to 2-cyano-

ethyl-4-*O*-(4,4'-dimethoxytrityl)butyl *N,N*-diisopropylphosphoramidite, slowly add 2-cyanoethyl-*N,N* diisopropylphosphonamidic chloride (237 mg, 1 mmol) to a solution of 4-*O*-(4,4'-dimethoxytrityl)-1,4-butanediol (390 mg, 1 mmol) and ethyldiisopropylamine (510 mg, 4 mmol) in dry methylene chloride (3 ml) under anhydrous conditions with cooling in an ice-bath.

9. Keep the mixture at room temperature for 1 h.

10. Evaporate the methylene chloride solvent using a rotary evaporator at a temperature of less than 40°C.

11. Add a few millilitres of benzene to the above residue and apply the crude product to a silica gel column (1.8 × 14 cm) and elute the product with a solvent of benzene:petroleum ether:triethylamine (6:3:1).

12. After evaporation of solvent from appropriate fractions (as determined by checking the silica gel TLC of UV-absorbing fractions and the presence of product with R_f = 0.7 in benzene:petroleum ether: triethylamine, 6:3:1), the product, 2-cyanoethyl-4-*O*-(4,4'-dimethoxytrityl)butyl *N,N*-diisopropylphosphoramidite, is obtained in about 98% yield (580 mg). Proton NMR (see step 7) (CDCl$_3$, 1% deuteriopyridine) δ (p.p.m.): 1.17 (t, J = 7.0 Hz, 12H, CH$_3$C); 1.70 (m, 4H, CH$_2$C); 2.60 (t, J = 6.5 Hz, 2H, CH$_2$CN); 3.08 (t, J = 5.7 Hz, 2H, CH$_2$O); 3.80 (m, 2H, CH); 6.80–7.49 (m, aromatic protons). ^{31}P NMR (CDCl$_3$, 1% deuteriopyridine) δ (p.p.m.): 147.6.

Protocol 2. Synthesis of N^6-benzoyl-5'-*O*-dimethoxytrityl-adenosine

Equipment and reagents
- Anhydrous magnesium sulfate
- Adenosine
- Kieselgel 60 (Fluka, 220–440 mesh)
- Triethylamine
- 4,4'-Dimethoxytrityl chloride
- 4-Dimethylaminopyridine
- Pyridine
- Trimethylsilyl chloride
- Benzoyl chloride
- Ammonium hydroxide

Method
1. Add DMAP (4-dimethylaminopyridine, 0.5 mmol, 61 mg), triethylamine (14 mmol, 1.9 ml), and 4,4'-dimethoxytrityl chloride (4.1 g, 12 mmol) to a suspension of anhydrous adenosine (2.67 g, 10 mmol) in dry pyridine (100 ml).

2. Stir the reaction mixture at room temperature for 2–3 h.

3. When the reaction is complete (TLC, chloroform:methanol, 95:5, using adenosine, dimethoxytrityl chloride, and DMAP as TLC standards), add trimethylsilyl chloride (7.7 ml, 60 mmol) slowly at 0°C.

Protocol 2. *Continued*

4. After 15 min, add benzoyl chloride (5.8 ml, 50 mmol).

5. Stir the mixture at room temperature for approximately 2 h.

6. Then cool the mixture in an ice-bath, and add cold water (20 ml) followed, after 5 min, by 20 ml of conc. ammonium hydroxide (the final concentration of ammonia should be approx. 2 M).

7. Continue stirring, for 30 min, then concentrate to about 20 ml on a rotary evaporator *in vacuo*. Then dissolve the residue in ethyl ether (700 ml).

8. Wash the ether solution with water and then dry over anhydrous magnesium sulfate.

9. Remove the ethyl ether by evaporation on a rotary evaporator.

10. Purify the product by fast silica gel column chromatography (5 × 20 cm, Kieselgel 60, 220–440 mesh), using 2% methanol in ethyl acetate containing 0.2% of pyridine as an eluent.

11. Combine fractions containing the product (as determined by running silica gel TLCs of UV-absorbing fractions, and comparing them for material with the same R_f as the product obtained in step 3), and evaporate solvent to dryness on a rotary evaporator.

12. Remove residual pyridine by several additions and evaporations of toluene. The product is obtained as a slightly yellow solid foam, with yields ranging from 50–70%. TLC, ethyl acetate:methanol, 95:5, R_f = 0.55. ^{1}H NMR: (CDCl$_3$) δ (p.p.m.): 3.34 (dd, 1H, H-5' or 5''); 3.49 (dd, 1H, H-5' or 5''); 4.39 (d, 1H, H-4'); 4.50 (dd, 1H, H-3'); 4.90 (t, 1H, H-2'); 6.13 (d, 1H, H-1'); 6.74–7.60 (m, aromatic protons); 8.02 and 8.04 (s, 1H, H-2 or H-8) (see *Protocol 1*, step 7).

Protocol 3. Preparation of N^6-benzoyl-5'-*O*-dimethoxytrityl-2'-*O*-*t*-butyldimethylsilyladenosine and N^6-benzoyl-5'-*O*-dimethoxytrityl-3'-*O*-*t*-butyldimethylsilyladenosine

Equipment and reagents
- N^6-Benzoyl-5'-*O*-dimethoxytrityladenosine
- Imidazole
- *t*-Butyldimethylsilyl chloride
- Dimethylformamide

Method

1. Prepare a solution of N^6-benzoyl-5'-*O*-dimethoxytrityl-adenosine (see *Protocol 2*) (1.75 g, 2.6 mmol), imidazole (544 mg, 8 mmol), and *t*-butyldimethylsilyl chloride (600 mg, 5 mmol) in anhydrous DMF (10 ml).

2. Maintain the solution at room temperature with continuous stirring for 2.5 h.

3. Run a TLC (silica gel, ethyl acetate:methanol, 95:5 or ethyl acetate: cyclohexane, 1:1) to check that all of the substrate has reacted. The silylated product runs with a faster R_f than starting material which should be consumed entirely.

4. When appropriate, terminate the reaction by the addition of 5% Na_2CO_3 (1.5 ml) at 0°C, and stir the resulting mixture for 15 min at a temperature below 10°C.

5. Evaporate the DMF on a rotary evaporator under vacuum.

6. Dissolve the resulting residue in methylene chloride (50 ml) and wash this organic solution with water.

7. Dry the organic layer over anhydrous $MgSO_4$, evaporate the CH_2Cl_2, dissolve the residue in approx. 10 ml of ethyl acetate:cyclohexane (1:1), containing 0.2% pyridine.

8. Apply this solution to a silica gel column (5 × 15 cm, Kieselgel 60, 220–440 mesh) and elute the column with the same solvent. The 2'-O-silylated isomer is eluted first (TLC).

9. After the 2'-O-silylated compound is eluted, increase the concentration of ethyl acetate to a 3:2 ratio to elute the 3'-O-silylated isomer.

10. For each of these two products, pool appropriate fractions in separate containers, and remove the solvent *in vacuo*, followed by addition and evaporation of toluene to remove traces of pyridine. The products are obtained as colourless solid foams, in the following yields: 2'-O-silylated isomer, 480 mg; 3'-O-silylated isomer, 660 mg. TLC, ethyl acetate:cyclohexane:pyridine, 1:1:0.004, R_f = 0.47 and 0.24, respectively. For this work, only the slower running 3'-silylated isomer is of interest. 3'-O-silylated isomer: ^1H NMR: (CDCl$_3$), δ (p.p.m.): 0.02 and 0.10 (ds, 6H, CH$_3$Si); 0.98 (s, 9H, t-butyl); 3.27 and 3.54 (dd, 1H, H-5' or H-5''); 3.77 (s, 3H, CH$_3$O); 4.21 (m, 1H, H-4'); 4.60 (t, 1H, H-3'); 4.80 (t, 1H, H-2'); 6.10 (d, 1H, H-1'); 8.03 and 8.05 (s, 1H, H-2 or H-8); 6.78–7.60, 8.28, 8.77 (m, aromatic protons) (see *Protocol 1*, step 7).

Protocol 4. Synthesis of N^6-benzoyl-5'-O-dimethoxytrityl-3'-O-t-butyl-dimethylsilyl-adenosine 2'-(N,N-diisopropyl-2-cyanoethyl)-phosphoramidite

Equipment and reagents

- N^6-Benzoyl-5'-O-dimethoxytrityl-3'-O-t-butyl-dimethylsilyl-adenosine
- 1H-tetrazole
- Phosphorus pentoxide
- Methylene chloride
- 2-Cyanoethyl-(N,N,N',N'-tetraisopropyl)-phosphordiamidite

Protocol 4. *Continued*

Method

1. Combine the N^6-benzoyl-5'-*O*-dimethoxytrityl-3'-*O*-*t*-butyldimethyl-silyladenosine (540 mg, 0.68 mmol, see *Protocol 3*) and 1*H*-tetrazole (48 mg, 0.68 mmol) in a two arm reaction flask (25 ml size), fitted with a rubber septum and a hypodermic needle (to allow evacuation of air and admission of dry nitrogen and reagents), and dry overnight in a vacuum desiccator in the presence of P_2O_5.

2. Admit dry nitrogen to the desiccator to fill.

3. Add a methylene chloride solution (5 ml) of 2-cyanoethyl-(*N,N,N',N'*-tetraisopropyl)-phosphordiamidite (206 mg, 0.68 mmol) through a rubber septum.

4. Stir the mixture for 4 h at room temperature and then leave overnight in a refrigerator at 4°C.

5. Concentrate the reaction mixture under vacuum on a rotary evaporator, and add a solution of benzene:cyclohexane:triethylamine (6:3:1) (2 ml) to the residue.

6. Apply this solution to a silica gel column (2 × 15 cm) using the same solvent system for elution. A partial separation of the two possible isomers can be achieved, but all product-containing fractions may be combined together, concentrated, and dried by addition and evaporation of toluene. The product, consisting of a mixture of both P-chiral diastereoisomers, is obtained as a colourless solid foam (500 mg). TLC (for the two diastereoisomers), benzene:cyclohexane:triethylamine, 6:3:1, $R_f = 0.61$ and 0.53. ^{31}P NMR: (CDCl$_3$, 1% C_5D_5N) δ (p.p.m.): 150.73 and 150.38 (see *Protocol 1*, step 7).

Protocol 5. Preparation of *bis*-(2-cyanoethyl)-*N,N*-diisopropyl-phosphoramidite

Equipment and reagents

- 2-Cyanoethyl *N,N*-diisopropylchlorophos-phoramidite
- 2-Cyanoethyanol
- 4-Dimethylaminopyridine
- Diisopropylamine

Method

1. Add 2-cyanoethyl *N,N*-diisopropylchlorophosphoramidite (1 g, 4.2 mmol) slowly from a syringe through a rubber septum into a flask containing a solution of 2-cyanoethanol (300 mg, 287 ml, 4.2 mmol), diisopropylamine (1 ml), 4-dimethylaminopyridine (10 mg) in anhydrous dichloromethane (5 ml), under an atmosphere of dry nitrogen.

2. Stir the mixture for 8 h at room temperature and then store it overnight in a refrigerator at 4°C.

3. Evaporate the solvent on a rotary evaporator.

4. Purify the residue with fast silica gel chromatography (1 × 10 cm, elution with benzene:petroleum ether:triethylamine (6:3:1).

5. Pool fractions containing product (determined by examining UV-absorbing fractions by TLC according to the conditions specified below are pooled, solvent is evaporated, and dry the residue by several additions and evaporations of benzene. The product is obtained as a slightly yellow, viscous oil (0.8 g, 70% yield). TLC, benzene:petroleum ether:triethylamine, 6:3:1, R_f = 0.70. NMR: ^{31}P (CDCl$_3$, 0.4% C$_5$D$_5$N) δ (p.p.m.): 148.8. ^{1}H (CDCl$_3$, 0.4% C$_5$D$_5$N) δ (p.p.m.): 1.19 (d, 12H, CH$_3$-C); 2.66 (t, 4H, CH$_3$-CN); 3.63 (m, 2H, CH); 3.86 (m, 4H, CH$_2$-O-P) (see *Protocol 1*, step 7).

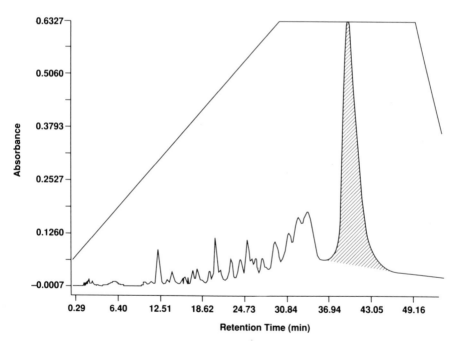

Figure 5. Anion exchange HPLC purification of 2–5A-antisense chimera on a Nucleogen DEAE 60–7 column (4 × 125 mm). The elution program was a linear gradient of 10–100% buffer B in buffer A in 30 min, and then isocratic at 100% buffer B (flow rate = 1 ml/min). Buffer A was 20 mM KH$_2$PO$_4$ pH 7.0 in H$_2$O/CH$_3$CN (8:2, v/v), and buffer B was 20 mM KH$_2$PO$_4$ in 1 M KCl. The cross-hatched section of the chromatogram indicates the collected fraction.

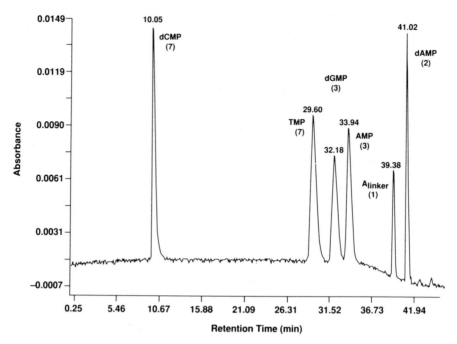

Figure 6. A typical reverse-phase HPLC of the products resulting from the snake venom phosphodiesterase digestion of the 2–5A-antisense chimera directed against the PKR mRNA. The sequence of this chimera is: p5'A2'p5'A2'p5'A2'p5'A2'pbupbup5'GTA CTA CTC CCT GCT TCT G3', where bu stands for 1,4-butanediol. The digestion conditions and HPLC analysis followed *Protocol 6*. A$_{linker}$ represents the product in which the terminal adenosine of 2–5A is still linked to the two butanediol linkers. This material is resistant to further digestion with the snake venom enzyme under these conditions.

Protocol 6. Oligonucleotide characterization by digestion with snake venom phosphodiesterase and product identification and quantitation

Equipment and reagents

- *Crotalus admantus* snake venom phosphodiesterase
- 50 mM Tris buffer pH 8.0, 0.5 mM MgCl$_2$
- Thermostated water-bath
- Nucleotide standards (dAMP, dCMP, dGMP, dTMP, AMP, A)
- HPLC with Ultrasphere ODS column, UV detector, and integrator

Method

1. Dissolve the oligonucleotide (0.1–0.2 OD units) to be analysed in 100 μl of a buffer of 50 mM Tris chloride (pH 8.0), 0.5 mM MgCl$_2$.

2. Add 0.05–0.1 U of snake venom phosphodiesterase.

3. Incubate the resulting solution for at least 3 h at 37°C.

4. Analyse the digestion products by injection onto an analytical Ultra-shere ODS column (Beckman, reversed-phase C18, 4.6 × 250 mm). Set the flow rate to 0.5 ml/min. Elute the column with 2% B in A for 20 min, then a linear gradient of 2–45% B in A for 15 min, followed by isocratic 45% B in A for 10 min, where A is 100 mM ammonium phosphate, and B is $CH_3OH:H_2O$ (1:1, v/v).

5. Identify products by comparison with authentic standards; e.g. 5'-dAMP, 5'-dCMP, 5'-dGMP, 5'-dTMP, 5'-AMP, and adenosine (*Figure 6*).

6. Integrate the area under each peak and determine the ratio of nucleotides (and adenosine, if present).

3.1 Reagents and chemicals for composite oligonucleotide synthesis

3.1.1 For initiation of synthesis on solid support

These DMT protected nucleosides are attached to controlled pore glass (CPG) through a succinyl group and a long chain alkyl amine linker; are contained in pre-packed columns; and are commercially available products of Applied Biosystems (Foster City, CA).

- 5'-*O*-dimethoxytrityl-N^6-benzoyl-2'-deoxyadenosine-3'-lcaa-CPG
- 5'-*O*-dimethoxytrityl-N^4-benzoyl-2'-deoxycytidine-CPG
- 5'-*O*-dimethoxytrityl-N^2-isobutyryl-2'-deoxyguanosine-3'-lcaa-CPG
- 5'-*O*-dimethoxytritylthymidine-3'-lcaa-CPG

Pre-packed columns all are 1 µmole scale.

3.1.2 For elongation of the DNA antisense chain

A total of 500 mg of each of the following phosphoramidites (Applied Biosystems) is dissolved in the indicated amount of anhydrous acetonitrile to make 0.1 M phosphoramidite solution:

- 5'-*O*-dimethoxytrityl-N^6-benzoyl-2'-deoxyadenosine-3'-(2-cyanoethyl-*N,N*-diisopropyl)phosphoramidite (5.6 ml)
- 5'-*O*-dimethoxytrityl-N^4-benzoyl-2'-deoxycytidine-3'(2-cyanoethyl-*N,N*-diisopropyl)phosphoramidite (5.9 ml)
- 5'-*O*-dimethoxytrityl-N^2-isobutyryl-2'-deoxyguanosine-3'-(2-cyanoethyl-*N,N*-diisopropyl)phosphoramidite (5.8 ml)
- 5'-*O*-dimethoxytrityl-2'-deoxythymidine-3'-(2-cyanoethyl-*N,N*-diisopropyl)phosphoramidite (6.6 ml)

3.1.3 Linker to join chimeric domains

The linker, 2-cyanoethyl-4-*O*-(4,4'-dimethoxytrityl)butyl *N,N*-diisopropyl-phosphoramidite was synthesized as described in *Protocol 1*. A 0.1 M solution is prepared by dissolving 100 mg linker in 1.5 ml of anhydrous acetonitrile.

3.1.4 For synthesis of 2',5'-oligoadenylate domain of the chimera

5'-*O*-dimethoxytrityl-N^6-benzoyl-3'-*O*-*t*-butyldimethylsilyl-adenosine-2'-*N,N*-di isopropylcyanoethylphosphoramidite. A 0.1 M solution is made by dissolving 500 mg of monomer in 2.5 ml of anhydrous acetonitrile.

3.1.5 Phosphorylation reagent for 5' terminus of 2',5'-oligoadenylate domain of chimera

Bis-(2-cyanoethyl)-*N,N*-diisopropylphosphoramidite, prepared according to *Protocol 5*, is used at a concentration of 0.2 M in anhydrous terazole:acetonitrile (ABI).

3.1.6 Other reagents

All other DNA synthesis reagents can be obtained from Applied Biosystems Inc. This includes diluent (acetonitrile), activator solution (tetrazole:acetonitrile), capping solutions (A: acetic anhydride solution and B: *N*-methylimidazole solution), deblock reagent (trichloroacetic acid solution), and oxidizer (iodine solution).

Tetrabutylammonium fluoride in tetrahyrofuran (Aldrich, Milwaukee, WI) is used to deblock the *t*-butyldimethylsilyl group for protecting the 3'-hydroxyls of (2',5')-oligoriboadenylate domain.

3.2 Assembly procedure: synthesis of oligonucleotides

(a) Syntheses are performed manually on DNA synthesis columns (1.5 cm, American Bionetics, Inc) loaded with approximately 1.5 μmole of CPG-bound 5'-*O*-dimethoxytrityl-thymidine, using adaptors and gas-tight syringes.

(b) Syntheses are controlled by quantitating spectrophotometrically the release of the dimethoxytrityl cation.

(c) 5'-Monophosphorylation of the assembled chimera is accomplished through the use of the reagent, *bis*(2-cyanoethoxy)(diisopropylamino) phosphine.

(d) The synthesized oligos are cleaved from the support with conc. ammonia: ethanol (3:1) by 2 h incubation at room temperature.

(e) N^6-benzoyl groups are removed by warming the resulting ethanolic solutions for 6 h at 55 °C.

(f) Finally, the 3'-*O*-*t*-butyldimethylsilyl protecting groups are removed by treatment with 1 M tetrabutylammonium fluoride in THF for at least 12 h at room temperature.

The exact procedures are shown in *Table 1*.

Table 1. Solid phase synthesis of oligonucleotides: coupling cycle

Step	Solvents/reagents	Time	Volume
Detritylation	3% DCA in CH_2Cl_2	90 sec	1 ml
Washing	2% Py in acetonitrile		1 ml
Washing	Acetonitrile		3 ml
Drying	Nitrogen	3 min	
Coupling	0.2 M monomer in 0.5 M tetrazole/acetonitrile	8 min for Ado 3 min for 5' end phosphitylation	0.15 ml
Washing	Acetonitrile		3 ml
Drying	Nitrogen	2 min	
Capping	A + B, 1:1 A: 30% Ac_2O in THF B: 0.6 M DMAP in Py/THF (3:2, v/v)	2 min	1 ml
Washing	Acetonitrile		3 ml
Drying	Nitrogen	2 min	
Oxidation	0.1 M I_2 in lutidine: THF:water, 20:80:1	45 sec	1 ml
Washing	Acetonitrile		3 ml
Drying	Nitrogen	3 min	

3.3 HPLC methods

The HPLC system which can be used consists of:

(a) Beckman System Gold software controlling an IBM PS/2 computer, two 110B solvent delivery modules, and a 167 UV/VIS variable wavelength detector (set to operate at 260 and 280 nm) (or the equivalent of this HPLC system).

(b) Anion exchange HPLC (see *Figure 5* for conditions) can be used for purification of synthetic oligos.

(c) An analytical Ultrasphere ODS (reversed-phase C18, 4.9 × 250 mm, flow rate 1 ml/min, linear gradients of B in 0.02 M ammonium phosphate, pH 7.0) is employed for analysis of purity, and to determine enzymic digests of the products (*Figure 6* and *Protocol 6*).

(d) Capillary electrophoresis is used to assess oligo purity (see *Figure 7* for example conditions).

4. Biological and biochemical evaluation of 2–5A-antisense chimeras

In the remaining sections, the analysis of the ability of 2–5A-antisense chimeras to induce selective decay of target mRNA species is described. The three systems involve purified RNase L, cell extracts, and intact cells. Each of

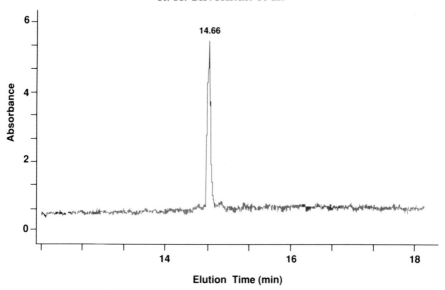

Figure 7. Gel capillary electrophoresis (CE) of the 2–5A-antisense chimera against the PKR protein. The CE was performed on an Applied Biosystems 270A-HT capillary instrument using MICRO-GEL$_{100}$-filled capilaries (Applied Biosystems, Inc, Foster City, CA). Capillary dimensions were 50 μm i.d., with an effective length of 27 cm. The running buffer was 75 mM Tris phosphate pH 7.6 with 10% methanol. UV detection was at 260 nm.

these has its particular value. The assays with cell extract or purified RNase avoid the barriers to uptake of oligos encountered with intact cells. In assays with purified RNase there is an absence of enzymes that can degrade 2–5A-antisense. Accordingly, much lower concentrations of 2–5A-antisense oligos, as little as one-hundredth compared to using cells, are required. Studies to determine kinetics and interactions of 2–5A-antisense with RNase L are best performed with the purified enzyme for obvious reasons. The cell-free systems have the advantage of more closely approaching conditions encountered in cells in comparison to assays with the purified RNase L. The advantages of using intact cells are that most types of antisense experiments, e.g. determining the functions of individual proteins in biological systems, such as in signal transduction and preventing synthesis of disease-causing proteins, requires the use of whole cells.

4.1 Targeted cleavage of mRNA species in cell-free systems

For cell-free system assays, we routinely use an extract of human lymphoblastoid Daudi cells as the source of RNase L activity. However, other types of human cells can be substituted depending on the amounts of RNase L and general RNase activities which are present. If murine cells are used, there may be a requirement for a 5′ diphosphate on the 2′,5′-oligoadenylate part of the chimeras. The human RNase L requires only a single 5′ phosphoryl group on the 2',5'-oligoadenylate for maximal activity (14) whereas the murine enzyme seems to require 2–5A with at least two 5' phosphoryl groups (40). Mammalian cell extracts are prepared by a modification of the method of Wreschner *et al.* (12) as desribed in *Protocol 7*.

Alternatively, purified recombinant RNase L can be used to perform the cleavage reactions. The precise sites of cleavage may be determined using primer extension assays (described in ref. 1).

Protocol 7. Preparation of cell extracts

Equipment and reagents

- Tissue culture hood
- T175 tissue culture flasks (Falcon, Cat. No. 3112)
- Tissue culture media and supplements from Gibco BRL:
- RPMI 1640 media (Cat. No. 31052)
- Fetal calf serum (Cat. No. 26140–038)
- L-Glutamine (Cat. No. 25030–016)
- Sodium pyruvate (Cat. No. 11360–13)
- Non-essential amino acids (Cat. No. 11140–019)

- Hepes buffer (Cat. No. 15630–015)
- Gentamycin (Cat. No. 15750–011)
- β-Mercaptoethanol (Sigma, Cat. No. M7522)
- Refrigerated centrifuge, such as a Beckman model GS-6R
- Ultracentrifuge, such as a Beckman model L-70/ SW41 Ti rotor
- Bradford protein assay reagent (Bio-Rad, Cat. No. 500–0006)

Method

1. Grow Daudi cells at 37°C in 5% CO_2 in 500 ml RPMI 1640 medium including 50 ml fetal calf serum, 5 ml of 200 mM L-glutamine, 5 ml 100 mM sodium pyruvate, 5 ml MEM non-essential amino acids, 2.5 ml Hepes buffer pH 7.4, 0.5 ml of 50 mg/ml gentamicin, 50 μl 2-mercaptoethanol (of a stock solution containing 250 μl RPMI + 10 μl 14.3 M 2-mercaptoethanol).

2. Harvest cells from eight T175 flasks each containing 100 ml of static cell suspension when the cell density is about 10^6 cells/ml. Collect cells, a total of about 2 ml of packed cell volume, by centrifugation at 1500 *g* at 4°C using a Beckman centrifuge model GS-6R or similar centrifuge.

3. Gently lyse cells in buffer containing 40 mM KCl, 10 mM Hepes pH 7.5, 2.5 mM magnesium acetate, 0.5 mM ATP, 2.5% (v/v) glycerol, 2.0 mM 2-mercaptoethanol, by several strokes in a sterile glass Dounce homogenizer, until 90% of cells are burst as determined by microscopy.

4. Centrifuge the homogenate at about 100 000 *g* in an ultracentrifuge for 1–2 h at 4°C.

5. Collect the supernatant and measure the protein concentration using Bradford protein assay reagent. Expected protein concentrations range from 1–3 mg/ml in a total volume of 1.5–1.8 ml. Pipette 50 μl aliquots, freeze on dry-ice, and store at –70°C.

In vitro synthesis and gel purification of the target RNA is described in Chapter 4. The 3' end-labelling of the RNA with [5'-^{32}P]pCp using T4 RNA ligase is performed by a modification of the method described by Silverman and Krause (41). RNA labelling reactions are performed as descibed in *Protocols 8* and *9*. Important note: care should be taken in handling RNA including the use of sterile DEPC treated water (see Chapter 7 for details on preparation).

Protocol 8. Radiolabelling of RNA at the 3' termini

Equipment and reagents

- [5'-^{32}P]pCp (3000 Ci/mmol) (New England Nuclear) in 10 mM Tricine pH 7.6 (Cat. No. NEG-019A)
- T4 RNA ligase (FPLC pure, cloned) (Pharmacia, Cat. No. 27–0883–01)

Method

1. Dry 10–15 µl (100–150 µCi) of [5'-^{32}P]pCp using a Speed Vac. Dissolve the dried [5'-^{32}P]pCp in 25 µl of RNA ligase buffer containing 100 mM Hepes pH 7.6, 15 mM MgCl$_2$, 6.6 mM dithiothreitol (DTT), and 20% dimethyl sulfoxide (DMSO).

2. While keeping the solution of [5'-^{32}P]pCp on ice add the following: 5 µl of 1.0 mM dTTP and 0.1 mM ATP pH 7.0, 50–100 µg RNA in 15 µl water (see Chapter 7), 3.5 µl water, 1.5 µl T4 RNA ligase (22.5 U).

3. Place the reaction tube in a small beaker containing ice in a refrigerator (2–4°C) for 16–18 h. The ice should melt during this time.

4. Extract RNA with an equal volume of phenol:chloroform (1:1). Remove RNA in upper aqueous phase and pass once through a Sephadex G50 Quick Spin column (*Protocol 9*).

5. Purify full-length RNA on a gel as described in Chapter 4.

Protocol 9. Radiolabelling of RNA at the 5' termini

Equipment and reagents

- Calf intestine alkaline phosphatase (CIP) and 10 × CIP buffer (Boehringer Mannheim, Cat. No. 713023)
- Proteinase K (Boehringer Mannheim, Cat. No. 161–519)
- 10 × polynucleotide kinase buffer (US Biochemical, Cat. No. 70083)
- Sephadex G50 columns (Boehringer Mannheim Quick Spin™ Columns for RNA, Cat. No. 100 411)
- T4 polynucleotide kinase (US Biochemical, Cat. No. 70031)
- [γ-^{32}P]ATP, 50 µCi (3000 Ci/mmol) (New England Nuclear, Cat. No. NEG-002H)

A. *Phosphatase treatment of RNA*

Phosphatase treatment of RNA is required to remove 5' phosphoryl groups prior to 5' end-labelling.

1. Mix the following and incubate for 30 min at 37°C: 21 μl RNA, 15–20 μg RNA from *in vitro* transcription reaction, diluted in water, 2.5 μl 10 × CIP buffer, 1.5 μl CIP (1 U/1 μl).

2. Digest CIP with proteinase K by adding the following to the reaction mixture: 2.5 μl 10% SDS, 2.5 μl 100 mM EDTA, 5 μl proteinase K (1 mg/ml), 15 μl water, and incubate at 56°C for 30 min. Then cool reaction mixture to room temperature.

3. Add 50 μl water, extract with an equal volume of phenol: chloroform:isoamyl alcohol (25:24:1), and precipitate in ethanol. Wash in 70% ethanol, dry, and redissolve in 20 μl water.

B. 5′ end-labelling of RNA

1. Perform 5′ end-labelling of RNA by mixing the following and incubating for 45 min at 37°C: 20 μl phosphatase treated RNA (about 15–20 μg) dissolved in water, 5 μl 10 × polynucleotide kinase buffer, 10 μl [γ-^{32}P]ATP, 50 μCi (3000 Ci/mmol), 2 μl T4 polynucleotide kinase (3 U/μl) (a 1:10 dilution of the stock using 50 mM Tris pH 8.0), 13 μl water.

2. Extract with an equal volume of phenol:chloroform:isoamyl alcohol (25:24:1). Pass aqueous phase through a Sephadex G50 Quick Spin column to remove unincorporated nucleotides. Ethanol precipitate, wash, dry, and redissolve RNA in 20 μl DEPC treated water. Gel purify the RNA as described in Chapter 4.

RNA cleavage assays are typically performed with different amounts of the 2–5A-antisense chimeras or control oligos. A variety of different control oligos can be used, e.g. 2–5A linked to the sense orientation oligo or a scrambled version of the antisense, 2–5A linked to antisense against a different RNA than the target RNA, or chimeras lacking the 5′ phosphoryl group which fail to efficiently activate RNase L (14).

Protocol 10. RNA cleavage in the cell-free system

Equipment and reagents

- Sequencing gel apparatus (Gibco BRL, Cat. No. 21105–010)
- Whatman 3MM filter paper (Cat. No. 3030917)
- Vacuum gel drier (Hoefer Scientific Instruments, model SE1160)
- PhosphoImager (Molecular Dynamics, Sunnyvale,California)

Method

1. Set-up reactions by mixing in the following order: 4.5 μl water, 2 μl radiolabelled RNA (25 000 c.p.m.; final conc. typically about 100 nM), 2 μl 2–5A-antisense (final conc. typically 25–100 nM), control oligo or

Protocol 10. *Continued*

water, 1.5 μl 1.0 M KCl (to a final concentration of 75 mM), 10 μl Daudi S100 (10–25 μg).

2. Incubate the reaction mixture at 30°C for desired time and stop re-action by extracting with equal volumes of phenol:chloroform (1:1), and chloroform:isoamyl alcohol (24:1). Ethanol precipitate, wash, dry, and resuspend RNA in 10 μl of water.

3. Add 10 μl of formamide RNA gel sample buffer, heat to 90–95°C for 3 min, and immediately place on ice.

4. Load RNA (in 5–8 μl) on 6% polyacrylamide–8 M urea gel (30 × 40 × 0.04 cm). Apply 45 W for about 2–3 h until the dye front of bromo-phenol blue reaches the bottom of the gel.

5. Transfer the wet gel to Whatman 3MM filter paper and dry on a vacuum gel drier.

6. Prepare an autoradiogram of the gel using Kodak X-OMAT film. You may quantitate radioactivity in the gel with a PhosphorImager or esti-mate relative intensities using the film and a densitometer.

Protocol 11. RNA cleavage in the assays with purified RNase L

Equipment and reagents

- Purified recombinant RNase L (prepared as described in ref. 14)
- Gel purified and ^{32}P end-labelled RNA (prepared as described in *Protocols 8* or *9*)

Method

1. Add in the following order: 14 μl buffer (25 mM Tris–HCl pH 7.4, 10 mM magnesium acetate, 8 mM β-mercaptoethanol, 100 mM KCl, 0.5 μg/ml leupeptin, and 50 μM ATP), 2 μl oligo solution (final conc. 25–100 nM), 2 μl ^{32}P end-labelled RNA (2500–5000 c.p.m., a final conc. 25–50 nM), 2 μl RNase L (2.5–15 mg/ml).

2. Incubate for 15 min at 37°C, terminate with the addition of an equal volume of gel sample buffer, and boil 5 min.

3. Analyse RNA by gel electrophoresis, autoradiography, and Phosphor-Image analysis as described in *Protocol 10*.

Results obtained from a Daudi cell-free system are shown in *Figure 8B* (from ref. 1). The target mRNA consisted of a modified HIV-1 *vif* mRNA into which was inserted a stretch of adenylyl residues (*Figure 8A*). Specific cleavage of the mRNA was observed in the Daudi cell-free system only in the

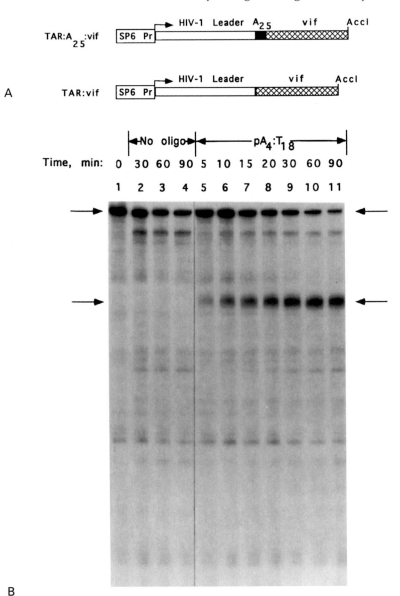

Figure 8. A cell-free system demonstrating the cleavage of a modified HIV-1 *vif* mRNA in response to the addition of a 2–5A-antisense. (A) *Acc*I digested plasmids containing TAR:A$_{25}$:vif and TAR:vif sequences. (B) Specific cleavage of TAR:A$_{25}$:vif RNA induced by the 2–5A-antisense, pA$_4$:T$_{18}$ at 100 nM in an extract of Daudi cells as a function of time. An autoradiogram of the dried gel is shown. Upper arrow indicates intact RNA; lower arrow indicates specific cleavage products as discussed previously (1). Reprinted from ref. 1 with permission.

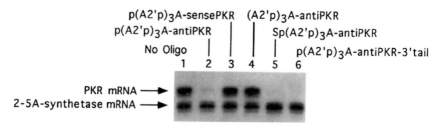

Figure 9. The 2–5A-antisense directs homogeneous, recombinant human RNase L to selectively cleave the mRNA target. The identities of the chimeric oligonucleotides and the positions of the PKR mRNA (2.0 kb) and 2–5A-synthetase mRNA (1.8 kb) are indicated. Reprinted from ref. 3 with permission.

presence of the 2–5A-antisense species, p5′A(2′p5′A)$_3$-(dT$_{18}$) (*Figure 8B*, lanes 5 to 11). We demonstrated through a series of controls that cleavage was due to RNase L (1). The rate of cleavage of the target mRNA was increased by more than fourfold in the presence of 2–5A-antisense.

RNA cleavage by RNase L typically uses recombinant human RNase L, expressed in insect cells from a baculovirus vector, and purified to homogeneity with fast protein liquid chromatography (Pharmacia). The methods involved in preparing the enzyme have been described in detail elsewhere and are therefore not included in this volume (14). For each preparation of RNase L, determine the efficiency of the cleavage reaction with different amounts of the ribonuclease (typically ranging from 5–30 ng per assay).

An assay performed with the purified RNase L is shown in *Figure 9* (from ref. 3). The target mRNA, encoding the protein kinase PKR, was added together with a non-target mRNA for 2–5A-synthetase. Addition of 2–5A-antisense species against PKR mRNA caused selective cleavage of the target mRNA (lane 2, 5, and 6). Neither of the control chimeras, at 25 nM, caused cleavage of the mRNAs. These consisted of 2–5A linked to the sense orientation sequence in PKR mRNA or the dephosphorylated form of 2–5A linked to antisense against PKR mRNA (lanes 3 and 4). The latter was inactive due to a requirement for a 5′ phosphoryl group on 2′,5′-oligoadenylate for efficient activation of RNase L (14). RNAs were present at 25 nM and 5 ng of RNase L was used. The incubations were at 37 °C for 15 min. The sequences of the oligos are in ref. 3.

4.2 Degradation of target RNA molecules in intact cells induced by 2–5A-antisense

Although we typically use human HeLa cells for these experiments, a variety of different types of human cells may be used. However, for murine cells, a 5′-diphosphate on the 2′,5′-oligoadenylate may be required (see Section 4.1).

Protocol 12. Cell culture and oligonucleotide treatment of cells

Equipment and reagents

- DMEM medium (Gibco BRL, Cat. No. 11995–032)
- Fetal bovine serum (Gibco BRL, Cat. No. 26140–020)
- 10 cm dishes (Falcon, Cat. No 3003)
- Penicillin–streptomycin (Gibco BRL, Cat. No. 15140–015)
- 24-well cell culture cluster (Costar, Cat. No. 3524)
- 2–5A-antisense and control oligos

Method

1. Culture HeLa cells in DMEM medium containing 10% fetal bovine serum and penicillin–streptomycin in either 24-well plates (with 250 µl of media for RT-PCR analysis) or in 10 cm dishes (with 5 ml of media for RNase protection assays).

2. One day prior to oligo treatment split the cell cultures so that cells will be about 80% confluent at the time of addition of oligos.

3. Aspirate medium and replace with fresh medium with or without 2–5A-antisense or control oligos (typically at about 2–10 µM final concentration). For example, dilute stock solution of oligo at 100 µM into media/serum.

4. Incubate cells at 37°C for different periods of time. Prepare RNA as described in *Protocol 15*.

Protocol 13. Preparation of RNA from intact cells

Equipment and reagents

- Dulbecco's phosphate-buffered saline (Gibco BRL, Cat. No. 14190–029)
- RNAzol reagent (Tel Test, Inc. cat no. CS-105)

Method

1. At the end of oligo treatment of cells, remove the medium, wash cells with ice-cold phosphate-buffered saline (PBS), and harvest by scraping into 1.0 ml of PBS, and pipette into 1.5 ml tubes. Centrifuge at 700 *g* for 10 min at 4°C and aspirate buffer.

2. Isolate RNA using RNAzol reagent as described in the instruction provided by the manufacturer (Tel-Test, Inc., Friendswood, TX). For RT-PCR analysis of RNA, add 500 µl of RNAzol and 100 µl of chloroform to the cell pellet.

3. Alternatively, for RNase protection assays use 4 ml of RNAzol reagent and 0.8 ml chloroform.

Protocol 14. Analysis of RNA by RT-PCR

Equipment and reagents

- DNase I, RNase-free (Boehringer Mannheim, Cat. No. 776785)
- rRNasin RNase inhibitor (Promega, Cat. No. 4346102)
- Oligo(dT)$_{12-18}$ (Pharmacia, Cat. No. 27–7841–01)
- Reverse transcriptase, M-MuLV cloned (Boehringer Mannheim, Cat. No. 1062603) and 5 × RT buffer (Cat. No. 1058495)
- DNA thermal cycler (Perkin Elmer Cetus)
- GeneAmp PCR reagent kit (Perkin Elmer Cetus, Cat. No. N801–0055)
- Nytran membranes (Schleicher & Schuell)
- Dot blot apparatus (Hybri.dot manifold, BRL Life Technologies, Inc., Cat. No. 1050 MM)
- Prime-a-Gene labelling system (Promega, Cat. No. U1100)
- Kodak X-ray film (XAR5, Cat. No. 1651454)

To avoid the possibility of DNA contamination, RNA preparations are treated with DNase as follows:

A. *DNase treatment of RNA preparations*

1. Suspend purified total RNA in 47 µl of buffer (100 mM Tris, 10 mM MgCl$_2$, 100 mM KCl), add 1 µl (40 U) rRNasin, and 2 µl (20 U) of RNase-free DNase.

2. Incubate for 15 min at 37 °C.

3. Extract with 50 µl of phenol:chloroform:isoamyl alcohol (25:24:1), collect the aqueous phase, and precipitate RNA with 2 vol. of 100% ethanol in the presence of 0.3 M sodium acetate, pH 5.1. Dissolve the RNA in 15 µl of water.

B. *Synthesis of cDNA*

1. Prepare cDNA as follows, making the desired amount of RT mix in the proportions indicated: 8 µl (20 U/µl) M-MuLV reverse transcriptase, 2.5 µl (40 U/µl) RNasin, 20 µl 5 × RT buffer, 2 µl oligo(dT) (5 A_{260} U/ml), 10 µl bovine serum albumin (1 mg/ml), 10 µl nucleotide mix (10 mM each of dATP, dGTP, dCTP, dTTP) pH 7, 1 µl 10 mM DTT.

2. Pipette 5 µl RT mix to 5 µl of RNA solution (3 µg) in water and incubate at 42 °C for 1 h.

C. *PCR*

1. Heat at 95 °C for 5 min and place on ice.

2. Prepare fourfold serial dilutions (typically three) of the cDNA solution, and perform PCR (usually for 30–40 cycles) as described in the information supplied with PCR reagent kit (Perkin Elmer Cetus).

3. For comparison, it is desirable to perform PCR on both the target RNA and on a control RNA, e.g. β-actin mRNA, using the same cDNA preparation.

D. *Dot blot hybridization*

1. Prepare ^{32}P-labelled DNA probe for hybridization as described in the information provided by the supplier of the random-priming kit (Promega).

2. Analyse the PCR products by dot blot hybridization as described in information provided by the supplier of the Nytran Membrane (*'Transfer and immobilisation of Nucleic acids to S&S solid supports'*, Schleicher & Schuell, p. 24) with modifications as shown below. To prepare sample for dot blot mix: 12 μl amplified cDNA, 33 μl 1 M Tris–HCl pH 7.6, 5 μl 3 M NaOH.

3. Heat at 85°C for 20 min to denature DNA, place on ice, and mix with 25 μl 20 × SSC.

4. Blot DNA on Nytran membranes using blotting apparatus. Perform pre-hybridization, hybridization, and washings as described (42).

5. Expose dried filter to X-ray film for autoradiography.

In the RT-PCR example shown in *Figure 10* (from ref. 3), PKR mRNA decay was induced in HeLa cells by the addition of three separate 2–5A-antisense species to the cultures at concentrations of 2 μM for 4 h. The antisense without 2–5A and 2–5A linked to irrelevant, unrelated sequences had no effect on PKR mRNA amounts. There was no effect on β-actin mRNA, monitored for comparison. Sequences of oligos are given in ref. 3. RNase protection assays are also routinely used to monitor RNA amounts in similar experiments (3).

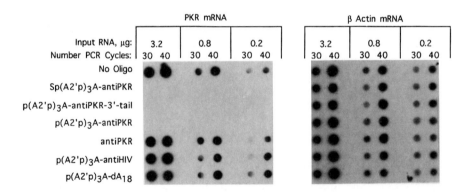

Figure 10. Selective ablation of PKR mRNA and activity in HeLa cells as demonstrated using an RT-PCR assay for PKR mRNA and β-actin mRNA. HeLa cells were incubated for 4 h in the absence or presence of 2 mM oligonucleotide (as indicated). DNA produced by RT-PCR amplification of (left panel) PKR mRNA and (right panel) β-actin mRNA was probed with radiolabelled cDNAs for PKR and β-actin. Reprinted from ref. 3 with permission.

References

1. Torrence, P. F., Maitra, R. K., Lesiak, K., Khamnei, S., Zhou, A., and Silverman, R. H. (1993). *Proc. Natl. Acad. Sci. USA*, **90**, 1300.
2. Lesiak, K., Khamnei, S., and Torrence, P. F. (1993). *Bioconjugate Chem.*, **4**, 467.
3. Maran, M., Maitra, R. K., Kumar, A., Dong, B., Xiao, W., Li, G., *et al.* (1994). *Science*, **265**, 789.
4. Kerr, I. M. and Brown, R. E. (1978). *Proc. Natl. Acad. Sci. USA*, **75**, 256.
5. Johnston, M. I. and Torrence, P. F. (1984). In *Interferon* (ed. R. M. Friedman), Vol. 3, pp. 189–298. Elsevier Science, New York.
6. Merlin, G., Chebath, J., Benech, P., Metz, R., and Revel, M. (1983). *Proc. Natl. Acad. Sci. USA*, **80**, 4904.
7. Marie, I. and Hovanessian, A. G. (1992). *J. Biol. Chem.*, **267**, 9933.
8. Williams, B. R. G., Kerr, I. M., Gilbert, C. S., White, C. N., and Ball, L. A. (1978). *Eur. J. Biochem.*, **92**, 455.
9. Johnston, M. I. and Hearl, W. G. (1987). *J. Biol. Chem.*, **262**, 8377.
10. Clemens, M. J. and Williams, B. R. G. (1978). *Cell*, **13**, 565.
11. Zhou, A., Hassel, B. A., and Silverman, R. H. (1993). *Cell*, **72**, 753.
12. Wreschner, D. H., McCauley, J. W., Skehel, J. J., and Kerr, I. M. (1981). *Nature*, **289**, 414.
13. Floyd-Smith, G., Slattery, E., and Lengyel, P. (1981). *Science*, **212**, 1030.
14. Dong, B., Xu, L., Zhou, A., Hassel, B. A., Lee, X., Torrence, P. F., *et al.* (1994). *J. Biol. Chem.*, **269**, 14153.
15. Maitra, R. K., Li, G., Xiao, W., Dong, B., Torrence, P. F., and Silverman, R. H. (1995). *J. Biol. Chem.*, **270**, 15071.
16. Cohen, J. (ed.) (1989). *Oligonucleotides: antisense inhibitors of gene expression*. Macmillan Press, London.
17. Uhlmann, E. and Peyman, A. (1990). *Chem. Rev.*, **90**, 543.
18. Imai, J., Johnston, M. I., and Torrence, P. F. (1982). *J. Biol. Chem.*, **257**, 12739.
19. Torrence, P. F., Imai, J., Lesiak, K., Jamoulle, J.-C., and Sawai, H. (1984). *J. Med. Chem.*, **27**, 726.
20. Torrence, P. F., Brozda, D., Alster, D., Charubala, R., and Pfleiderer, W. (1988). *J. Biol. Chem.*, **263**, 1131.
21. Kitade, Y., Nakata, Y., Hirota, K., Maki, Y., Pabuccuoglu, A., and Torrence, P. F. (1991). *Nucleic Acids Res.*, **19**, 4102.
22. Imai, J., Lesiak, K., and Torrence, P. F. (1985). *J. Biol. Chem.*, **260**, 1390.
23. Kovacs, T., Pabuccuoglu, A., Lesiak, K., and Torrence, P. F. (1993). *Bioorg. Chem.*, **21**, 192.
24. Kitade, Y., Alster, D. K., Pabuccuoglu, A., and Torrence, P. F. (1991). *Bioorg. Chem.*, **19**, 283.
25. Jamoulle, J.-C. and Torrence, P. F. (1986). *Eur. J. Med. Chem.*, **21**, 517.
26. Jamoulle, J.-C. and Torrence, P. F. (1984). *Biochemistry*, **23**, 3063.
27. Jamoulle, J.-C., Lesiak, K., and Torrence, P. F. (1987). *Biochemistry*, **26**, 376.
28. Lesiak, K. and Torrence, P. F. (1986). *J. Med. Chem.*, **29**, 1015.
29. Lesiak, K. and Torrence, P. F. (1987). *J. Biol. Chem.*, **262**, 1961.
30. Torrence, P. F., Xiao, W., Li, G., and Khamnei, S. (1994). *Curr. Med. Chem.*, **1**, 176.
31. Uznanski, B., Koziolkiewicz, M., and Stec, W. J. (1990). *Chem. Scr.*, **26**, 221.

32. Pond, R. T., Usman, N., and Ogilvie, K. K. (1988). *Biotechniques*, **6**, 768.
33. Letsinger, R. L. and Lunsford, W. B. (1976). *J. Am. Chem. Soc.*, **98**, 3655.
34. Beaucage, S. L. and Caruthers, M. H. (1981). *Tetrahedron Lett.*, **22**, 1859.
35. Ogilvie, K. K., Usman, N., Nicoghosian, K., and Cedergren, R. J. (1988). *Proc. Natl. Acad. Sci.* USA, **85**, 5764.
36. Usman, N., Ogilvie, K. K., Jiang, M.-Y., and Cedergren, R. J. (1987). *J. Am. Chem. Soc.*, **109**, 7845.
37. Scaringe, S. A., Franclyn, C., and Usman, N. (1990). *Nucleic Acids Res.*, **18**, 5433.
38. Ogilvie, K. K., Beaugage, S. L., Schifman, A. L., Theriault, N. Y. M., and Sadana, K. L. (1978). *Can. J. Chem.*, **56**, 2768.
39. Eckstein, F. (ed.) (1991). *Oligonucleotides and analogues: a practical approach.* Oxford University Press, Oxford.
40. Martin, E. M., Birdsall, N. J. M., Brown, R. E., and Kerr, I. M. (1979). *Eur. J. Biochem.*, **95**, 295.
41. Silverman, R. H. and Krause, D. (1987). In *Lymphokines and interferons: a practical approach* (ed. M. J. Clemens, A. G. Morris, and A. J. H. Gearing), pp. 149–93. IRL Press, Oxford.
42. Sambrook, J., Fritsch, E. F., and Maniatis, T. (ed.) (1989). *Molecular cloning: a laboratory manual*, 2nd edn. Cold Spring Harbor Laboratory Press, NY.

7

In vitro applications of antisense RNA and DNA

STEPHEN H. MUNROE

1. Introduction

In vitro studies with antisense RNAs and DNAs have served a number of different purposes:

(a) Analysis of mechanisms of antisense inhibition in naturally occurring systems such as those responsible for control of plasmid replication (1, 2).

(b) Evaluation of the effectiveness of antisense RNAs and DNAs as potential modulators of gene expression *in vivo* (3, 4).

(c) Screening of cloned cDNAs to identify ones encoding a particular protein by inhibition of *in vitro* translation (5).

(d) Identification of RNAs required for complex genetic processes such as pre-mRNA splicing (6).

(e) Characterization of the structure and accessibility of sites within an RNA molecule (7–9).

Despite these diverse applications, there have been relatively few studies with antisense nucleic acids in cell-free systems in comparison with the large number of investigations carried out in intact cells. It is anticipated that *in vitro* analysis will play an increasingly prominent role in future investigations as new antisense systems are discovered and efforts are directed towards understanding their mechanisms.

In this chapter I will discuss methods which are common to many investigations involving the use of complementary RNAs and DNAs to block gene expression and probe RNA structure and interactions in cell-free systems. Some of these methods, and a number of related procedures, are further discussed in other compilations of methods (10) and in another volume in this series (11).

2. Synthesis of RNA *in vitro*

In vitro transcription of cloned DNA sequences provides the most widely used source of specific RNAs for *in vitro* studies as described in Chapter 4.

Transcripts ranging in length from a dozen to more than 1000 nucleotides (nt) are readily prepared by *in vitro* transcription as described in Chapter 4. In many cases, shorter transcripts can be used directly after extraction and ethanol precipitation in the presence of 1 M ammonium acetate, which removes unincorporated nucleotides. Longer transcripts usually require further purification to remove substantial amounts of incomplete or partially degraded transcripts as described in Chapter 4, *Protocol 10*.

Vectors containing opposing promoters for two different phage polymerases flanking the cloning site are particularly useful for antisense studies since one cloned insert may serve as a template for both sense and antisense RNAs. Typical vectors of this type include pBlueScript (Stratagene) and pGEM vectors (Promega). A variation on this strategy is provided by the Litmus vectors (New England Biolabs) in which either of two opposing T7 promoters is selectively inactivated by restriction endonuclease cleavage prior to transcription. In studies where the exact sequence of the 5' or 3' end of the transcript is critical for interactions between complementary antisense molecules, a variety of strategies exist for construction of a specific plasmid insert or for modifying the transcript *in vitro*, after transcription. These approaches are described in Chapter 1 of ref. 11.

3. Working with RNA under RNase-free conditions

Consistent with its role as a transient cellular messenger, RNA is less intrinsically stable than DNA. In the laboratory, isolated RNA is often subject to assault from adventitious contamination with RNases from a variety of sources. Precautions must be taken in handling RNA to minimize exposure to conditions and sources of RNase that will rapidly degrade RNA samples. The most important of these precautions are summarized below.

(a) Use only reagent grade water which has been deionized and treated to remove organic contaminants. Water must be collected and stored in an RNase-free container. As a further precaution, water may be treated with diethyl pyrocarbonate (DEPC) as described below.

(b) Glassware used in handling RNA should be treated with DEPC, a reactive alkylating agent which is effective in inactivating RNases. Glassware is allowed to sit for 30 min with 0.02% DEPC in water, then drained, wrapped, and autoclaved. This concentration of DEPC is also used to remove residual RNase contamination in reagent water. Autoclaving glassware and water subsequent to treatment degrades DEPC. Because of its high reactivity, dialysis tubing, Tris, and other common buffers should not be treated directly with DEPC. As an alternative to DEPC, clean glassware can be baked for 4 h at 250°C to remove RNase activity, but autoclaving by itself is not sufficient.

(c) Sterilize all solutions to eliminate microbial growth as a source of RNase contamination. Sterilization of small volumes is most conveniently accomplished by filtration through 0.2 μm membranes in disposable filter units (Nalge or Schleicher & Schuell) with the aid of vacuum or centrifugation.

(d) Use pre-packaged, sterile plasticware for preparing and storing solutions. Most manufacturers' products are reliably RNase-free.

(e) Microcentrifuge tubes are siliconized for use with dilute solutions of RNA since RNA is efficiently absorbed at low ionic strength. Rinse tubes with 5% dichloromethylsilane in ethanol and rinse with reagent water. Commercially prepared siliconized tubes are available from several suppliers, but often display distressing batch-to-batch variations in quality.

(f) Latex gloves should be worn at all times when handling RNA samples. This protects tube caps, pipette tips, etc., from inadvertent exposure to RNases on hands and fingers.

Although RNases are rather resistant to denaturation, many standard reagents (including sodium dodecyl sulfate, phenol, formamide, EDTA, and borate buffers) either reduce the activity of RNases or minimize microbial contaminants. A small RNA transcript, radiolabelled to a high specific activity, can be used as a substrate for detecting persistent RNase activity.

4. Protein catalysed RNA:RNA annealing

The co-operative pairing of complementary polynucleotide strands is central to all antisense processes. The rate of pairing may be critical for regulation as shown for antisense interactions which regulate replication of ColE1 plasmids (1). The rate of annealing of RNAs *in vitro* provides a quantitative measure of efficiency of annealing and accessibility of target sequences. There is particular interest in characterizing proteins which affect the rate of annealing or stability of base paired structures. Some proteins may facilitate annealing either by altering the conformation of individual strands in a manner favourable for annealing (12, 13) or by increasing transient interactions which lead to productive nucleation (14). Many RNA binding proteins have recently been found to enhance the rate of RNA:RNA annealing *in vitro* (12, 13, 15–18). It has been proposed that RNA:RNA annealing proteins play a general role as molecular chaperones in facilitating folding and conformational transitions essential for biological activity of RNA (19, 20). These proteins have diverse structures which include several different RNA binding domains (*Figure 1*). These annealing proteins enhance annealing by as much as several thousand-fold *in vitro* (12, 16) and include hnRNP proteins such as A1 and A2, which are among the most abundant proteins in the HeLa cell nucleus.

Other nuclear proteins have been shown to destabilize RNA:RNA base pairing. These include helicases, which unwind base paired RNAs in concert

Stephen H. Munroe

hnRNP Proteins

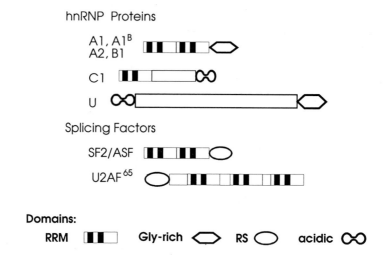

Figure 1. RNA:RNA annealing proteins. The domain structure of six hnRNP proteins and two splicing factors is represented schematically.

with the hydrolysis of ATP (21), and double-stranded RNA deaminases (22, 23), which convert adenosine residues to inosine in double-stranded RNA, and thereby destabilize A:U base pairing. Formation (or dissociation) of RNA:RNA base pairing in the presence of purified proteins or unfractionated cell extracts can be analysed using an RNase protection assay (*Protocol 2*) or native gel electrophoresis (*Protocol 4*). As shown in *Figure 2* (lower panel), this assay is carried out with two RNAs of unequal length to produce a shift in size when paired duplex molecules are trimmed with RNase. The longer RNA (the sense strand) is labelled with [^{32}P]UTP. The shorter RNA (antisense strand) is labelled to a low specific activity with [^{3}H]UTP and added in five-fold molar excess. In *Protocol 1*, low concentrations of sense and antisense RNA are used to permit analysis of the rate of annealing seen in the presence of hnRNP A1 protein. Substantially higher concentrations of RNA (0.8 nM sense strand, 8 nM antisense) are used to examine the stoichiometry and extent of duplex formation as shown in *Figure 2*.

Protocol 1. RNA:RNA annealing assay

Equipment and reagents

- Buffer DN: 20 mM Hepes–KOH pH 8.0, 100 mM NaCl, 5% glycerol, 0.1 mM EDTA, 0.5 mM DTT
- 2 × buffer M: 5 mM MgCl$_2$, 0.2 mM EDTA pH 7.5, 1 mM DTT
- Protein, diluted in buffer DN

- Antisense RNA, 5.0 fmole/μl (^{3}H-labelled)[a]
- Sense strand RNA, 2.5 fmol/μl (^{32}P-labelled)[a]
- Heating block or water-bath adjusted to 30°C

Method

1. Prepare the annealing mix as follows for 10 × 25 µl reactions and store on ice:
   - 50 µl 2 × buffer M
   - 3.0 µl RNase inhibitor
   - 2.0 µl sense strand RNA
   - 40.0 µl H_2O

2. Dilute the protein with buffer DN to 1.67 × final concentration and distribute 15 µl aliquots of protein into individual microcentrifuge tubes for annealing reaction. Include one tube with buffer DN only (minus protein).

3. Pre-incubate tubes with protein and annealing mix (step 1) at 30 °C for 5 min.

4. Add 5 µl antisense RNA to the annealing solution, step 1, mix, and immediately add 10 µl of the RNA mix to aliquots of diluted protein. To obtain a time course of annealing, 150 µl protein stock is added directly to 95 µl pre-warmed annealing mix after adding antisense RNA. 25 µl aliquots are withdrawn at intervals of 0.1–32 min.

5. Stop annealing as indicated in *Protocols 2* or *4*.

[a] Concentration of labelled RNA is calculated from the specific activity of the labelled nucleotide triphosphate (NTP) used for *in vitro* transcription and the c.p.m./µl of labelled RNA as follows:
(a) Specific activity of label equals total c.p.m. in sample at step 2 divided by total moles of NTP in reaction (moles labelled NTP + moles unlabeled NTP).
(b) Specific activity of transcript is equal to specific activity of label times moles labelled NMP residues/mole transcript.
(c) Concentration of RNA is equal to c.p.m./µl divided by specific activity of transcript (c.p.m./mole transcript).

Protocol 2. RNase protection assay of RNA:RNA annealing

Equipment and reagents

- RNase T1, 2500 U/ml in buffer DN (Sigma Chemical or Boehringer Mannheim)
- 2 × PK buffer: 0.2 M Tris–HCl pH 7.5, 25 mM EDTA, 0.3 M NaCl, 2% SDS
- Proteinase K stock, 10 mg/ml H_2O (E. Merck)
- tRNA, 10 mg/ml (R7125, Sigma Chemical)
- Phenol, equilibrated with buffer to pH 7.8 (Gibco BRL or AMRESCO)
- PK mix (for ten reactions prepare fresh): 1 ml 2 × PK buffer, 50 µl proteinase K stock, 10 µl tRNA, 0.63 ml H_2O
- Phenol (pH 7.8):chloroform:isoamyl alcohol (25:24:1) (P:C mix)
- 95% ethanol
- 5% urea–polyacrylamide gel (*Protocol 3*)
- Radioanalytical imager (Ambis 1000 Radioscanner)

Method

1. Terminate annealing reactions (*Protocol 1*) by adding 6.25 µl of RNase T1 solution. Mix and incubate 10 min at 30 °C.

Protocol 2. *Continued*

2. Add 169 μl PK mix, incubate 30 min at 30°C.

3. Extract once with an equal volume of P:C mix, vortex for 1 min, or shake vigorously on oscillating shaker for 1–2 min.

4. Spin for 2–3 min in a microcentrifuge to separate phases. Remove top, aqueous layer to new tube.

5. Add 2.5 vol. ethanol (1.0 ml) to aqueous layer. Mix and store on dry or wet ice for at least 15 min to precipitate RNA.

6. Spin the sample for 15 min at 4°C to recover RNA. Carefully remove and discard supernatant in shielded container for radioactive waste.

7. Dry pellet in a vacuum centrifuge.

8. Dissolve in 6 μl FSB and load on gel as described in *Protocol 3*. The gel is run at 600 V for 45 min.

9. Autoradiograph gel (*Protocol 3*) or scan gel with a radioanalytical imager for quantitative analysis.

Protocol 3. Purification of RNA by gel electrophoresis

Equipment and reagents

- Glass plates (about 20 × 20 cm) to match gel apparatus: one plate square, one plate with 16 cm long, 2 cm deep notch on one edge
- Plastic spacers (0.4 mm thick)
- Comb (18–22 teeth for forming wells 0.55 cm wide and 0.8 cm deep)
- Tape, 1.5 inch wide electrical tape, Scotch type 56 (3M Company)
- Vertical gel apparatus
- Binder clips
- 10 × TBE stock: 108 g Tris, 55 g boric acid, 9.3 g disodium EDTA, H₂O to 1 litre (store at 4°C)
- 20% acrylamide stock (20:1 acrylamide: methylene *bis*-acrylamide) in 1 × TBE, 7 M urea (Bio-Rad, ICN, or Gibco BRL)
- 1 × TBE, 7 M urea (store at 4°C)
- Formamide sample buffer (FSB): 4.8 ml deionized formamide,[a] 95 μl 1% bromophenol blue, 95 μl 1% xylene cyanol, 10 μl 0.5 M EDTA pH 7.5
- 10% ammonium persulfate (APS) made fresh every two weeks
- Tetramethylethylenediamine (TEMED) (Sigma Chemical Co.)
- Autoradiography film (XAR film, Kodak)
- Constant voltage power supply

Method

1. Set-up gel mould by inserting spacers at each side and sealing all but the top, notched edge of the plates with tape. After taping, clamp plates and spacers firmly together with a binder clip on either side.

2. Prepare solution for a 5% gel as follows:[b]

   - Acrylamide stock   5 ml
   - TBE–urea   14.75 ml
   - 10% APS   0.25 ml

   Mix and add 5–10 μl TEMED to initiate polymerization.

3. Pour the gel immediately. Fill mould to the top, clearing any bubbles from the gel by tapping on plates with the mould held upright. Insert comb, clamp the top with two additional clips, and lay the gel almost flat with the top edge propped up 1–3 cm.

4. Remove the comb after 20–30 min. The gel should polymerize in 10 min. If allowed to sit for an extended period the comb may be difficult to remove without distorting the wells.

5. Remove tape, rinse excess gel from area near wells, and clamp the gel in place on the apparatus, with the notched edge fitting against the notched face on the upper buffer chamber. Add 1 × TBE buffer to the upper and lower chambers.

6. Pre-electrophorese at 450 V for 15 min.

7. Prepare samples by dissolving RNA directly in FSB, or by diluting RNA dissolved in H_2O or low salt buffer with 4 vol. of FSB. Allow samples to sit on ice for 5–10 min with occasional mixing.

8. Squirt buffer into the bottom of each well several times with a Pasteur pipette before loading the samples. This procedure flushes out dense urea which collects in the wells.

9. Heat the samples at 85 °C for 1–2 min immediately before loading.

10. Load the samples into wells covered with buffer. This is done with an adjustable pipettor fitted with a small tip. The tip is placed directly over the centre of the well resting on the rim of the inner plate and angled slightly against the outer plate. The sample is gently expelled and allowed to settle into the well. A 0.55 cm well will hold up to 8 μl of sample.

11. Run samples at 450 V for 1 h or until the dyes migrate an appropriate distance into the gel. Bromophenol blue comigrates with oligonucleotides of 25, 15, or 7 nucleotides on 5%, 8%, or 20% gels respectively. Xylene cyanol runs with oligos of 130, 75, or 25 nucleotides, respectively, on the same gels.

12. When the run is complete, remove the gel from the apparatus and pry off one plate. Wrap the exposed gel with plastic wrap. Mark the gel with phosphorescent stickers or ink.

13. Store the gel with the film pressed flat at −70 °C for an appropriate length of time to expose autoradiogram. Develop film.

[a] Formamide is deionized by mixing 100 ml of formamide for 30 min with 10 g of mixed bed resin (Bio-Rad, AG 501-X8, 20–50 mesh).
[b] A 5% gel is used for resolving RNAs 75–500 nucleotides in length. Gels of 4–20% acrylamide are used for resolving shorter or longer RNAs.

The extent of annealing is determined by including control samples which are extracted without prior T1 digestion. The per cent of labelled RNA

Stephen H. Munroe

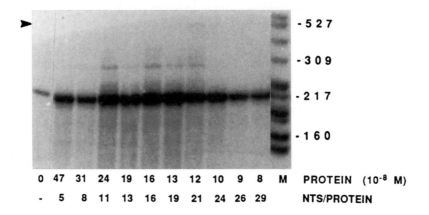

0	47	31	24	19	16	13	12	10	9	8	M	**PROTEIN (10⁻⁸ M)**

PROTEIN $(10^{-8}$ M)

NTS/PROTEIN

-	5	8	11	13	16	19	21	24	26	29

RNA-RNA ANNEALING ASSAY

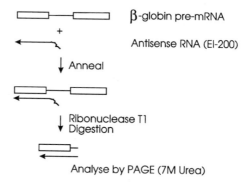

Figure 2. RNA:RNA annealing assay. Top panel: RNA annealing activity with hnRNP A1 protein (14). Under these conditions a single molecule of A1 binds to 14–18 nt. Results indicate that annealing is optimal at protein concentrations which saturate the binding sites of the RNA. The position of the labelled transcript is indicated by arrowhead at left. Bottom panel: summary of procedure for measuring RNA:RNA annealing with RNase protection assay.

annealed is calculated after correcting for size and base composition of protected RNA. It is important to note that RNase T1 remains active in SDS solutions, but it is efficiently degraded by treatment with proteinase K. Native gel electrophoresis provides another means of assaying RNA:RNA annealing (16). In this assay annealing is observed as a shift in mobility as the labelled RNA is converted from single-stranded to double-stranded molecules. It is necessary to use RNAs which give discrete, resolvable bands in both single-strand and double-stranded forms. The larger RNA substrate shown in *Figure 2* migrates diffusely in polyacrylamide gels prepared without urea which probably reflects the presence of multiple conformations. Smaller RNAs, 150–200 nucleotides long, are resolved on native gels prepared without urea (*Protocol 4*).

Protocol 4. Native gel assay of RNA:RNA annealing

Reagents

- Native gel sample buffer: 2% SDS, 25% glycerol, 5 μg/μl tRNA, 0.2% bromophenol blue, 0.2% xylene cyanol
- 10 mg/ml proteinase K (PK) stock in H_2O (E. Merck)
- 20% acrylamide stock: 19% acrylamide, 1% methylene *bis*-acrylamide in 1 × TBE (see *Protocol 3*)

Method

1. Add 0.1 vol. PK stock to native gel sample buffer just before use.
2. Stop annealing by adding 3 μl sample buffer with PK to 25 μl reactions.
3. Incubate the samples for 10 min at 30°C.
4. Load the samples directly on the gel (without heating) and run as described in *Protocol 3*.
5. Analyse by autoradiography or radioanalytical scanning (*Protocol 3*).
6. Identify bands corresponding to single- and double-stranded DNA by comparison with control sample containing ^{32}P-labelled sense strand RNA mock annealed without antisense RNA.

---

5. Antisense inhibition of pre-mRNA splicing

5.1 Introduction

Antisense strategies have played an important role in characterizing pre-mRNA splicing. Antisense RNAs and DNAs complementary to pre-mRNA have been used to examine changes in the accessibility of different sites on the pre-mRNA during assembly of the spliceosome (7–9). Directed digestion with RNase H has provided an important technique for examining the role of small nuclear RNAs in splicing (6).

5.2 mRNA splicing *in vitro*

Splicing is carried out *in vitro* with nuclear extracts prepared as described by Dignam *et al.* (24). *Protocol 5* is a modification of this preparation which was originally introduced by Krainer *et al.* (25). This extract has been used to splice many different pre-mRNAs as described in *Protocol 6* with only minor adjustments of $MgCl_2$, KCl, and ATP concentrations to optimize splicing activity. To obtain active preparations of extract it is critical to start with a rapidly growing cell culture. HeLa cells grown in spinner culture should double once every 20–24 hours. The following procedure requires 8 litres of culture, but the preparation can be carried out on a larger or much smaller scale (26). The extract can be used directly for splicing (*Protocol 6*) or it can serve as a starting point for isolation of specific factors involved in nuclear mRNA processing (see ref. 11, Chapter 3).

Protocol 5. Preparation of nuclear extract[a]

Equipment and reagents

- Dounce glass tissue grinders (one each, 40 ml and 15 ml capacities with tight pestles) (Kontes): rinse with DEPC and autoclave
- Centrifuge bottles, 250–1000 ml capacity (DuPont/Sorvall)
- Centrifuge tubes, 40 ml Oak Ridge type (DuPont/Sorvall)
- Centrifuge tubes, 50 ml disposable (Falcon)
- Small beaker (30–50 ml) and stirring bar, rinse with DEPC, and autoclave together
- Dialysis tubing (pre-rinse and autoclave in distilled water)
- PBS: 0.8 g NaCl, 0.2 g KCl, 1.44 g Na_2HPO_4, 0.24 g KH_2PO_4 in 1 litre H_2O, adjust to pH 7.4 with HCl

- Phase microscope
- Buffer A (200 ml): 10 mM Hepes, 1.5 mM $MgCl_2$, 10 mM KCl, 0.5 mM DTT[b]
- Buffer C (20 ml): 20 mM Hepes, 25% glycerol, 0.42 M NaCl, 1.5 mM $MgCl_2$, 0.2 mM EDTA, 0.5 mM DTT, 0.5 mM phenylmethylsulfonyl fluoride (PMSF)[b]
- Buffer D (1 litre): 20 mM Hepes, 5% glycerol, 100 mM KCl, 0.2 mM EDTA, 0.5 mM DTT, 0.5 mM PMSF[b]
- HeLa cells spinner culture, grown in 8 litres Joklik's medium with 7% horse serum, approx. 5×10^5 cells/ml (about 16 ml packed cells)

Method

Note: all glassware, centrifuge tubes, and rotor should be cooled to 4°C. Step 6 and all subsequent steps are carried out at 4°C in a cold room.

1. Centrifuge cells at 100–300 *g* for 3–10 min.

2. Decant carefully and discard supernatants. Resuspend cell pellets in 5 vol. PBS.

3. Pool the cell suspensions in two 50 ml centrifuge tube and hold on ice until all cells are collected.

4. Centrifuge cells at 400 *g* for 5 min. Measure packed cell volume for calculating volumes of additions in steps 4, 5, and 8. Remove supernatant and resuspend cells in 5 vol. buffer A.

5. Centrifuge cells at 400 *g* for 5 min. Carefully pipette off supernatant and gently resuspend pellets in 2 vol. of buffer A.

6. Transfer cell suspension to large Dounce homogenizer. Disrupt cells with 15 strokes of pestle and check cell lysate with phase microscope to determine extent of cell lysis. If fewer than 95% are broken, homogenize cells with 5–15 additional strokes until lysis is nearly complete.

---

7. Centrifuge homogenate at 4000 *g* for 10 min in 40 ml Oak Ridge type tube to pellet nuclei. Carefully pipette off the dense supernatant. Centrifuge pellet a second time at 31 000 *g* for 20 min.

8. Resuspend pelleted nuclei in 0.7 ml buffer C per millilitre of initial packed cell volume. Use about ten strokes of the small Dounce homogenizer to obtain a uniform suspension.

9. Transfer suspension to small beaker and stir gently on a magnetic stirrer for 30 min at 4°C.

10. Centrifuge extracted nuclear suspension at 31 000 *g* for 30 min.

11. Transfer supernatant to dialysis tubing, and dialyse for 4–8 h against two changes of buffer D, 500 ml each.

12. Centrifuge the dialysate for 30 min at 31 000 *g*.

13. Remove supernatant (nuclear extract), split into small aliquots of 0.2–1.0 ml each, and store below –70°C. The extract is stable for more than one year.

[a] Based on ref. 24 and a protocol provided by A. R. Krainer.
[b] These buffers are made with sterile water using the following sterile stock solutions: 1.0 M Hepes-KOH, pH 8.0 (filter sterilized); 2.0 M KCl; 1.0 M $MgCl_2$; 5.0 M NaCl; 0.5 M EDTA-NaOH, pH 7.5; 1 M DTT; and 0.10 M PMSF in ethanol. DTT and PMSF are added to ice cold solutions immediately before use.

Protocol 6. Antisense inhibition of pre-mRNA splicing

Equipment and reagents

- Nuclear extract (*Protocol 5*)
- ^{3}H-labelled antisense RNA, 250 fmole/μl
- ^{32}P-labelled, capped pre-mRNA, 25 fmoles/μl
- 125 mM ATP
- 400 mM $MgCl_2$
- Polyvinyl alcohol, 13% (w/v) in H_2O

- 0.5 M phosphocreatine (Tris salt) (Sigma Chemical)
- Placental RNase inhibitor (RNasin™) 40 U/μl (Promega)
- 5% urea–polyacrylamide gel (*Protocol 3*)
- Heating block or water-bath adjusted to 30°C

Method

1. Prepare the following splicing mix for ten 25 μl splicing reactions:

   - 2.0 μl $MgCl_2$
   - 1.0 μl   ATP
   - 10 μl   phosphocreatine
   - 2.5 μl   RNase inhibitor
   - 50 μl   polyvinyl alcohol
   - 14.5 μl H_2O

2. Add 8 μl aliquots of splicing mix to a series of microcentrifuge tubes on ice.

Protocol 6. *Continued*

3. Add 1 μl ³H-labelled antisense RNA and 1 μl ³²P-labelled RNA to each tube.

4. Start splicing reaction by adding 15 μl nuclear extract, mix gently, and incubate at 30°C for 0.5–3 h.

5. Terminate splicing by adding 175 μl proteinase K mix (*Protocol 2*), and incubating an additional 20–30 min at 30°C.

6. Extract splicing reactions with P:C mix, and analyse samples by gel electrophoresis as described in *Protocol 2*, steps 3–9.

The time of addition of antisense RNA relative to other components of the splicing reaction affects the efficiency of inhibition as shown in *Figure 3*. If antisense is added 15 min after splicing is initiated a substantial reduction in the level of inhibition is observed, even when the reaction is continued for a total of 4 h (*Figure 3*, lane 5). This reduction of inhibition is seen when the antisense is added only a few minutes after nuclear extract (8) and reflects the rapid sequestering of pre-mRNA at an early stage of spliceosome assembly. In contrast, the most efficient inhibition of splicing is observed when pre-mRNA and antisense RNA are pre-incubated with nuclear extract in the absence of ATP (*Figure 3*, lane 4). This reflects the requirement for ATP in the sequestering of pre-mRNA early in the splicing reaction. Under the conditions described in *Protocol 6*, more than 80% of splicing of β-globin and erbAα pre-mRNAs is inhibited with 5–10 nM antisense RNA (8, 27). This inhibition is strikingly more efficient than that observed with oligos 15–20 nucleotides in length where much higher concentrations are required.

5.3 Role of snRNPs in pre-mRNA splicing

Antisense oligodeoxyribonucleotides have been extensively used to direct cleavage with RNase H. In this method a complementary oligo, typically 6–15 nucleotides in length, is incubated with the RNA target in the presence of RNase H which digests only RNA that is base paired to DNA. An important application of this technique is its use to analyse requirements for snRNA molecules in pre-mRNA splicing (6). This approach is also used to evaluate the accessibility of specific regions of snRNAs and pre-mRNAs during spliceosome assembly (7, 9).

The usefulness of directed RNase H cleavage derives both from its specificity and its convenience. Cleavage of RNA by RNase H requires as few as four consecutive base pairs between DNA and RNA (28) and requires no cofactors except Mg^{2+}. RNase H is also present in many types of cells and extracts. Although *E. coli* RNase H is often added, HeLa cell nuclear extracts contain sufficient activity to efficiently cleave nanomolar concentrations of pre-mRNA during a short pre-incubation. For targeted digestion of snRNAs,

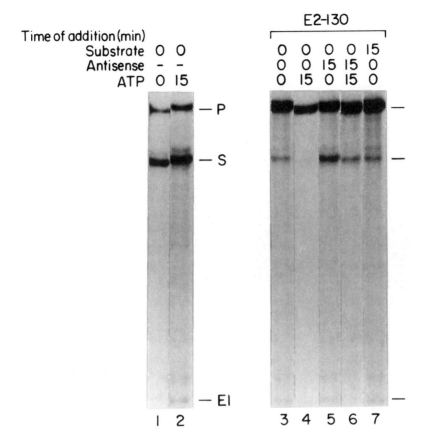

Figure 3. Antisense inhibition of splicing. Effect of delayed addition of ATP or antisense RNA. β-globin pre-mRNA was incubated for 4 h in the absence (lanes 1, 2) or presence (lanes 3–7) of RNA E2–130, complementary to the 3′ exon (8). P, position of unspliced RNA; S, spliced RNA; E1, the 5′ exon intermediate.

incubation may be carried out in the presence of other components of the splicing reaction (*Protocol 6*). The concentration of the antisense oligo used (0.2–100 μM) depends on the length of the oligo, the stability of base pairing, and the concentration and accessibility of the target. 10–20 μM oligo is optimal for complete digestion of U1 and U2 snRNA in HeLa cell nuclear extract and avoids non-specific degradation. These concentrations represent a 50- to 200-fold excess over the concentration of these abundant snRNAs (29).

RNase H degradation of U1 is described in *Protocol 7* (6, 30). The same approach is also used to target cleavage of pre-mRNA by use of a suitable complementary oligo. Addition of RNase H is not required for efficient degradation of low concentrations of pre-mRNA in HeLa cell nuclear extract in the presence of 0.3–1 μM oligo.

Protocol 7. Oligonucleotide-directed cleavage of U1 snRNA[a]

Equipment and reagents

- HeLa cell nuclear extract and other components of the *in vitro* splicing reaction (*Protocol 6*)
- Antisense oligodeoxyribonucleotide: 5′ TGCCAGGTAAGTAT 3′ complementary to 5′ end of U1 snRNA, 500 pmol/μl (Integrated DNA Technologies)

- Control oligodeoxyribonucleotide: 5′ GGT-GAATTCTTTGC 3′, 500 pmol/μl
- Ribonuclease H (*E. coli*), 2000 U/ml (Boehringer Mannheim)
- Ultraviolet transilluminator (Fotodyne)

Method

1. Prepare splicing mix as indicated in *Protocol 6*, step 1.

2. Add 5 μl oligo, 150 μl nuclear extract, 5 μl RNase H, and incubate for 30 min at 30 °C.

3. Remove 48 μl sample and set aside for analysis of snRNAs.

4. Distribute remaining sample into 24 μl aliquots for splicing reactions. Add 1 μl of [^{32}P]pre-mRNA substrate to each, and carry out splicing reactions and subsequent analysis as described in *Protocol 6*, steps 4–9.

5. Treat sample of digested extract (step 3) in parallel with untreated extract as described in *Protocol 6* (steps 5–8) to evaluate cleavage of U1 snRNA. Run snRNAs on 12% urea–polyacrylamide gel (*Protocol 3*) and stain with 1 μg/ml ethidium bromide to visualize snRNAs with ultraviolet transillumination.

[a] Adapted from ref. 30.

---

6. Antisense inhibition of mRNA translation

One of the first uses of antisense inhibition was to block *in vitro* translation (5). Hybrid-arrest of translation was used to screen cloned cDNAs and identify those which encoded a particular *in vitro* translation product. In the presence of complementary cDNA translation is selectively inhibited. Inhibition of translation usually results from cleavage of the mRNA base paired to cDNA by endogenous RNase H present in cell extracts used for translation (31, 32). While more direct techniques are now routinely used for identifying cDNAs corresponding to a particular protein, inhibition of translation by DNA-directed digestion of mRNA remains a useful method for blocking expression of a specific protein *in vitro* (33).

Techniques for oligo-directed RNase H cleavage of mRNA *in vitro* are similar to those described for inhibition of pre-mRNA splicing (*Protocols 5–7*). Wheat germ extract contains relatively high levels of endogenous RNase

H which are usually sufficient for efficient digestion of added mRNA (31). Almost no RNase H is present in reticulocyte lysates, but efficient cleavage of target RNA can be obtained by adding exogenous enzyme (31, 32).

In studies where the mechanism of a particular *in vivo* process is under investigation, it is often desirable to adapt an approach whereby a specific step is inhibited in a manner that does not irreversibly alter the RNA component. 2'-*O*-methyl oligoribonucleotides are particularly useful for this purpose since these modified oligoribonucleotides form base paired duplexes with RNA which are both more stable than the corresponding RNA:RNA duplex and are resistant to degradation with nucleases *in vivo*. Thus these derivatized antisense RNA molecules provide an important complement to antisense approaches based on DNA oligos which irreversibly destroy the target being studied. The synthesis and use of 2'-*O*-methyl oligoribonucleotides are described in ref. 11, Chapter 4.

Acknowledgements

I am grateful to Barbara DeNoyer for expert assistance in preparing this manuscript. The author's research was supported by a grant from Marquette University and awards GM44044 and DK48034 from the National Institutes of Health.

References

1. Eguchi, Y. and Tomizawa, J. I. (1990). *Cell*, **60**, 199.
2. Persson, C., Wagner, E. G. H., and Nordstrom, K. (1990). *EMBO J.*, **9**, 3767.
3. Bertrand, E., Pictet, R., and Grange, T. (1994). *Nucleic Acids Res.*, **22**, 293.
4. Dominski, Z. and Kole, R. (1993). *Proc. Natl. Acad. Sci. USA*, **90**, 8673.
5. Paterson, B. M., Roberts, B. E., and Kuff, E. L. (1977). *Proc. Natl. Acad. Sci. USA*, **74**, 4370.
6. Kramer, A., Keller, W., Appel, B., and Luhrman, R. (1984). *Cell*, **38**, 299.
7. Ruskin, B. R., Greene, J. M., and Green, M. R. (1985). *Cell*, **41**, 833.
8. Munroe, S. H. (1988). *EMBO J.*, **7**, 2523.
9. Eperon, I. C., Ireland, D. C., Smith, R. A., Mayeda, A., and Krainer, A. R. (1993). *EMBO J.*, **12**, 3607.
10. Sambrook, J., Fritsch, E. F., and Maniatis, T. (ed.) (1989). *Molecular cloning: a laboratory manual*, 2nd edn. Cold Spring Harbor Laboratory Press, NY.
11. Higgins, S. J.and Hames, B. D. (ed.) (1994). *RNA processing: a practical approach*, Volume I. IRL Press, Oxford.
12. Munroe, S. H. and Dong, X. (1992). *Proc. Natl. Acad. Sci. USA*, **89**, 895.
13. Portman, D. S. and Dreyfuss, G. (1994). *EMBO J.*, **13**, 213.
14. Pontius, B. W. (1993). *Trends Biochem. Sci.*, **18**, 181.
15. Krainer, A. R., Conway, G. C., and Kozak, D. (1990). *Genes Dev.*, **4**, 1158.
16. Pontius, B. W. and Berg, P. (1990). *Proc. Natl. Acad. Sci. USA*, **87**, 8403.
17. Lee, C.-G., Zamore, P. D., Green, M. R., and Hurwitz, J. (1993). *J. Biol. Chem.*, **268**, 13472.

18. Mayeda, A., Munroe, S. H., Caceres, J. F., and Krainer, A. R. (1994). *EMBO J.*, **13**, 5483.
19. Herschlag, D., Khosla, M., Tsuchihashi, Z., and Karpel, R. L. (1994). *EMBO J.*, **13**, 2913.
20. Herschlag, D. (1995). *J. Biol. Chem.*, **270**, 20871.
21. Wassarman, D. A. and Steitz, J. A. (1991). *Nature*, **349**, 463.
22. Wagner, R. W., Smith, J. E., Cooperman, B. S., and Nishikura, K. (1989). *Proc. Natl. Acad. Sci. USA*, **86**, 2647.
23. Bass, B. L. and Weintraub, H. (1988). *Cell*, **55**, 1089.
24. Dignam, J. D., Lebovitz, R. M., and Roeder, R. G. (1983). *Nucleic Acids Res.*, **11**, 1475.
25. Krainer, A. R., Maniatis, T., Ruskin, B., and Green, M. R. (1984). *Cell*, **36**, 993.
26. Lee, K. A. and Green, M. R. (1990). In *Methods in enzymology* (ed. J. E. Dahlberg and J. N. Abelson), Vol. 181, pp. 20–30. Academic Press, San Diego.
27. Munroe, S. H. and Lazar, M. A. (1991). *J. Biol. Chem.*, **266**, 22083.
28. Donis-Keller, H. (1979). *Nucleic Acids Res.*, **7**, 179.
29. Black, D. L., Chabot, B., and Steitz, J. A. (1985). *Cell*, **42**, 737.
30. Krainer, A. R. and Maniatis, T. (1985). *Cell*, **42**, 725.
31. Minshull, J. and Hunt, T. (1986). *Nucleic Acids Res.*, **14**, 6433.
32. Walder, R. Y. and Walder, J. A. (1988). *Proc. Natl. Acad. Sci. USA*, **85**, 5011.
33. Minshull, J., Blow, J. J., and Hunt, T. (1989). *Cell*, **56**, 947.

Antisense applications in *Dictyostelium*: a lower eukaryotic model system

MARTIN HILDEBRANDT

1. General considerations

The cellular slime mould *Dictyostelium discoideum* is a single-cell haploid amoeba, which, upon starvation, enters a well characterized developmental cycle. During this process, multicellular aggregates, also called pseudo-plasmodia are generated, which later form a fruiting body and differentiate into a tube of non-viable stalk cells and a sorocarp of encysted spore cells. Since the many *Dictyostelium* proteins involved, for example, in signal trans-duction or cell movement are very similar to their mammalian homologues, antisense RNA experiments in *Dictyostelium* are an easy and inexpensive model system to generate and study mutants lacking a protein of general interest.

Since several attempts to establish antisense RNA-mediated gene inactiva-tion in yeast have failed (1), *Dictyostelium* is so far the only lower eukaryote where antisense approaches are a more or less routine procedure. A detailed analysis of the data reported from *Dictyostelium* will not only help with appli-cations in this system but will also provide a valuable source of information for the design of antisense experiments in other fungi or amoebae.

1.1 History of antisense-mediated gene inactivation in *Dictyostelium*

Soon after the establishment of DNA-mediated transformation for *Dictyo-stelium* (2), and after the first application of antisense strategies in a few other organisms (3, 4), Crowley and colleagues inactivated a *Dictyostelium* gene by antisense expression (5). Thus, *Dictyostelium* became the first eukaryote, where antisense expression had been applied for gene repression in a whole organism. As a target, the *discoidin I* gene family (involved in cell–substrate interaction) was chosen. The phenotype of the transformants did not only confirm previous observations on chemically-induced discoidin minus mutants, but also demonstrated that a whole gene family could be silenced by antisense expression.

In 1987, Knecht and Loomis (6) reported on antisense inactivation of the myosin II heavy chain gene (*MIIHC*). In parallel, the same gene had been disrupted by homologous recombination (7), and the efficiency of both techniques could be compared. Neither the mutant nor the antisense silenced strain expressed any detectable levels of MIIHC protein, and both showed the same phenotype: an inability to grow in suspension culture, multinucleated cells, and an impairment in cytokinesis.

1.2 Advantages of antisense-mediated gene inactivation

Although gene disruption by homologous recombination has become a commonly used technique to knock-out *Dictyostelium* genes, it is in many cases preferable to choose the antisense strategy for gene inactivation. A comparison of both techniques is given in *Table 1*.

First, the construction of an antisense vector is usually easy, since typically only a fragment of the target gene needs to be cloned in an antisense orientation into one of the available *Dictyostelium* expression vectors. Transformation results in clones that have randomly integrated tandem arrays of 20 to 200 vector copies into the genome. Due to different copy numbers and genomic integration sites, clones with different levels of antisense RNA expression are obtained, and thus different levels of gene inactivation can be studied. For genes where a complete loss of function is lethal, the maximal, still viable inactivation can be obtained. Antisense effects can be enhanced, by increasing selective pressure; this often results in increased plasmid copy

---

Table 1. Comparison of antisense expression and gene replacement

Antisense expression	Gene replacement
+ Easy protocol	− More difficult procedure
+ High frequency	− In some cases low frequency
+ Partial gene sequence sufficient	− Requires essential parts of target gene
+ Clones with different levels of inactivation	− Complete inactivation
+ Partial inactivation of essential genes possible	− Not possible with essential genes
+ Inactivation of gene families possible	− Inactivation of single genes only
+ Regulated inactivation possible (e.g. inducible promoter, increased selective pressure)	− No regulated dosage possible
− Transformants may be unstable	+ Usually no revertants
− Residual amounts of gene product not excluded	+ Complete loss of gene product
− Not applicable for late developmental genes	+ Inactivation of late genes possible

---

numbers and, simultaneously, increased antisense RNA expression. Depending on the extent of DNA sequence similarity (see also Chapter 9), expression of an antisense RNA directed against one member of a gene family can result in inactivation of the entire family. In addition to discoidin I (5), this has also been seen for the gp138A and gp138B gene products (8). Moreover, depending on the promoter, the timing of antisense inactivation can be chosen between vegetative phase and early/mid-development (see below).

There are two major problems with antisense experiments. First, a low, undetectable residual level of the gene product can never be excluded and cells may fail to show a phenotype. Secondly, at least in *Dictyostelium*, genomic instability of antisense transformants has been observed in a few cases: although constantly grown under selection, some strains tend to lose copies of the integrated vector DNA, and/or produce less and less antisense RNA. Finally, this may result in a loss of gene inactivation and the mutant phenotype (9, 10).

1.3 Mechanism of antisense-mediated gene inactivation

The mechanisms of antisense RNA-mediated gene silencing are far from being understood. It has been shown that antisense expression does not affect mRNA transcription of the targeted gene (5, 11). Since in most cases the targeted mRNA does not accumulate to detectable levels, antisense RNA therefore seems to destabilize the mRNA. It has been proposed that mRNA and antisense RNA form hybrids which are rapidly degraded by a double-strand RNA-specific, sequence non-specific nuclease. Investigations on *PSV-A* gene regulation, which is mediated by an endogenous antisense RNA (11), and the discovery of a dsRNA-specific cytoplasmatic nuclease activity (Nellen and co-workers, unpublished data) support this hypothesis.

Surprisingly, overexpression of partial sense RNAs, at least in the cases of the *MIIHC* gene (9, 12), and the *V4* gene (13), may also result in post-transcriptional inactivation of the target genes. This co-suppression effect has previously been observed in plant systems (14, 15). Another puzzling observation is, that even within populations of clonally grown cells, a small percentage of the cells is still capable of expressing the targeted gene at wild-type levels (5, 12).

2. Construction of vectors expressing antisense RNAs

2.1 Examples of antisense-mediated gene inactivation

Since the mechanisms of antisense RNA-mediated gene inactivation are not understood, it is impossible to set-up a general strategy for antisense experiments. Therefore, the best approach is the comparison of different strategies which have previously been employed. *Table 2* describes several vectors

Table 2. Frequently used vectors for *Dictyostelium* transformation

Name	Promoter	Cloning sites	Terminator	Examples
pA6NPTII (6)	A6Pr–NPTII	*Hpa*I–*Bam*HI	SV40T (NPTII–antisense fusion)	1.1, 2.1, 3.1, 4.1
pA15XPTI (9)	A15Pr NPTI	*Bam*HI	A15T (NPTI–antisense fusion)	1.2
pDNeoII (32)	A6Pr	MCS	A8T	1.3, 3.2, 5, 6, 7, 8
pDexRH (33)	A15Pr	*Eco*RI–*Hind*III	A8T	10
PVEII (25)	Disclγipr	MCS (inducible expression upon depletion of folate)	A8T	18, 19, 20

frequently used for antisense inactivation. *Table 3* summarizes the results of all available data on antisense experiments in *Dictyostelium*.

It should be noted that there is not a single example for antisense inactivation of genes in late *Dictyostelium* development. Since both sense RNA and antisense RNA are readily detectable in late development (16), it appears that the mechanisms of antisense RNA-mediated gene silencing are impaired in these developmental stages.

2.2 Choice of the appropriate vector

With one exception (8), only integrating plasmids have been used for antisense experiments although extrachromosomal, self-replicating vectors are available for *Dictyostelium* transformation (17, 18). Except for their large size, which may cause problems in *in vitro* manipulation, there is no specific reason why these vectors could not be used with equal efficiencies.

Integration of vector DNA into the genome mostly occurs at high copy number simultaneously resulting in high levels of antisense expression. Integration sites appear to be random with some preference for non-coding, AT-rich regions (19). It is therefore unlikely that 'artificial' phenotypes, caused by disruption of an unrelated gene, will be generated.

In a few cases, antisense transformation has produced a high number of clones disrupted in the target gene by homologous recombination (10, 12, 20). These clones may express truncated gene products and should be clearly distinguished from true antisense transformants.

Several different markers for selection of antisense transformant cells may be used. However, the ura (21), thyl (18) and Bs^R (22) selection systems usually result in single copy vector integration and thus presumably in low antisense expression levels. Among the different selection markers, G418 resistance is the most common and so far the only one used for antisense expression.

In some experiments, antisense gene fragments have been cloned directly 3' to the G418 resistance gene (Tn5 or Tn903 neomycin phosphotransferase gene), generating hybrid RNAs containing both the phosphotransferase mRNA and the antisense transcript (6). Although here antisense expression is directly under selective pressure, similar effects of gene silencing have been obtained with separate resistance and antisense transcription units.

Cotransformation of a resistance plasmid and a non-selectable antisense vector result in independent integration of both vectors into the same genome. Cotransformed cells usually maintain both DNAs over many generations unless the antisense vector confers, e.g. a growth disadvantage.

2.3 Choice of the appropriate promoter

In most cases, an excess of antisense RNA expression relative to the level of mRNA is required. Therefore, the most commonly used promoters *actin6* (2), *actin15* (23), or *discoidinI*γ (5) are strong and usually produce sufficient levels of antisense transcripts (see also Chapter 9). Problems may arise targeting genes such as *gip17* (encoding NDP kinase) which are regulated by strong promoters. Possibly for this reason, *gip17* could only be partially inactivated by antisense expression (Hildebrandt and Veron, unpublished data). Successful antisense experiments have also been done by using the promoter of the target gene, e.g. for the G-protein subunit α1 gene (24) and the discoidin I family genes (5).

For timed antisense inactivation, several developmentally regulated promoters and the inducible *discoidinI*γ promoter are available (*Protocol 1*) (25). With the latter procedure, transcription of an antisense construct can be suppressed during transformation and reactivated later during growth. The protocol gives a modification of the general transformation procedure which is described in detail in ref. 39.

A fortuitous ATG and an open reading frame may promote the formation of a translational initiation complex on the (probably capped) antisense transcript. Indeed, a 'masking effect' has been described for antisense inactivation of the *P8A7* gene when the antisense fragment was inserted into the open reading frame of the *E. coli* tetracycline repressor gene (26). Similarly, constructs using an *actin15* promoter fragment including the ATG initiation codon have raised some problems (24). Pollenz and co-workers (28) had to delete the translation initiation site for successful antisense experiments. There are, however, also examples where a putative translation initiation site apparently does not interfere with antisense effects, since a defined developmental phenotype was obtained (27).

Finally, for the improvement of antisense effects, it is recommended to introduce a polyadenylation signal ('terminator') at the 3' end of the antisense sequence. The common *actin8*, *actin15*, and *2H3T* 'terminator' fragments are sufficiently tight. Most 'terminators' function in both orientations and thus prevent synthesis of potential read-through transcripts from adjacent promoters.

Table 3. Antisense transformation in *Dictyostelium*

Gene:	**Myosin II heavy chain (6, 9, 12)**
Constructs:	1. A6Pr–NPTII–SV40T–AS (various fragments)
	2. A15Pr–NPTI–AS (3.7 kb genomic) –A15T
	3. A6Pr–AS (3.7 kb genomic) –A8T
Phenotype	Multinucleate cells, blocked in cytokinesis (only observed with 1 and 2)
Protein level	Undetectable
mRNA level	Undetectable
Antisense level	Excess
Remarks	No inhibition in 1 when cells are grown on bacteria (A6Pr off). Detailed analysis with various gene fragments: 3′ end most efficient, a 0.4 kb fragment has no effect, variability of effects within a clonal isolate (assayed by immunofluorescence), co-suppression observed by overexpression of sense fragments.

Gene:	**Cyclic AMP receptor I, cARI (34)**
Constructs:	1. A6Pr–NPTII–SV40T–AS (complete 2 kb cDNA)
	2. A15Pr–AS (complete 2 kb cDNA) –2H3T
Phenotype	Blocked in aggregation (with both constructs)
Protein level	Undetectable
mRNA level	1. Undetectable
	2. Not significantly reduced
Antisense level	Excess

Gene:	**Conditioned media factor, CMF (35)**
Constructs:	1. A6Pr–NPTII–SV40T–AS (0.5 kb partial cDNA)
	2. A6Pr–AS (1.4 kb complete cDNA) –A8T
Phenotype	Blocked in aggregation
Protein level	Residual activity, almost undetectable
mRNA level	Undetectable
Antisense level	Undetectable

Gene:	**rasG (G. Weeks, with permission)**
Constructs:	1. A6Pr–NPTII–SV40T–AS (complete cDNA)
	2. A15Pr–AS (complete cDNA) –2H3T
Phenotype	Wild-type
Protein level	nd
mRNA level	Wild-type
Antisense level	Detectable in 1, undetectable in 2

Gene:	**Clathrin heavy chain (36)**
Constructs:	A6Pr–AS (2.8 kb partial cDNA) –A8T
Phenotype	Defects in endocytosis and osmoregulation
Protein level	Undetectable
mRNA level	Undetectable
Antisense level	Excess, degradation products detected

Gene:	**NDP kinase (M. Hildebrandt, M. Veron, unpublished)**
Constructs:	A6Pr–AS (0.5 kb complete cDNA including two exon junctions) –A8T
Phenotype	Minor temporal delay in streaming

Table 3. *Continued*

Protein level	Reduced to 50%
mRNA level	Reduced to 50%
Antisense level	Excess in some clones
Remarks	Target gene has a strong promoter

Gene: **Prespore vesicle protein A, PSV-A (20)**

Constructs:	A6Pr–AS (0.7 kb genomic including intron) –A8T
Phenotype	Wild-type
Protein level	nd
mRNA level	Wild-type
Antisense level	Excess
Remarks	Endogenous AS RNA destabilizes mRNA in vegetative cells

Gene: **dutA, non-coding cDNA (37)**

Constructs:	A6Pr–AS (1.3 kb cDNA fragment) –A8T
Phenotype	Wild-type
Protein level	Not applicable (non-coding)
mRNA level	nd
Antisense level	nd
Remarks	Extensive secondary structure predicted

Gene: **gp24 (38)**

Constructs:	A6Pr–AS (0.5 kb complete cDNA including two exon junctions) –A8T
Phenotype	Blocked in EDTA-sensitive cell adhesion
Protein level	Undetectable
mRNA level	Undetectable
Antisense level	Excess until 6 h of development

Gene: **Profilin I (10)**

Constructs:	A15Pr–AS (0.4 kb complete cDNA) –A8T
Phenotype	Wild-type
Protein level	Undetectable
mRNA level	Undetectable
Antisense level	Undetectable
Remarks	Redundant gene, only disruption of both profilin I and II (antisense inactivation not obtained) resulted in defects in cytokinesis, development, and F-actin content.

Gene: **rap1 (G. Weeks, with permission)**

Constructs:	A15Pr–AS (cDNA without 5′ end) –2H3T
Phenotype	Wild-type
Protein level	50–90%
mRNA level	90%
Antisense level	Fivefold excess

Gene: **G-protein α1 (24)**

Constructs:	1. A15Pr–AS (1 kb cDNA fragment) –2H3T
	2. Gα1Pr–AS (1 kb cDNA fragment) –2H3T
Phenotype	Wild-type

Table 3. *Continued*

Gene: **G-protein α1 (24)**

Protein level	Wild-type in 1, undetectable in 2
mRNA level	nd
Antisense level	nd

Gene: **gp138 (8)**

Constructs:	A15Pr–AS (complete cDNA) –2H3T (on extrachromosomal plasmid DDP1)
Phenotype	Repression of sexual cell fusion
Protein level	Strongly reduced
mRNA level	Strongly reduced (but wild-type levels are already very low)
Antisense level	Low but in excess compared to mRNA
Remarks	gp138 is encoded by two genes

Gene: **Essential myosin light chain (28)**

Constructs:	A15Δ3'Pr–AS (0.5 kb 5' cDNA fragment) –A6Pr (function as T)
Phenotype	Multinucleate cells, blocked in cytokinesis
Protein level	Less than 0.5% of wild-type
mRNA level	Approx. 10% of wild-type
Antisense level	Excess
Remarks	A15Δ3'Pr = deletion of translation initiation site in the A15 promoter fragment

Gene: **Discoidin I (5)**

Constructs:	DisclγPr–AS (0.3 kb partial cDNA), no selection marker, cotransformation with G418 resistance plasmid
Phenotype	Complete block of streaming
Protein level	Less than 10% of all discoidin I isoforms
mRNA level	Undetectable
Antisense level	nd
Remarks	Discoidin I is encoded by three closely related genes, mRNA and antisense transcription shown by run-on assays, variability in protein expression within clonal isolates.

Gene: **D2-esterase (27)**

Constructs:	DisclγPr–AS (0.6 kb genomic fragment containing intron) –2H3T
Phenotype	Incomplete aggregate formation
Protein level	nd
mRNA level	Undetectable until 12 h of development
Antisense level	Excess
Remarks	No mutant phenotype if cells are grown on bacteria, there, DisclγPr is off but activated with the onset of development, i.e. antisense effect has to occur before the mutant phenotype is observed.

Gene: **P8A7, putative transmembrane protein (26)**

Constructs:	1. ALFPr–5' CAT–AS (0.4 kb cDNA fragment incl. two exon junctions) –3' CAT–A8T

Table 3. *Continued*

Gene:	**P8A7, putative transmembrane protein (26)**
	2. DisclγPr–TetR–AS (0.4 kb cDNA fragment incl. two exon junctions) –A8T
	3. DisclγPr–AS (0.4 kb cDNA fragment incl. two exon junctions) –A8T
Phenotype	Feed-back induction of a second, upstream promoter
Protein level	nd
mRNA level	Undetectable until 8–10 h of development, then reaccumulation
Antisense level	Undetectable until 10 h of development, then reaccumulation
Remarks	Masking effect: reduced antisense effect when CAT or TetR coding sequence preceeds the antisense fragment.

Gene:	**Calmodulin (29)**
Constructs:	1. DisclγPr–AS (87 bp 5′ cDNA fragment) –A8T
	2. DisclγPr–AS (complete cDNA) –A8T
	3. A6Pr, D19Pr, or CP2Pr–AS (complete cDNA or cDNA without 5′ end) –A8T
Phenotype	1 and 2 multinucleate cells, block of cytokinesis
Protein level	50% of wild-type with 1 and 2
mRNA level	50% of wild-type with 1 and 2
Antisense level	nd
Remarks	DisclγPr repressed by growing cells at very low densities, induction of antisense expression occurs when cells reach higher densities, high antisense expression presumably lethal. This may explain why A6Pr failed; D19Pr and CP2Pr probably failed because they are only expressed in late development.

Gene:	**Vacuolar ATPase β-subunit (J. Gross, with permission)**
Constructs:	DisclγPr–AS (87 bp 3′ cDNA fragment) –A8T
Phenotype	Development blocked at slug stage
Protein level	30% of wild-type under stringent selection (200 μg G418/ml)
mRNA level	nd
Antisense level	nd

Gene:	**PKA catalytic subunit (C. Reymond, with permission)**
Constructs:	DisclγPr–AS (80 bp internal fragment) –A8T
Phenotype	Wild-type
Protein level	nd
mRNA level	Wild-type
Antisense level	Excess

Gene:	**V4 gene (13)**
Constructs:	V18Pr–AS (complete 1 kb cDNA) –CAT–A15T, no selection marker, cotransformation with G418 resistance plasmid
Phenotype	Blocked in aggregation
Protein level	nd
mRNA level	Undetectable
Antisense level	Excess

Protocol 1. Transformation at repressed antisense RNA levels (25)

Equipment and reagents
- pVE II vector (contains the folate-repress-ible *discoidinIγ* promoter to drive antisense expression)
- 100 mM folate solution (made up in water, pH adjusted with NaOH to bring folate into solution)

Method

1. Grow cells for three days in medium containing 1 mM folate. Keep cell density below 10^6/ml by diluting once a day with medium containing 1 mM folate.

2. Perform transformation as described in ref. 39.

3. After transformation, plate cells at a lower density than usual (1–5 × 10^6/9 cm Petri dish) in medium containing 1 mM folate and the appropriate antibiotic (in most cases G418).

4. Exchange medium daily by carefully pipetting off the edge of the tilted plate, take special care not to disturb cells which are loosely attached to the plate, add fresh medium (containing folate and the appropriate antibiotic) slowly to the edge of the plate.

5. When clones appear, wash cells off the plate and distribute onto two or three plates. Keep one plate at low density in the presence of folate to maintain the repressed state, refill the other plate with fresh medium not containing folate.

6. Grow to confluence (will strongly induce expression from the discoidin promoter) or keep at low density in the absence of folate to obtain intermediate antisense expression. Full antisense expression will be reached within two days.

7. Depending on the target gene, antisense effects on, for example, growth or cytokinesis may be observed directly on the culture plate. In other cases, analyse by Northern or Western blot or phenotypic screening (see *Protocol 2*).

2.4 Choice of sequence for antisense constructs

An intriguing problem for the design of antisense expression vectors is the choice of the gene fragment with optimal effects in gene silencing. The systematic studies of Scherczinger and Knecht demonstrate that different fragments of the same gene may vary significantly in efficiency (12): phenotypic variation ranging from wild-type to a complete block of development have been seen using different fragments of the *MIIHC* gene for antisense expression. However, a general algorithm for the best choice of sequence can so far not be suggested (see also Chapter 5). Similarly, the relative position of an optimal target sequence cannot be predicted (see also Chapter 9). A fragment overlapping the 5' end seems to be essential for inactivation of the calmodulin gene (29), whereas for the *MIIHC* gene, a 3' fragment was found to be most effective (12).

The size of the antisense transcript also seems not to be a crucial parameter: a 0.4 kb internal cDNA fragment of the *P8A7* gene (26), a 0.3 kb fragment of the *discoidinI* gene (5), and an 87 bp fragment of the calmodulin gene (29) were highly efficient whereas a 0.4 kb fragment of the *MIIHC* was not (12).

cDNA fragments containing exon junctions (26) as well as genomic fragments containing 5' or 3' non-coding sequences as well as introns (27) have all been used successfully. However, since introns and non-coding regions in *Dictyostelium* are extremely AT-rich, special care should be taken to avoid accidental AAUAAA polyadenylation signals which may lead to processing and premature termination of the antisense transcripts (28).

Scherczinger and Knecht (9, 12) have suggested that the most likely reasons for variations in antisense efficiency conferred by different parts of the target gene, are due to secondary structures in the antisense transcripts:

(a) Intrinsic instability of the antisense RNA may prevent gene inactivation. Combined strand-specific run-on and Northern blot analysis could identify this potential problem.

(b) The potential of an antisense transcript and/or the corresponding target to form intramolecular secondary structures, may inhibit or favour the generation of intermolecular hybrids. In addition, there are proteins such as RNA helicases and hybrid promoting proteins which may modulate secondary structures within the cell and make them quite different from those predicted by computer programs (see also Chapter 2). Consequently, the *in vivo* formation of RNA secondary structures can not be reliably predicted or controlled. Thus, small changes in the length of an antisense RNA may significantly alter secondary structure and lead to dramatic changes in the gene silencing capacity (30).

(c) Antisense RNA may also hybridize to heterologous mRNAs with a certain degree of homology to the targeted sequence. The 5' part of the *MIIHC* gene shows some homology to other myosin genes and is almost ineffective in antisense experiments. This suggests that the antisense RNA may be titrated out by binding to many other myosin mRNAs. For the essential myosin light chain gene, despite sharing most phenotypes, antisense transformants have a more severe developmental defect than targeted gene disruptions of the same gene. This may be due to a cross-hybridization of the antisense RNA to other myosin mRNAs (Chisholm, personel communication). Thus, when choosing fragments for the design of antisense constructs, one should always consider sequence properties and homologies with related genes.

Antisense transformation and selection

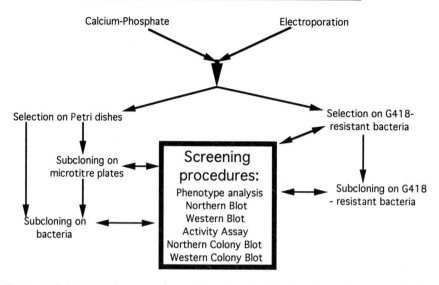

Figure 1. Schematic diagram of possible strategies for transformation and selection. Arrows indicate the different possibilities in order to generate clones expressing high levels of antisense RNA. The different time points, where screening procedures have to be performed, are indicated by double-headed arrows.

Due to its high genomic AT content, *Dictyostelium discoideum* has a very characteristic codon usage (31), which results in an extreme preference for certain codons, and thus in a high degree of homology in sequences encoding conserved domains from different proteins (e.g. ATP binding sites). Therefore, it may be important to avoid such sequences as antisense targets because other genes could also be affected.

Finally, sequences with internal repeats should be avoided to minimize secondary structure formation in the antisense RNA and the likelihood of recombination events.

3. Methods for antisense transformation

Transformation of *Dictyostelium* is inexpensive and easy. The procedures, calcium phosphate precipitation (39) or electroporation (40), are similar to those used for mammalian tissue culture. Transient transfections are not sufficiently effective (transformation efficiency 10^{-4} to 10^{-5}) to obtain reliable results but stably transformed clones are readily selected within 10–14 days.

Clones are obtained by plating limited dilutions of transformants on bacterial lawns (without selective pressure) and allowing for the formation of plaques (see *Protocol 2*). Alternatively, clones are directly picked from the transformation plate into microtitre wells and kept under permanent selection. However, clones isolated in this way may not be pure. Additional methods have been described to select clones on G418-resistant *E. coli* (41) or *Micrococcus luteus* (42).

Usually, the first screen to analyse antisense effects is for mutants displaying an aberrant developmental phenotype. The plaques that are formed upon spreading of cells on a bacterial lawn, contain all developmental stages from fruiting bodies in the centre to vegetative cells at the rim of the plaque. Therefore, subcloning on bacteria already allows for a visual screening of potential mutants. Lawns of *Klebsiella* on SM plates provide high amounts of food bacteria and consequently high densities of *Dictyostelium* cells. Frequently however, antisense inacitivation of a specific gene only results in subtle phenotypes. A more detailed analysis is done by growing cells on a more limiting food source such as *E. coli B/r* on normal agar plates. Cells reach lower densities and the different developmental stages are more spread out than on the rich *Klebsiella*/SM plates.

Protocol 2. Phenotypic screening

Equipment and reagents

- Neubauer counting chamber
- Stereo microscope
- SM agar: 20 g Bacto agar (Difco), 10 g Peptone (Oxoid),10 g glucose, 1 g yeast extract, 1 g $MgSO_4.7H_2O$, 2.2 g KH_2PO_4, 1 g K_2HPO_4, water to 1 litre, autoclave

- Normal agar (NA): 15 g Bacto agar (Difco), 1 g Peptone (Oxoid), 1 g glucose, 2.2 g KH_2PO_4, 1 g Na_2HPO_4, water to 1 litre, autoclave
- Soerensen buffer: 14.5 mM KH_2PO_4, 2.5 mM Na_2HPO_4 pH 6

Method

1. Grow a thick lawn of *Klebsiella aerogenes* on SM agar plates (two days at 37 °C).

2. Wash cells off the plates with 30 ml Soerensen buffer (resulting in a cell density of 10^8–10^{10} cells/ml).

3. Harvest *Dictyostelium* cells from a transformation plate with small (approx. 2 mm wide) clones by pipetting up and down several times.

4. Count *Dictyostelium* cells in a Neubauer counting chamber.

5. Mix aliquots containing 50–100 cells with approx. 150 µl of the *Klebsiella* suspension, spread a maximum of 200 µl of the mixture per SM agar plate.

6. Grow at 22 °C, until plaques become visible and examine with a stereo microscope.

7. The same procedure is used for plating on *E. coli*/NA plates except that *E. coli* cells are grown in suspension culture (NB or LB).

After the isolation of stable transformants, clones expressing the highest level of antisense RNA may be isolated. This can be done by enzyme activity measurements (if the targeted protein has a defined enzymatic activity, see *Protocol 3*), by phenotype, or by Northern and Western blot analysis.

Protocol 3. Preparation of cell lysates for screening by activity assay

This is a rapid and easy method to prepare cell lysates and has so far been successfully applied for measuring CAT, NDP kinase, and NPTII activity.

Equipment and reagents
- Eppendorf centrifuge
- Heating block
- Dry ice
- Appropriate buffer for the respective enzyme: 0.1 M Tris pH 7 for CAT, phosphate buffer for NDP kinase

Method
1. Wash 10^5–10^7 cells once in 17 mM phosphate buffer (Soerensen buffer) and spin at 500 g at 4°C.
2. If cells have been grown in the presence of bacteria, wash cells twice in Soerensen buffer, spin at 350 g at 4°C to pellet *Dictyostelium* cells, and keep the majority of bacteria in the supernatant, then wash once again as in step 1.
3. Resuspend cells in 100 μl of the appropriate enzyme buffer.
4. Freeze the sample for several minutes on dry ice or for at least 20 min at −20°C.
5. Thaw sample for 10 min at 37°C, for heat stable enzymes (e.g. CAT) heating at 60°C denatures other proteins without impairing enzyme activity and results in a cleaner preparation.
6. Repeat steps 4 and 5 at least twice.
7. For soluble, cytoplasmic proteins, spin the sample for 5 min (4°C) at maximum speed in an Eppendorf centrifuge.
8. Transfer supernatant to a fresh tube and measure enzyme activity.

Western blots can be performed with an equivalent of as little as 1–2×10^5 cells per lane. Therefore, it is sufficient to harvest cells directly from microtitre plates or from the growing edge of a clone on a bacterial lawn. In order to load comparable amounts of protein, the cell suspension has to be counted prior to lysis in protein sample buffer. Alternatively, cells may be lysed by repeated freezing on dry ice and thawing, and the protein concentration of the cell lysate can be determined.

Large numbers of clones can be screened for the expression of a specific protein by Western colony blots described in *Protocol 4*. The method is sufficiently sensitive to allow for a semi-quantitative determination of proteins.

Protocol 4. Western colony blot (43, 44)

Equipment and reagents

- Dry ice
- Metal plate (aluminium) 30 × 20 × 3 mm for six filters
- Plastic tray to fit the plate
- Round nitrocellulose filters
- NCP buffer (pH 8.0): 12.1 g Tris–HCl, 87 g NaCl, 5 ml Tween 20, 2.0 g sodium azide, water to 10 litres

Method

1. With a toothpick, transfer colonies in an ordered array from a bacterial lawn in duplicate onto two fresh SM plates with a lawn of *Klebsiella aerogenes*. Grow colonies to the desired size of approx. 1 cm diameter.

2. Cool a metal plate in a tray on dry ice to –70°C.

3. Blot the colonies from one duplicate plate carefully for 5 min onto a nitrocellulose membrane.

4. Label the orientation of the filter with a pen on the agar plate.

5. Place membranes for 10 min on the pre-cooled metal plate (colony side-up) to lyse the cells.

6. Thaw filters for 5 min at room temperature.

7. Wash the filters six times for 20 min in NCP while shaking constantly (all bacteria should be washed off).

8. Denature proteins for 5 min at 80°C in 0.1% SDS.

9. Wash twice for 30 min in NCP.

10. Incubate filters with antibodies in the same way as for Western blots.

Note: colony blots can also be performed directly on randomly distributed clones as obtained, for example, from phenotypic screening. After blotting of cells to the membrane, the plates are stored at 4°C. There are sufficient cells left on the plate to allow for re-picking onto fresh plates once an interesting clone has been identified.

Screening of transformants by conventional Northern blot analysis is advantageous, because it directly provides information on the levels of both antisense and target mRNA. Hybridization with strand-specific riboprobes (see Chapter 4) is recommended to discriminate between mRNA and antisense RNA even if they both can be distinguished by size. Additional, unexpected transcripts can thus be directly attributed to the sense or the antisense orientation. For *Dictyostelium*, it is usually not necessary to prepare poly(A) RNA. Due to the small genome size, most transcripts are readily detected by Northern blots made with total RNA. *Protocol 5* has been optimized for RNA isolation from *Dictyostelium*.

Protocol 5.   RNA isolation for Northern blot analysis (45)

Equipment and reagents
- Microcentrifuge
- Heating block
- Sol D: 4 M guanidinium isothiocyanate, 25 mM sodium citrate pH 7.0, 0.1 M β-mercaptoethanol, 0.5% Sarcosyl
- Phenol:chloroform:isoamyl alcohol
- Isopropanol
- 2 M sodium acetate pH 4.7
- DEPC (diethyl pyrocarbonate) treated water

Method

1. Harvest 5×10^7 cells by centrifugation at 500 *g*, 4°C.
2. Resuspend in 650 μl Sol D and transfer to a 1.6 ml Eppendorf tube (wear gloves!).
3. Add 75 μl 3 M sodium acetate pH 4.7.
4. Extract once with 750 μl phenol:chloroform:isoamyl alcohol.
5. Precipitate with 750 μl isopropanol 30 min at –20°C.
6. Resuspend in 300 μl DEPC treated water.
7. Add 30 μl sodium acetate pH 4.7.
8. Extract once with 660 μl phenol:chloroform:isoamyl alcohol.
9. Extract once with 330 μl chloroform:isoamyl alcohol.
10. Precipitate with 660 μl ethanol.
11. Resuspend in 100 μl DEPC treated water.
12. In case the RNA pellet cannot be resuspended by pipetting up and down, the tubes may be heated at 65°C for 5 min.
13. Determine RNA concentration by mesuring the OD_{260} and perform electrophoresis of 10 μg RNA following standard procedures.

As with a Western colony blot, a Northern colony blot can be performed to screen, for example, large numbers of clones for antisense RNA expression. As described in ref. 46, this method is applicable not only for *Dictyostelium* but also for yeast, and should work for all cells which can be lysed by guanidinium isothiocyanate. A detailed protocol for Northern colony blots is given in Chapter 2. (See *Figure 1* for overview.)

Acknowledgements

I would like to thank all the colleagues who generously contributed their unpublished results.

References

1. Atkins, D., Arndt, G. M., and Izant, J. G. (1994). *Biol. Chem. Hoppe-Seyler*, **375**, 721.
2. Nellen, W., Silan, C., and Firtel, R. A. (1984). *Mol. Cell. Biol.*, **4**, 2890.
3. Izant, J. G. and Weintraub, H. (1984). *Cell*, **36**, 1007.
4. Coleman, J., Green, P. J., and Inouye, M. (1984). *Cell*, **37**, 429.
5. Crowley, T. E., Nellen, W., Gomer, R. H., and Firtel, R. A. (1985). *Cell*, **43**, 633.
6. Knecht, D. A. and Loomis, W. F. (1987). *Science*, **236**, 1081.
7. DeLozanne, A. and Spudich, J. A. (1987). *Science*, **236**, 1087.
8. Fang, H., Aiba, K., Higa, M., Urushihara, H., and Yanagisawa, K. (1993). *J. Cell Sci.*, **106**, 785.
9. Scherczinger, C. A. and Knecht, D. A. (1992). *Ann. N. Y. Acad. Sci.*, **660**, 45.
10. Haugwitz, M., Noegel, A. A., Karakesisoglou, J., and Schleicher, M. (1994). *Cell*, **79**, 303.
11. Hildebrandt, M. and Nellen, W. (1992). *Cell*, **69**, 197.
12. Scherczinger, C. A. and Knecht, D. A. (1993). *Antisense Res. Dev.*, **3**, 207.
13. McPherson, C. E. and Singleton, C. K. (1992). *Dev. Biol.*, **150**, 231.
14. Napoli, C., Lemieux, C., and Jorgensen, R. (1990). *Plant Cell*, **2**, 279.
15. van der Krol, A. R., Mur, L. A., Beld, M., Mol, J. N. M., and Suite, A. R. (1990). *Plant Cell*, **2**, 291.
16. Sadiq, M., Hildebrandt, M., Maniak, M., and Nellen, W. (1994). *Antisense Res. Dev.*, **4**, 263.
17. Leiting, B. and Noegel, A. A. (1988). *Plasmid*, **20**, 241.
18. Dynes, J. L. and Firtel, R. A. (1989). *Proc. Natl. Acad. Sci. USA*, **86**, 7966.
19. Maniak, M. and Nellen, W. (1989). *Nucleic Acids Res.*, **17**, 4894.
20. Hildebrandt, M., Saur, U., and Nellen, W. (1991). *Dev. Genet.*, **12**, 163.
21. Kalpaxis, D., Zündorf, I., Werner, H., Reindl, N., Boy-Marcotte, E., Jaquet, M., *et al.* (1990). *Mol. Gen. Genet.*, **225**, 492.
22. Suton, K. (1993). *Plasmid*, **30**, 150.
23. Knecht, D. A., Cohen, S., Loomis, W. F., and Lodish, H. (1986). *Mol. Cell. Biol.*, **6**, 3973.
24. Kumagai, A., Hadwiger, J. A., Pupillo, M., and Firtel, R. A. (1991). *J. Biol. Chem.*, **266**, 1220.
25. Blusch, J., Morandini, P., and Nellen, W. (1992). *Nucleic Acids Res.*, **20**, 6235.
26. Maniak, M. and Nellen, W. (1990). *Nucleic Acids Res.*, **18**, 5375.
27. Rubino, S., Mann, S. K. O., Hori, R. T., Pinko, C., and Firtel, R. A. (1989). *Dev. Biol.*, **131**, 27.
28. Pollenz, R. S., Chen, T. L. L., Trivinos-Lagos, L., and Chisholm, R. L. (1992). *Cell*, **69**, 951.
29. Liu, T., Williams, J. G., and Clarke, M. (1992). *Mol. Biol. Cell*, **3**, 1403.
30. Rittner, K., Burmester, C., and Sczakiel, G. (1993). *Nucleic Acids Res.*, **21**, 2809.
31. Warrick, H. M. and Spudich, J. A. (1988). *Nucleic Acids Res.*, **16**, 6617.
32. Witke, W., Nellen, W., and Noegel, A. A. (1987). *EMBO J.*, **6**, 4143.
33. Faix, J., Gerisch, G., and Noegel, A. A. (1992). *J. Cell Sci.*, **102**, 203.
34. Sun, T. J., van Haastert, P. J. M., and Devreotes, P. N. (1990). *J. Cell Biol.*, **110**, 1549.
35. Clarke, M., Dominguez, N., Yuen, I. S., and Gomer, R. H. (1992). *Dev. Biol.*, **152**, 403.

36. O'Halloran, T. J. and Anderson, R. G. W. (1992). *J. Cell Biol.*, **118**, 1371.
37. Yoshida, H., Kumimoto, H., and Okamoto, K. (1994). *Nucleic Acids Res.*, **22**, 41.
38. Loomis, W. F. (1991). In *Gene regulation: biology of antisense RNA and DNA* (ed. R. Ericksen and J. Izant), pp. 197–208. Raven Press, New York.
39. Nellen, W. and Saur, U. (1988). *Biochem. Biophys. Res. Commun.*, **154**, 54.
40. Howard, P. K., Ahern, K. G., and Firtel, R. A. (1988). *Nucleic Acids Res.*, **16**, 2613.
41. Podgorsky, G. and Welker, D. (1992). *Plasmid*, **28**, 46.
42. Wilczynska, Z. and Fisher, P. R. (1994). *Plasmid*, **32**, 182.
43. Wallraff, E., Schleicher, M., Modersitzki, M., Rieger, D., Isenberg, G., and Gerisch, G. (1986). *EMBO J.*, **5**, 61.
44. Wallraff, E. and Gerisch, G. (1991). In *Methods in enzymology*, Vol. 196 (ed. Vallee), pp. 334–48. Academic Press, London.
45. Chomczynzki, P. and Sacci, N. (1987). *Anal. Biochem.*, **162**, 156.
46. Maniak, M. and Nellen, W. (1989). *Anal. Biochem.*, **176**, 78.

9

Applications of antisense technology in plants

ANJA G. J. KUIPERS, EVERT JACOBSEN, and
RICHARD G. F. VISSER

1. Introduction

In 1984, Izant and Weintraub (1) published the first experiments on the antisense effect of gene constructs in eukaryotic cells. Since then, the inhibition of gene expression via antisense RNA has developed into a technique that is widely applied in molecular biology. In the field of plant biotechnology and plant molecular biology Ecker and Davis (2) were the first to describe a transient assay system in which carrot protoplasts were simultaneously electroporated with sense and antisense CAT genes. This resulted in the inhibition of CAT activity of up to 95% as compared to the control which only contained a sense CAT gene. These initial studies were followed by experiments with various plant species, in which the expression of a wide variety of genes has been effectively inhibited, either transiently with *in vitro* transcribed antisense RNAs, or by the introduction of antisense genes via transformation (reviewed in refs 3–5).

In this chapter, the application of antisense RNA technology in plants is discussed. A comparison is made between aspects of antisense RNA inhibition in plants and in other systems, and the application of the antisense approach is compared with the generation of mutants. An overview of various applications of antisense RNA technology in plants is followed by a description of factors that affect the effectiveness of antisense inhibition, such as composition of the antisense gene, promoter choice, and gene dosage. The chapter concludes with examples of current applications of antisense technology in plants.

2. Characteristics of antisense RNA in plants

The preferred approach for the application of antisense technology in plants is the endogenous expression of antisense genes that are stably integrated in the plant genome. Endogenous expression of these genes results in the forma-

Table 1. Antisense approaches in eukaryotes

Approach	Applicability in animal systems	Applicability in plant systems
Introduction of synthetic antisense oligonucleotides	+	–
Introduction of *in vitro* transcribed antisense RNAs	+	–
Endogenous expression of antisense genes in transgenic organisms	+	+

tion of *in vivo* transcribed antisense RNAs, that are supposed to interact with their target mRNAs, resulting in suppressed gene expression.

This approach is essentially different from the strategies that are usually followed to achieve antisense inhibition in other organisms (*Table 1*). As addressed in other chapters of this book, in mammalian systems much attention has been paid to the use of short (11–19 bp), specific synthetic oligonucleotides to control gene expression (6, 7). These oligos, either with or without chemical modification, are taken up actively via receptors on the cell surface (8), or are microinjected into cultured cells (9). Inside the cells, antisense oligos are targeted to interact with complementary regions of a specific mRNA. Targeting of oligos to form triple helices with genomic DNA has also been studied (6). Since they cannot be efficiently introduced into plant cells, antisense oligos are of limited use for the inhibition of gene expression in plants. Another approach is the application of *in vitro* transcribed antisense RNA molecules. In, for example, *Xenopus* (10, 11) and *Drosophila* (12) cells, microinjected antisense RNAs resulted in the transient inhibition of the expression of several target genes. As for oligos, the injection of *in vitro* transcribed antisense RNAs is not very suitable in plants.

The endogenous expression of antisense genes stably integrated into the genome of higher plants is dependent on the availability of a transformation system. In general, *A. tumefaciens*-mediated transformation is used to incorporate the antisense gene into the plant DNA. This system exploits a natural genetic transformation system: *A. tumefaciens* harbours a tumour-inducing or Ti plasmid. During infection of wounded plant tissue, in nature, a portion of this Ti plasmid is transferred to and integrates into the plant nuclear genome. The transferred DNA (T-DNA) is expressed resulting in tumour formation. This system has been modified in such a way that transfer of DNA still occurs, however without the formation of tumours. The *A. tumefaciens* transformation system has been optimized for an increasing number of plant species (13). Since efficient *A. tumefaciens* transformation is restricted to dicotyledonous species, other techniques, such as electroporation and particle bombardment, have been developed for monocots (13). Electroporation has been used quite frequently in initial experiments studying the effectiveness of antisense RNA in plant cells using transient expression of, for example, sense and antisense chloramphenicol acetyltransferase (CAT) genes (2). Recently, an antisense granule-bound starch synthase (GBSS) gene has been introduced into rice protoplasts by electroporation. The antisense gene was expressed in grains of plants regenerated from transgenic calli (14). To allow the selection of transgenic clones, the antisense gene is always co-integrated with a selectable marker gene. In the case of *A. tumefaciens* transformation, both the antisense gene and the marker genes are located on the T-DNA that is inserted into the plant genome. Common marker genes are neomycin phosphotransferase (NPTII) (15), hygromycin phosphotransferase (HPT) (16), and bialaphos (*bar*) (17).

3. Alternative for creating mutants

Antisense technology has become a valuable alternative for creating mutants. Mutant plant lines can be obtained via several procedures, such as chemical mutagenesis, transposon tagging, T-DNA insertion, and targeted gene disruption by homologous recombination. Compared with these methods, the antisense approach has some advantages. Most importantly, antisense inhibition is gene-specific and can be directed against a known target gene. Secondly, there is no need for gene targeting by homologous recombination, which works for lower eukaryotes such as yeast (18) and filamentous fungi (19). Due to the very low recombination rates that are achieved (20), homologous recombination so far has not been successful in plants. Other characteristics of antisense inhibition are summarized in *Table 2*. Due to its characteristics, antisense technology is an asset for plant breeding, since it can reduce the time involved in a breeding programme, which would be necessary to express the mutant phenotype in plants with a higher ploidy level. In addition, the possibility of incomplete antisense inhibition allows the analysis and identification of genes.

Table 2. Comparison of the antisense approach and mutation induction in plants

Antisense approach	Mutation induction
Gene-specific	Undirected
Variable effect on target gene expression	Complete block of gene expression
Dominant suppressor	Recessive in most cases
Independent from ploidy level	Influenced by ploidy level
Inactivation of gene families	Not applicable for gene families
DNA sequences required	DNA sequences not required
Regeneration/transformation system required	Tissue culture procedures not necessary
Acceptance by society unknown	Accepted by society

4. Inhibition of gene expression in plants

At present, antisense RNA-mediated inhibition has been applied to suppress the expression of a wide variety of genes in plants. An overview is given in *Tables 3* and *4*. As can be seen in these tables, the chimeric antisense genes usually consist of the full-length cDNA of the target gene which is cloned in reversed orientation between the 35S cauliflower mosaic virus (CaMV) promoter and the nopaline synthase (nos) transcription terminator. Several studies have demonstrated that antisense inhibition can also be achieved with differently composed antisense genes (see below). Expression of the antisense genes in transgenic plants in most cases results in a combination of the reduction of the mRNA steady state level, the protein steady state level, and the enzyme activity, and phenotypic changes. The mRNA steady state level in clones expressing an antisense gene can be examined in various plant organs by extracting mRNA from each organ (*Protocol 1*), followed by Northern hybridization. When using strand-specific probes, it is possible to discriminate between antisense RNA and target mRNA.

Protocol 1. RNA extraction method for plant tissue

Equipment and reagents

- RNA extraction buffer: 50 mM Tris pH 9.0, 10 mM EDTA, 2% SDS
- Phenol
- Phenol:chloroform
- Isopropanol
- Distilled water
- 4 M LiCl
- 3 M NaAc
- 96% ethanol
- Pestle and mortar
- Spatula
- Liquid nitrogen
- 10 ml tubes with screw cap
- 1.5 ml Eppendorf tubes
- Centrifuge for 10 ml tubes
- Microcentrifuge
- Vacuum desiccator

To prevent RNase contamination, all aqueous solutions are autoclaved before use. Glassware is heat sterilized overnight at 180°C. See Chapter 7 for detailed procedures in RNA handling.

Method

1. Grind 1 g frozen plant tissue to a fine powder in liquid N_2 using a pestle and mortar.[a]

2. Transfer powder with an N_2-cooled spatula to a 10 ml tube containing 3 ml of RNA extraction buffer and 3 ml phenol.

3. Vortex vigorously and leave at room temperature until all samples are ground.

4. After centrifugation (8000 *g*, 5 min), transfer aqueous phase to a new tube.

5. Add 3 ml phenol:chloroform (1:1), vortex vigorously, centrifuge (8000 *g*, 5 min), and transfer aqueous phase to a new tube.

---

6. Add 3 ml of isopropanol, precipitate nucleic acids at –80°C for 15 min, and centrifuge (8000 *g*, 5 min).

7. Remove supernatant and completely dissolve pellet in 750 μl of distilled water. Transfer solution to a 1.5 ml Eppendorf tube.

8. Add 750 μl of 4 M LiCl and keep on ice for 3 h.

9. After centrifugation (9000 *g*, 10 min), dissolve RNA pellet in 400 μl distilled water.

10. Add 40 μl of 3 M NaAc and 1 ml 96% ethanol, precipitate nucleic acids at –80°C for 15 min, and centrifuge (9000 *g*, 5 min).

11. Vacuum dry the pellet and dissolve in 50 μl of distilled water.[b]

[a] For potato leaf tissue, this procedure yields 400–500 μg of RNA.
[b] The presence of target mRNA and/or antisense RNA can be determined by Northern hybridization using strand-specific probes.

Using antisense inhibition, the expression of the target gene can be almost completely suppressed. Reductions of over 99% of the mRNA, protein and/or enzyme activity level have been reported (e.g. 22, 39, 44). However, the response to the introduced antisense gene can vary greatly in different transgenic plants which carry the same antisense gene. The variability of inhibition seems to be a general feature of the antisense technology, and has been ascribed to position effects that are assumed to be caused by influences of adjacent plant genomic DNA sequences, or by the chromosomal structure at the integration site (72).

With respect to the target genes for antisense inhibition, two classes can be discerned. First, non-plant genes have been targeted to evaluate the applicability of antisense inhibition in plants. These genes, often bacterial or fungal marker genes such as chloramphenicol acetyltransferase (CAT) (22), β-glucuronidase (*uidA*) (25), nopaline synthase (*nos*) (21), or phosphinothricin acetyltransferase (PAT) (24), had been introduced into the plant genome by stable transformation prior to the introduction of the antisense gene. The second class consists of a large variety of endogenous plant genes, which are involved in, e.g. flower pigmentation (35), fruit ripening (33), starch metabolism (44), photosynthesis (32), fatty acid metabolism (47). To date, the antisense technology has been successfully applied in at least ten different plant species: alfalfa, arabidopsis, carrot, gerbera, petunia, potato, rape, rice, tobacco, and tomato (*Table 4*).

5. Factors affecting the effectiveness of antisense inhibition

5.1 Composition of the antisense gene

The influence of construct composition on the effectiveness of antisense inhibition has been studied for the suppression of the chalcone synthase (CHS)

Table 3. Successful inhibition of marker gene expression in plants by antisense RNA[a]

Target gene	Target gene promoter	Antisense gene promoter	Antisense gene length	Plant	Reference
Chloramphenicol acetyltransferase (CAT)	35S CaMV, pal, nos	35S CaMV, pal, nos	Full	Carrot protoplasts	2
Nopaline synthase (nos)	nos	35S CaMV	5' end (65%)	Tobacco	21
Chloramphenicol acetyltransferase (CAT)	19S CaMV	35S CaMV	Full, 5' end (172 bp)	Tobacco	22
Nopaline synthase (nos)	nos	Cab22R (petunia)	Full, fragments	Tobacco	23
Phospinotricin acetyltransferase (PAT)	TR'2	35S CaMV	Full	Tobacco	24
β-Glucuronidase (uidA)	35S CaMV	35S CaMV	Full	Tobacco	25
β-Glucuronidase (uidA)	35S CaMV	ca/b (A. thaliana)	Full, 5' end (41 bp)	N. plumbaginifolia	26
Chalcone synthase (chs)	Endogenous (heterol. bean gene)	35S CaMV	Full; partial	Alfalfa protoplasts	27
Chloramphenicol acetyltransferase (CAT)	35S CaMV	35S CaMV, tRNA^{met} (soybean)	Full, 5' end (154 bp), 5' end (432 bp), 3' end (341 bp)	Carrot protoplasts	28
β-Glucuronidase (uidA)	35S CaMV	β-conglycinin	Full	Tobacco	29
β-Glucuronidase (uidA)	35S CaMV	35S CaMV	Full, 5' end (647 bp), 5' end (583 bp), 5' end (68 bp), 3' end (1239 bp), 3' end (723 bp), int. (516 bp)	Petunia protoplasts	30

[a] Due to the rapid progress in this field of research, this table may not be complete.

Table 4. Successful inhibition of the expression of endogenous and viral genes in plants by antisense RNA[a]

Target gene	Antisense gene promoter	Antisense gene length	Plant	Reference
Cucumber mosaic virus (CMV)	35S CaMV	Full	Cucumber	30a
Potato virus X (PVX) coat protein gene	35S CaMV	Full (almost)	Tobacco	31
Ribulose bisphosphate carboxylase (RUBISCO) small subunit (rbcS)	35S CaMV	Full	Tobacco	32
Polygalacturonase (pg)	35S CaMV	Full	Tomato	33
Polygalacturonase (pg)	35S CaMV	5' end (730 bp)	Tomato	34
Chalcone synthase (chs)	35S CaMV	Full	Petunia and tobacco	35
Tobacco mosaic virus (TMV) coat protein gene	35S CaMV	Full	Tobacco	36
pTOM13 (ACC oxidase)	35S CaMV	5' end	Tomato	37
Polygalacturonase (pg)	35S CaMV	5' end (730 bp)	Tomato	38
10 kDa protein of photosystem II	35S CaMV	Full	Potato	39
Chalcone synthase (chs)	35S CaMV	5' end (704 bp), 3' end (628 bp), int. (328 bp), 3' end (292 bp)	Petunia	40
pTOM5 gene	35S CaMV	5' end (794 bp)	Tomato	41
Tomato golden mosaic virus (TGMV) AL1 gene	35S CaMV (enhanced)	Full, + 5' ends of AL2 and AL3	Tobacco	42
Potato leafroll virus (PLV) coat protein gene	35S CaMV	Full	Potato	42a
1-aminocyclopropane-1-carboxylate synthase (ACC synthase)	35S CaMV	Full	Tomato	43
Granule-bound starch synthase (gbss)	35S CaMV	Full	Potato	44
Vacuolar H+ ATPase A	35S CaMV	Full, 5' end (150 bp)	Carrot	45

Table 4. *Continued*

Target gene	Antisense gene promoter	Antisense gene length	Plant	Reference
Class I patatin	35S CaMV	Full	Potato	46
Stearoyl–ACP desaturase	Napin, ACP	Full	Rape	47
Prosystemin	35S CaMV	Full	Tomato	48
ADP glucose pyrophosphorylase (subunit B)	35S CaMV	Full (almost)	Potato	49
E8 gene	35S CaMV	Full	Tomato	50
Pectin methylesterase (PME)	35S CaMV	Full	Tomato	51
Chalcone synthase (chs)	35S CaMV + anther box	(genomic DNA) Full	Petunia	52
Ankyrin repeat-containing gene (AKR)	35S CaMV	3' end (1051 bp)	*Arabidopsis*	53
Triose phosphate translocator (TPT)	35S CaMV	Full	Potato	54
Chalcone synthase (chs)	35S CaMV	Full (almost)	*Gerbera hybrida*	55
Pectin esterase	35S CaMV	5' end (428/756 bp)	Tomato	56
NADH hydroxypyruvate reductase	35S CaMV	Full (heterol. cucumber gene)	Tobacco	57
Granule-bound starch synthase (gbss)	35S CaMV	Full (heterol. cassava gene)	Potato	58
A. rhizogenes rolC gene	35S CaMV	Full, 5' end (192 bp)	Tobacco	59
Granule-bound starch synthase (waxy gene)	35S CaMV	Int. (genomic, 1.0 kb)	Rice	14
Glutamine synthase (gs1)	35S CaMV	Full (heterol. alfalfa gene)	Tobacco	60

Gene	Promoter	Region	Plant	Ref.
UDP glucose pyrophosphorylase	35S CaMV	Full	Potato	61
Polyphenol oxidase (ppo)	35S CaMV, GBSS, patatin I	Full, 5' end (800 bp)	Potato	62
Tomato golden mosaic virus (TGMV) AL1 gene	35S CaMV, (enhanced)	Full, + 5' ends of AL2 and AL3	Tobacco	63
Leucine aminopeptidase (LAP)	35S CaMV	Full	Potato	65
Fructose-1,6-bisphosphatase (FBPase)	35S CaMV	Full	Potato	66
Granule-bound starch synthase	35S CaMV, GBSS	cDNA: full, int. (0.7kb), genomic: full, 5' end (1.8 kb), int. (1.1 kb), 3' end (1.4 kb/ 0.6 kb)	Potato	67
Carbonic anhydrase	35S CaMV	Full	Tobacco	68
Caffeic acid O-methyltransferase (COMT)	35S CaMV	int. (0.43 kb; heterol. lucerne gene)	Tobacco	69
Sucrose carrier	35S CaMV	Full	Potato	70
Chitinase	35S CaMV	Full, (enhanced)	Arabidopsis	71
Pyrophosphate:fructose-6-phosphate phosphotransferase (PFP)	35S CaMV	a-subunit full, b-subunit full	Potato	64

[a] Due to the rapid progress in this field of research, this table may not be complete.

gene in *Petunia* (40), the granule-bound starch synthase (GBSS) gene in potato (67), and the CAT enzyme activity in carrot protoplasts that transiently expressed both sense and antisense CAT genes introduced simultaneously by co-electroporation (28). Although relatively small numbers of transgenic clones have been evaluated, these studies indicate the occurrence of differences in the effectiveness of the constructs analysed. For petunia, inhibition of CHS gene expression, resulting in inhibited flower pigmentation, has been achieved with antisense genes based on either the full-length CHS cDNA or 3' cDNA fragments (40). Antisense genes based on 5' fragments of the CHS cDNA were shown not to affect flower pigmentation. Effective inhibition of flower pigmentation has been achieved with antisense genes driven by the 35S CaMV or the endogenous CHS promoter, which were found to be of comparable strength (40).

In potato, antisense inhibition of granule-bound starch synthase (GBSS), involved in the biosynthesis of amylose, has been used to study the role of antisense gene composition in the effectiveness of inhibition. Suppression of GBSS gene expression results in a reduced amount of amylose in tuber starch, which can easily be screened by iodine staining of tuber starch granules (67). Nine antisense constructs derived from the full-length GBSS cDNA, the genomic GBSS coding region (gDNA), or fragments of each of these sequences, were analysed with respect to their inhibitory effect by means of iodine staining of tuber starch from transgenic potato plants (*Table 5*).

For convenience, the number of clones giving a certain degree of inhibition is usually taken as a measure for antisense effectiveness and not the average degree of inhibition obtained with a certain construct. Using this as a criterion, the best inhibition was found in plants carrying the full-length cDNA constructs, full-length gDNA constructs yielded a lower percentage of clones with complete inhibition. This difference was suggested to be caused by a difference in hybridization to mRNA and pre-mRNA. Complete inhibition of GBSS gene expression could also be achieved with a construct based on a 0.7 kb internal fragment of the GBSS cDNA, and with a construct based on the 0.6 kb 3' end of the GBSS gDNA. The latter fragment comprises transcribed GBSS sequences and contains the 3' end of the GBSS coding region (0.3 kb) including one intron.

In carrot protoplasts, reduced CAT activity has been observed after co-electroporation of a sense CAT gene and an antisense gene based on either a 5' fragment, a 3' fragment, or the complete CAT gene (28). The inhibition was found to be highest after introduction of an antisense gene based on the 3' CAT gene fragment.

The various studies in which differently composed antisense genes have been used, demonstrate that effective inhibition can be obtained with constructs based on the full-length cDNA or the genomic coding region, as well as 5', 3' and internal cDNA, and genomic fragments of varying size. The presence of a specific sequence or domain, such as the ribosomal binding site in

Table 5. Inhibition of GBSS[a] gene expression assessed by iodine staining of tuber starch from *A. tumefaciens* transformed potato clones carrying different antisense constructs

Construct	Promoter	Antisense origin	Fragment size[b]	No. transf. with complete inhibition[c]	No. transf. with incomplete inhibition[d]
pGBA10	35S CaMV	Genomic coding region	f.l. (3.0 kb)	1 (3%)	23 (64%)
pKGBA10	GBSS	Genomic coding region	f.l. (3.0 kb)	1 (3%)	17 (49%)
pGBA20	35S CaMV	5' genomic fragment	1.8 kb	–	1 (3%)
pKGBA20	GBSS	5' genomic fragment	1.8 kb	–	2 (7%)
pKGBA25	GBSS	Internal genomic fragment	1.1 kb	–	6 (12%)
pKGBA31	GBSS	3' genomic fragment	0.6 kb	2 (5%)	21 (51%)
pGB50	35S CaMV	GBSS cDNA	f.l. (2.4 kb)	3 (12%)	21 (80%)
pKGBA50	GBSS	GBSS cDNA	f.l. (2.4 kb)	8 (25%)	14 (44%)
pKGBA55	GBSS	Internal cDNA fragment	0.7 kb	2 (4%)	2 (4%)

[a] GBSS: granule-bound starch synthase.
[b] f.l.: full-length.
[c] Complete inhibition: tuber starch granules showing red staining starch with a small blue staining core after iodine staining, which indicates amylose-free starch.
[d] Incomplete inhibition: tuber starch granules showing a medium sized or large blue staining core and a red staining outer part of the granule after iodine staining. This indicates a reduced amount of amylose as compared to control plants.

antisense inhibition in prokaryotes does not seem to be required for effective inhibition of plant gene expression. Furthermore, the presence of intron sequences in antisense genes derived from genomic DNA does not greatly influence the efficiency of antisense inhibition. Although antisense inhibition can be obtained with different types of antisense genes, the effectiveness of inhibition, i.e. the percentage of transgenic clones with an antisense effect, varies greatly between constructs that are derived from different antisense sequences. This may be due to the secondary structure of the antisense RNA, which may determine the occurrence of interactions between the antisense RNA and the target mRNA.

5.2 Heterologous sequences

Several studies have demonstrated that a 100% homology between the anti-sense gene and the target gene is not always required to obtain antisense inhibition (*Table 6*). Van der Krol *et al.* (35, 73) have demonstrated that the petunia antisense CHS-A gene inhibits CHS gene expression in flowers of petunia and tobacco. However, the efficiency of antisense inhibition was found to differ. For petunia, an antisense effect was observed in 52% of the transgenic clones, whereas only 10% of the transgenic tobacco clones showed an antisense effect. In tobacco, glutamine synthase gene expression was suppressed after introduction of an antisense GS_1 gene from alfalfa (60). The alfalfa and tobacco GS genes have an overall sequence homology of 81% and share five highly conserved regions.

A lower overall sequence homology is found between the GBSS genes of potato and maize (55%) and potato and cassava (74%) (58). The potato and maize GBSS genes share 35 complementary regions of at least 10 nucleotides with a sequence similarity of more than 75% (74). The introduction of anti-sense constructs based on either the maize gene or the cassava cDNA into potato resulted in transgenic clones with reduced GBSS expression. However, using the maize gene, the frequency of inhibition was very low.

These studies demonstrate that antisense inhibition can be achieved with heterologous genes, but that the efficiency tends to be lower than with homo-logous genes. The lower efficiency seems to be determined by the overall sequence homology between the heterologous antisense gene and the target gene, and by the presence of regions with homologous sequences that will allow the formation of double-strands. The possible role of clusters of homo-logy in antisense inhibition is also suggested for the inhibition of the AL1 gene of tomato golden mosaic virus (TGMV) (63), which is discussed in Section 6.

5.3 Promoter choice

Most antisense genes that are used for the inhibition of gene expression in plants are driven by the 35S cauliflower mosaic virus (CaMV) promoter (see *Table 4*). This promoter is widely used for the expression of transgenes in

Table 6. Antisense inhibition with heterologous sequences

Antisense gene	Source plant	Target plant	Homology (%)	Maximum inhibition (%)[a]	Reference
Chalcone synthase	Petunia	Tobacco	80	Complete[b]	35
Glutamine synthase	Alfalfa	Tobacco	81	40	60
Granule-bound starch synthase	Cassava	Potato	74	97	58
NADH-hydroxypyruvate reductase	Cucumber	Tobacco	79	50	57
Caffeic acid O-methyl-transferase	Lucerne	Tobacco	77	80	69

[a] Reduction of enzyme activity.
[b] Complete inhibition of flower pigmentation.

plants, because of its strength and almost constitutive expression pattern. However, it can be useful to direct the antisense gene expression by using an organ- or tissue-specific promoter. For the inhibition of flower pigmentation in petunia it was found that the 35S CaMV-driven antisense CHS gene suppressed pigmentation in the corolla but not in anthers. Insertion of a so-called anther box, a promoter element found in genes expressed in anthers, modified the expression pattern of the 35S CaMV promoter (52) and resulted in inhibited pigment synthesis in anthers. The experiment shows that transcription of the antisense gene in the appropriate cell type is required to obtain antisense effects.

In *Brassica rapa* and *Brassica napus* the stearoyl–acyl carrier protein desaturase gene expression in seeds has been inhibited by the expression of antisense genes driven by embryo-specific promoters. These promoters assured a high expression level of the antisense genes in developing seed during storage lipid biosynthesis, resulting in a reduction of the desaturase enzyme activity of up to 100%, and the accumulation of stearate in seeds (47). The use of specific promoters also demonstrated that a large excess of antisense RNA compared to target mRNA is not required for effective antisense inhibition. Similarly, expression of an antisense CHS gene in petunia from the endogenous CHS promoter, resulted in the complete inhibition of flower pigmentation in one of the obtained transgenic clones (40). In potato, GBSS gene expression could be completely inhibited after the introduction of GBSS antisense genes driven by the endogenous promoter. Expression of antisense GBSS genes from the GBSS promoter even resulted in a higher percentage of clones with complete inhibition than expression from the 35S CaMV promoter (67). This difference was especially clear during development of field grown tubers (75), and may be explained by the three- to tenfold higher expression level of the promoter in tuber tissue as compared to the 35S CaMV promoter (76). The more pronounced inhibition by a GBSS promoter-driven antisense gene suggests that the effectiveness of inhibition is influenced by the type and the strength of the promoter used.

All promoters described above are transcribed by RNA polymerase II. Bourque and Folk (28) have reported the use of a promoter for RNA polymerase III. Electroporation of antisense CAT genes, fused to a soybean tRNA^{met} gene, into carrot protoplasts resulted in suppressed CAT enzyme activity. The inhibition of CAT gene expression was fivefold higher than when the 35S CaMV promoter was used. The relatively short length of the RNA polymerase III transcripts was suggested to improve the accessibility and hybridization to target sequences, whereas the attached tRNA structure may improve stability. The study indicates that the expression of antisense genes from tRNA promoters may be advantageous.

5.4 T-DNA copy number

The copy number of the antisense gene may also affect the extent of antisense inhibition. Since genes are transferred to the plant genome as part of the T-

DNA, the copy number of antisense genes is also given by the T-DNA copy number. Rodermel *et al.* (32) analysed the antisense inhibition of the ribulose bisphosphate carboxylase (Rubisco) enzyme level in five transgenic tobacco clones, and showed that the lowest Rubisco RNA and protein content were found in a clone that carried at least four T-DNA copies. Clones with lower antisense effects contained fewer T-DNA copies. For tomato, selfings have been carried out with transgenic lines containing one T-DNA copy with an antisense polygalacturonase gene (77) or an antisense gene from a cDNA clone involved in ethylene synthesis (37). Analysis of the progeny showed almost complete inhibition of gene expression in homozygous plants (two T-DNA copies) and incomplete inhibition in hemizygous plants (one T-DNA copy).

In potato, the T-DNA copy number of primary transformants with varying degrees of inhibition of GBSS gene expression was determined by Southern hybridization of *Sst*I digested total genomic DNA isolated by a miniprep procedure (*Protocol 2*) (67). When a 1.1 kb 5' fragment of the GBSS cDNA was used as a probe, individual T-DNA inserts are detected as hybridizing bands of varying size. Comparison of the extent of antisense inhibition and of T-DNA copy number in individual clones revealed a positive correlation between these characteristics: one or two inserts did not always result in antisense inhibition, whereas three or more inserts always generated a certain degree of antisense inhibition (67). Multiple copies may increase the chances of insertion into a favourable genomic position and/or lead to an additive effect in antisense RNA accumulation. The latter can be elucidated by the analysis of progeny plants in which the T-DNA copies are segregated.

Protocol 2. DNA miniprep for plant tissue

Equipment and reagents

- Extraction buffer (94): 100 mM Tris pH 8.0, 50 mM EDTA, 500 mM NaCl
- Phenol
- Isopropanol
- TE: 10mM Tris pH 8.0, 1 mM EDTA
- 0.1 μg/ml RNase
- Phenol:chloroform
- 3 M NaAc
- 96% ethanol
- 80% ethanol
- Glass rod
- 2 ml Eppendorf tubes with screw cap
- Microcentrifuge
- Vacuum desiccator

Method

1. Put 250 mg leaf tissue in a 2 ml Eppendorf tube with screw cap and freeze in liquid nitrogen.[a]

2. Grind frozen tissue with a cooled glass rod.

3. Mix 0.75 ml of extraction buffer and 0.5 ml of phenol with the powder.

4. After centrifugation (9000 *g*, 10 min), transfer supernatant to a new Eppendorf tube and add 0.75 ml of isopropanol.

Protocol 2. *Continued*

   5. Precipitate nucleic acids at –80°C for 15 min.

   6. Pellet nucleic acids by centrifugation (9000 *g*, 5 min).

   7. Dissolve pellet in 0.3 ml of TE with 0.1 μg/ml RNase, and incubate at 37°C for 30 min.

   8. After phenol:chloroform extraction, precipitate the DNA with 30 μl of 3 M NaAc and 600 μl of 96% ethanol at –20°C.

   9. Pellet the DNA by centrifugation (9000 *g*, 5 min), wash pellet with 500 μl of 80% ethanol.

   10. After centrifugation (9000 *g*, 5 min), vacuum dry the pellet, and dissolve in 50 μl of TE.[b]

[a] For potato leaf tissue, this procedure yields 30–40 μg of DNA.
[b] The number of inserted T-DNAs can be determined by Southern hybridization using suitable restriction enzyme–probe combinations, or by PCR analysis.

5.5 External factors affecting the antisense effect

Several studies demonstrated that antisense effects can be influenced by physiological conditions or external factors. Antisense inhibition of chalcone synthase gene expression results in reduced flower pigmentation in petunia. For some of the transgenic clones, the flower phenotype varied within plant cuttings derived from the same clone. Experimentally, flower pigmentation can be influenced by treatment with gibberellic acid or extra light during early flower development (71a). Samac and Shah (71) studied disease resistance in transgenic *A. thaliana* plants expressing an antisense RNA targeted against the class I chitinase. Prior to infection, reduced chitinase levels compared to control plants were found. However, upon inoculation with the fungal pathogen *Botrytis cinerea*, induced chitinase expression was only slightly reduced by the antisense gene. In this case, infection with *Botrytis* reduced antisense efficiency, similar effects may occur with other antisense constructs and have to be considered when applying antisense technology.

6. Specificity of antisense inhibition

The inhibition of plant gene expression by means of antisense inhibition is generally considered to be gene-specific. However, several reports have demonstrated the possibility of inhibiting the expression of heterologous genes (see above) or genes of a multigene family. The latter is exemplified by van der Krol *et al.* (40), who demonstrated that an antisense *chsA* gene was also capable of inhibiting expression of the *chsJ* gene, which shares 86% sequence homology.

The inhibition of related or heterologous genes requires a certain minimal sequence similarity. In tomato, it was shown that antisense inhibition of the E8 protein gene (50) did not affect expression of the EFE (ethylene-forming enzyme) gene, which shares 51% sequence similarity.

The expression of an antisense RNA targeted against the AL1 gene of tomato golden mosaic virus (TGMV) resulted in a strong reduction of symptoms and viral DNA accumulation upon TGMV infection in transgenic tobacco (63). The same construct also reduced the accumulation of beet curly top virus (BCTV) DNA, but not of African cassava mosaic virus (ACMV) DNA. TGMV, BCTV, and ACMV are all geminiviruses and show an overall sequence homology of 63% and 64%, respectively in the AL1 gene. The presence of a 280 nucleotide region of high homology (82%) in BCTV only, may account for the observed antisense effect on BCTV and not on ACMV accumulation.

The specificity of antisense RNAs depends on the overall homology between genes, apparently, the presence of regions with a sequence homology high enough to allow the interactions involved in antisense inhibition is crucial. The specificity of antisense RNAs may thus be optimized by using relatively short sequences directed against gene-specific target sequences sharing little or no homology with related genes.

7. Stability of antisense inhibition

Several studies have demonstrated stable inheritance of the antisense trait in subsequent generations of plants. Clonal progeny of potato clones in which AGPase B gene expression was inhibited, showed a similar degree of inhibition as the primary transformants. Tubers from both generations were found to be devoid of starch, and accumulated comparable amounts of sucrose and glucose (49). The sexual transmission of antisense genes has been studied by backcrossing of transgenic petunia clones (40) and selfing of tomato (37, 38, 51). Progeny analysis showed stable inheritance, and thus the stable integration of the antisense construct into the plant genome.

With respect to environmental conditions (see Section 5.5) field grown and greenhouse grown tomato plants expressing an antisense polygalacturonase (PG) gene were compared (78). This study demonstrated that the 10–50% reduction of PG activity observed in greenhouse grown plants was maintained in the field. For potato, antisense inhibition of granule-bound starch synthase (GBSS) gene expression appeared to be stably maintained in five successive generations of transgenic clones grown in the field (R. G. F. Visser, unpublished data).

8. Optimization of antisense inhibition

In most experiments described here, complete or almost complete inhibition of target gene expression can be achieved with antisense constructs composed

Anja G. J. Kuipers et al.

of a strong promoter and an appropriate antisense sequence. Although this approach is sufficient for most purposes, two aspects may require further optimization of the antisense system. First, application of antisense technology in plant breeding requires a high percentage of transgenic clones in which the antisense gene is expressed at an optimal level. This will allow for the selection of clones in which agronomic traits, such as yield, are preserved. Secondly, optimization of effectiveness will be useful for applications using heterologous genes for antisense inhibition.

Several strategies may be followed to improve the effectiveness of antisense inhibition. One of these is the incorporation of a ribozyme sequence in the antisense gene (see Chapter 5). Antisense ribozymes are composed of a catalytic domain, the so-called 'hammerhead' structure, and flanking antisense sequences required for the specific binding of the ribozyme to the target RNA (reviewed in refs 79 and 80). After binding, the target RNA is cleaved and the ribozyme and target RNA fragments may dissociate, thus allowing for cleavage of additional targets. Thus, inhibition of gene expression may be more efficient by antisense ribozymes than by ordinary antisense genes (see Chapter 5). The use of ribozymes in gene suppression has been exploited in several studies. Steinecke *et al.* (81) cotransfected tobacco protoplasts with the *nptII* gene and a gene encoding a ribozyme directed against *nptII* mRNA. The activity of the ribozyme was demonstrated by the presence of cleaved substrate mRNA. Ribozymes have also been used successfully to generate virus-resistance in plants. Upon targeting to the positive strand of tobacco mosaic virus (TMV), site-specific cleavage of the target was observed *in vitro*, and inhibition of virus replication was observed *in vivo* (82). Expression of the ribozyme construct and a control antisense construct without ribozyme, reduced the accumulation of TMV by 90% or 20%, respectively.

Another factor that may influence the effectiveness of antisense inhibition, is the secondary structure of the antisense RNA. Natural antisense RNAs in prokaryotes have a high degree of secondary structure, mainly consisting of stem-loops (83). The analysis of mutants has revealed that mutations influencing the inhibitory effect, are frequently located in or near the loops. Binding rates are greatly influenced by the three-dimensional structure of stem-loops of both antisense and target RNA (83). In chimeric antisense genes in eukaryotes, the sequence and the length of the fragment will influence its secondary structure and may thus change its capacity to bind the target RNA. By using various gene fragments, the antisense construct can be optimized for effective inhibition of target gene expression. Moreover, understanding of the mechanisms of antisense-mediated inhibition will increase insights into the structural requirements for optimal antisense RNA.

Finally, the stability of transgene expression may be improved by the inclusion of matrix attachment regions (MARs) flanking the transgene. MARs are thought to define DNA loops on which genes are located, and are considered to interact with nuclear organizing elements. By flanking a transgene with MAR elements, and thus creating a discrete transcriptional domain, the expression level of transgenes has been increased, whereas the variability of expression was reduced. This was demonstrated by Mlynerova *et al.* (84), who found a reduced variability of β-glucuronidase (GUS) gene expression and a higher average GUS activity in transgenic tobacco plants carrying T-DNA constructs in which MAR elements flanked either the GUS gene or the GUS gene and a selection gene. With respect to antisense inhibition, this approach may be useful to stabilize and increase antisense gene expression.

9. Application of antisense technology in plants

9.1 Applications in crop improvement

Antisense technology has a large potential for crop improvement. So far, several monogenic qualitative traits have been modified by suppressing specific genes (*Table 7*). The most advanced application is the inhibition of the polygalacturonase (PG) gene in tomato (33, 34). PG is involved in cell wall degradation; inhibition of its expression results in retarded fruit softening and thus in an improvement of shelf-life. The stability of inhibition was confirmed in a field trial, which also demonstrated that introduction and expression of the antisense PG gene did not have detrimental effects on agronomic characteristics (78). In 1994, FLAVR SAVR™ tomatoes with strongly reduced levels of PG were the first genetically modified crop to be introduced on the US market. The requirements for the commercial introduction such as variety development, regulatory considerations, product specifications, and consumer acceptance have been extensively assessed (86).

An important advantage of antisense technology over conventional plant breeding, is the direct applicability in established varieties. This is best illustrated by employing the antisense approach in potato, which is a tetraploid crop. Introduction of antisense PPO genes controlled by the tuber-specific GBSS promoter, strongly inhibited the expression of PPO and prevented discoloration after bruising of tubers from transgenic, field grown plants of the varieties Diamant and Van Gogh (62).

The enzyme GBSS is involved in the biosynthesis of amylose in potato tubers. Together with amylopectin, amylose constitutes the main portion of starch. To inhibit GBSS gene expression, antisense genes based on the full-length GBSS cDNA and driven either by the 35S CaMV promoter or the potato GBSS promoter, were introduced into a tetraploid potato variety (75). Suppression of GBSS resulted in the production of amylose-free tuber starch, which, because of its modified characteristics (i.e. reduced retrogradation), will give rise to novel applications in the starch industry (87). Field analysis of the transgenic clones indicated that GBSS inhibition could be achieved without significantly affecting the starch and sugar content, the expression levels of other genes involved in starch and tuber metabolism, and agronomic

Table 7. Applications of antisense inhibition: improvement of qualitative characteristics in crops

Gene	Crop	Trait	Transgen. clones (No.)	Plants with detectable effect (No.)	Maximum inhibition[a]	Reference
Polygalacturonase	Tomato	Ripening	10	9	93	33
Chalcone synthase	Petunia	Flower colour	25	13	Complete[b]	35
ACC-synthase[c]	Tomato	Ripening	34	3	99.5	43
Stearoyl-ACP desaturase	Rape	Fatty acid composition	22	20	100	47
Pectin methylesterase	Tomato	Ripening	18	10	> 90	51
Polyphenol oxidase	Potato	Discoloration	1400	72–74%	100	62
Granule-bound starch synthase	Potato	Amylose formation	58	46	97	67

[a] Maximum inhibition of enzyme activity (% reduction compared to control).
[b] Complete inhibition of flower pigmentation.
[c] 1-aminocyclopropane-1-carboxylate synthase.

characteristics such as yield and dry matter content. Antisense plants have several advantages over the use of the amylose-free (*amf*) potato mutant, which carries a recessive mutation in the GBSS gene (88). Since the recessive *amf* mutation had been induced in a monohaploid potato clone, the isolated mutant had to be introduced into a breeding programme to develop a cultivar in which the *amf* character will be combined with appropriate agronomic traits (89). In contrast, the dominant character of the inserted antisense GBSS gene(s) allows the application of antisense technology in existing cultivars, which will significantly reduce the time required to develop amylose-free potato cultivars.

9.2 Virus-resistance by antisense RNA

The successful use of antisense RNA to inhibit expression of endogenous plant genes has led to using the same approach to suppress the accumulation of viruses in plants (90). Several studies have demonstrated that it is indeed possible to achieve a certain degree of virus protection in transgenic plants expressing an antisense gene targeted against the viral coat protein gene of plant RNA viruses (*Table 8*). For potato virus X (PVX) (31), tobacco mosaic virus (TMV) (36), and cucumber mosaic virus (CMV) (30a), expression of antisense coat protein genes in transgenic potato or tobacco plants, respectively, resulted in virus protection at low inoculum concentrations. In transgenic potato plants expressing an antisense coat protein gene of potato leafroll luteovirus (PLRV) (42a), the level of protection was found to be considerably higher: the amount of accumulated virus was found to be less than 10% of the controls.

A drawback in the use of antisense RNA to generate virus-resistance in plants, is the fact that 98% of the plant viruses contain RNA genomes (see the examples mentioned above). These viruses are supposed to replicate in the cytoplasm, whereas the antisense RNA is expressed in the cell nucleus. Hence, the compartmentalization of viral and antisense RNA may account for the relatively low efficiency of suppression. Antisense RNA has also been applied to suppress the accumulation of plant DNA viruses (Caulimo- and Geminiviruses), for which this approach is more suitable (63). In contrast to RNA viruses, the replication of DNA viruses occurs in the plant cell nucleus, which may make them a better target for antisense suppression. This is exemplified by the study of Day *et al.* (42), who demonstrated that the expression of antisense RNA effectively suppressed the replication of the geminivirus, tomato golden mosaic virus (TGMV) in tobacco. The introduced antisense gene was driven by the 35S CaMV promoter and directed against three TGMV genes, which are required for viral replication. Transgenic tobacco plants showed a high level of virus-resistance upon agro-inoculation with TGMV. The accumulation of the related beet curly top virus (BCTV) DNA was also reduced in infected transgenic tobacco plants (63). In general,

Table 8. Applications of antisense inhibition for virus-resistance

Gene	Crop	Transgenic clones (No.)[a]	Plants with detectable effect (No.)	Level of protection upon virus infection	Reference
Cucumber mosaic virus (CMV) coat protein gene	Tobacco	n.r.	2	Protection at low inoculum concentrations	30a
Potato virus X (PVX) coat protein gene	Tobacco	n.r.	1	No symptoms, reduced DNA replication	31
Tobacco mosaic virus (TMV) coat protein gene	Tobacco	3	3	Low level of protection against TMV	36
Tomato golden mosaic virus (TGMV) AL1 gene	Tobacco	8	6	No symptoms, reduced DNA replication	42
Tomato golden mosaic virus (TGMV) AL1 gene	Tobacco	n.r.	1	TGMV: no symptoms, reduced DNA replication, BCTV: reduced symptoms and DNA replication, ACMV: no effects	63

[a] n.r.: not reported.

sequences that are essential for viral replication are likely to be better antisense targets than the coat protein gene, because of their direct involvement in accumulation of virus. The more effective antisense-mediated suppression of plant DNA virus replication (as compared to plant RNA viruses) suggests that the antisense effect is exerted in the cell nucleus rather than in the cytoplasm.

9.3 Characterization of gene expression and dissection of metabolic pathways

The role of individual enzymes in plant metabolism can be dissected by creating mutant phenotypes via antisense inhibition. The resulting clones can then be analysed for effects of the suppressed target enzyme on plant metabolism. The power of antisense technology in dissecting complex metabolic pathways is demonstrated by the study of sink-source relationships with respect to carbohydrate metabolism in potato (*Table 9*) (reviewed in ref. 85). In carbohydrate metabolism, three major processes can be discerned: production, transport, and utilization of photoassimilates. The role of several enzymes in carbohydrate metabolism has been assayed by means of antisense inhibition.

Antisense inhibition of the enzymes ribulose-1,5-bisphosphate carboxylase-oxygenase (RUBISCO) (32), fructose-1,6,bisphosphatase (FBPase) (66), and aldolase (85), all involved in the Calvin cycle, only had effects after a relatively strong suppression (60–90%) as compared to the wild-type. To analyse the transport of photoassimilates, antisense experiments have been performed to suppress the triose phosphate translocator (TPT) (54), pyrophosphate fructose-6-phosphate 1-phosphotransferase (PFP) (64), and sucrose carrier (70). In general, inhibition of these genes results in the accumulation of photosynthetic products in leaf tissue, indicating a distortion of metabolite transport. The utilization of assimilates in sink organs could also be modified. Suppression of ADP glucose pyrophosphorylase (AGPase) resulted in an almost complete block of starch formation and an accumulation of soluble sugars (49). This finding proved that AGPase is the major regulatory step in the biosynthesis of tuber starch. The antisense inhibition of GBSS gene expression demonstrated this enzyme to be exclusively involved in amylose formation (44).

9.4 Assignment of gene function

For the characterization of isolated cDNA sequences, antisense inhibition can be a useful tool in addition to sequence analysis and expression studies. By introducing an antisense construct derived from the unknown sequence and controlled by a strong, constitutive promoter, expression of the gene of interest can be down-regulated. Characterization of transgenic plants expressing the antisense gene can elucidate the biochemical function of the protein

Table 9. Applications of antisense inhibition: dissection of sink/source relationships in carbohydrate metabolism in potato

Gene	Metabolic function	Maximum inhibition[a]	Phenotypic effect of antisense inhibition	Reference
Ribulose bisphosphate carboxylase (RUBISCO) small subunit	Calvin cycle	63	Reduced photosynthesis and plant growth	32
Fructose-1,6 bisphosphatase	Calvin cycle	90	Reduced photosynthesis and plant growth	66
Triose phosphate translocator	Transport of photoassimil.	20–30	Increased starch accumulation in chloroplasts	54
Pyrophosphate:fructose-6-phosphate phosphotransferase	Transport of photoassimil.	94	No phenotypic effects	64
Sucrose carrier	Transport of photoassimil.	n.r.	Retarded growth, reduced photosynthesis, increased carbohydrate content in leaves, reduced root system, reduced tuber yield	70
ADP glucose pyrophosphorylase (subunit B)	Starch formation	98.5	Block of starch formation, accumulation of soluble sugars in tubers	49
Granule-bound starch synthase	Amylose formation	97	Block of amylose formation, no effect on total starch content	67

[a] Maximum inhibition of enzyme activity or protein level (% reduction compared to control). n.r.: not reported.

encoded by the cDNA. This is demonstrated by the characterization of pTOM5, a tomato cDNA which had been shown to encode a ripening-related 46.7 kDa protein of unknown function (41). Antisense inhibition of the pTOM5 gene resulted in transgenic plants with yellow ripening fruit and pale coloured flowers. Additional biochemical and molecular analyses indicated that the pTOM5 gene is crucial for carotenoid biosynthesis. The function of the tomato E8 gene was elucidated by expressing an appropriate antisense gene (50). Inhibition of E8 protein accumulation stimulated the production of ethylene during ripening. Therefore, the authors suggested an involvement of E8 in the suppression of ethylene production, probably by limiting the amount of the ethylene precursor 1-aminocyclopropane-1-carboxylic acid (ACC).

Both studies show that functions of isolated cDNAs can be identified by antisense inhibition of the corresponding genes. Since antisense inhibition can be achieved with constructs derived from cDNA fragments, gene function studies may be performed with incomplete cDNA sequences. Furthermore, antisense inhibition in a heterologous system can be used to characterize cDNAs isolated from plant species that are not yet amenable to transformation.

10. Antisense inhibition and co-suppression

The success of the antisense approach in plant science is proven by the numerous examples described in this chapter. Despite extensive studies, the mechanism of antisense inhibition is still not fully understood. In some cases, transcription of antisense RNA has been shown by Northern blot analysis or run-on transcription assays (33, 35, 39). This may point to a mechanism based on interactions between antisense and target RNA, resulting in the breakdown of double-stranded (ds) RNA hybrids. However, the presence of dsRNA has never been demonstrated in plant tissues (however, see indications in Chapter 14), and the amount of transcribed antisense RNA does not always correlate with the extent of inhibition observed in transgenic plants (26, 35, 39). It may be that antisense genes do not exclusively act by dsRNA formation, but also by RNA:DNA or DNA:DNA interactions.

Several mechanisms that have been postulated for gene suppression by antisense genes are also supposed to be involved in gene silencing by sense genes (co-suppression) (91). Co-suppression often results in similar phenotypes in transgenic plants as compared to antisense inhibition (35, 92). Some of the mechanisms that are suggested to be involved in co-suppression (readthrough transcription resulting in antisense RNA, methylation, and ectopic pairing) may also be involved in antisense inhibition (91). At present, much effort is made to elucidate the mechanisms underlying gene suppression (93). A better understanding of these mechanisms will be useful in those cases where further improvement of the extent of suppression is required.

References

1. Izant, J. G. and Weintraub, H. (1984). *Cell*, **36**, 1007.
2. Ecker, J. R. and Davis, R. W. (1986). *Proc. Natl. Acad. Sci. USA*, **83**, 5372.
3. Van der Krol, A. R., Mol, J. N. M., and Stuitje, A. R. (1988). *BioTechniques*, **6**, 958.
4. Watson, C. F. and Grierson, D. (1993). In *Transgenic plants, fundamentals and applications* (ed. A. Hiatt), pp. 255–81. Marcel Dekker, Inc. New York, NY.
5. Tabler, M. (1993). In *Morphogenesis in plants* (ed. K. Tran Thanh Van), pp. 237–58. Plenum Press, New York, NY.
6. Hélène, C. and Toulmé, J-J. (1990). *Biochim. Biophys. Acta*, **1049**, 99.
7. Miller, P. (1991). *Bio/Technology*, **9**, 358.
8. Loke, S., Stein, C., Zhang, X., Mori, K., Nakanishi, M., Subasinghe, C., et al. (1989). *Proc. Natl. Acad. Sci. USA*, **86**, 3474.
9. Shuttleworth, J. and Colman, A. (1988). *EMBO J.*, **7**, 427.
10. Harland, R. and Weintraub, H. (1985). *J. Cell. Biol.*, **101**, 1094.
11. Melton, D. (1985). *Proc. Natl. Acad. Sci. USA*, **82**, 144.
12. Rosenberg, U. B., Preiss, A., Seifert, E., Jäckle, H., and Knipple, D. C. (1985). *Nature*, **313**, 703.
13. Potrykus, I. (1991). *Annu. Rev. Plant Physiol. Plant Mol. Biol.*, **42**, 205.
14. Shimada, H., Tada, Y., Kawasaki, T., and Fujimura, T. (1993). *Theor. Appl. Genet.*, **86**, 665.
15. Bevan, M. W., Flavell, R. B., and Chilton, M.-D. (1983). *Nature*, **304**, 184.
16. Van den Elzen, P. J. M., Townsend, J., Lee, K. Y., and Bedbrook, J. R. (1985). *Plant Mol. Biol.*, **5**, 299.
17. De Block, M., Botterman, J., Vandewiele, M., Dockx, J., Thoen, C., Gosselé, V., et al. (1987). *EMBO J.*, **6**, 2513.
18. Hinnen, A., Hicks, J. B., and Fink, A. R. (1978). *Proc. Natl. Acad. Sci. USA*, **75**, 1929.
19. Timberlake, W. E. and Marshall, M. A. (1989). *Science*, **244**, 1313.
20. Offringa, R., van den Elzen, P. J. M., and Hooykaas, P. J. (1992). *Transgenic Res.*, **1**, 114.
21. Rothstein, S. J., DiMaio, J., Strand, M., and Rice, D. (1987). *Proc. Natl. Acad. Sci. USA*, **84**, 8439.
22. Delauney, A. J., Tabaeizadeh., Z., and Verma, D. P. (1988). *Proc. Natl. Acad. Sci. USA*, **85**, 4300.
23. Sandler, S. J., Stayton, M., Townsend, J. A., Ralston, M. L., Bedbrook, J. R., and Dunsmuir, P. (1988). *Plant Mol. Biol.*, **11**, 301.
24. Cornelissen, M. and Vandewiele, M. (1989). *Nucleic Acids Res.*, **17**, 833.
25. Robert, L. S., Donaldson, P. A., Ladaique, C., Altosaar, I., Arnison, P. G., and Fabijanski, S. F. (1989). *Plant Mol. Biol.*, **13**, 399.
26. Cannon, M., Platz, J., O'Leary, M., Sookdeo, C., and Cannon, F. (1990). *Plant Mol. Biol.*, **15**, 39.
27. Choudhary, A. D., Kessmann, H., Lamb, C. J., and Dixon, R. A. (1990). *Plant Cell Rep.*, **9**, 42.
28. Bourque, J. E. and Folk, W. R. (1992). *Plant Mol. Biol.*, **19**, 641.
29. Fujiwara, T., Lessard, P. A., and Beachy, R. N. (1992). *Plant Mol. Biol.*, **20**, 1059.
30. De Lange, P., de Boer, G.-J., Mol, J. N. M., and Kooter, J. M. (1993). *Plant Mol. Biol.*, **23**, 45.

30a. Cuozzo, M., O'Connell, K. M., Kaniewski, W., Fang, R.-X., Chua, N.-H., and Turner, N. E. (1988). *Bio/Technology*, **6**, 549.

31. Hemenway, C., Fang, R.-X., Kaniewski, W. K., Chua, N.-H., and Turner, N. E. (1988). *EMBO J.*, **7**, 1273.

32. Rodermel, S. R., Abbott, M. S., and Bogorad, L. (1988). *Cell*, **55**, 673.

33. Sheehy, R. E., Kramer, M., and Hiatt, W. R. (1988). *Proc. Natl. Acad. Sci. USA*, **85**, 8805.

34. Smith, C. J., Watson, C. F., Ray, J., Bird, C. R., Morris, P. C., Schuch, W., *et al.* (1988). *Nature*, **334**, 724.

35. Van der Krol, A. R., Lenting, P. E., Veenstra, J., van der Meer, I. M., Koes, R. E., Gerats, A. G., *et al.* (1988). *Nature*, **333**, 866.

36. Powell, P. A., Stark, D. M., Sanders, P. R., and Beachy, R. N. (1989). *Proc. Natl. Acad. Sci. USA*, **86**, 6949.

37. Hamilton, A. J., Lycett, G. M., and Grierson, D. (1990). *Nature*, **346**, 284.

38. Smith, C. J., Watson, C. F., Morris, P. C., Bird, C. R., Seymour, G. B., Gray, J. E., *et al.* (1990). *Plant Mol. Biol.*, **14**, 369.

39. Stockhaus, J., Höfer, M., Renger, G., Westhoff, P., Wydrzynski, T., and Willmitzer, L. (1990). *EMBO J.*, **9**, 3013.

40. Van der Krol, A. R., Mur, L. A., de Lange, P., Mol, J. N. M., and Stuitje, A. R. (1990). *Plant Mol. Biol.*, **14**, 457.

41. Bird, C. R., Ray, J. A., Fletcher, J. D., Boniwell, J. M., Bird, A. S., Teulieres, C., *et al.* (1991). *Bio/Technology*, **9**, 635.

42. Day, A. G., Bejarano, E. R., Bulk, K. W., Burrell, M., and Lichtenstein, C. P. (1991). *Proc. Natl. Acad. Sci. USA*, **88**, 6721.

42a. Kawchuk, L. M., Martin, R. R., and McPherson, J. (1991). *Mol. Plant Microbe Int.*, **4**, 247.

43. Oeller, P. W., Min-Wong, L., Taylor, L. P., Pike, D. A., and Theologis, A. (1991). *Science*, **254**, 437.

44. Visser, R. G. F., Somhorst, I., Kuipers, G. J., Ruys, N. J., Feenstra, W. J., and Jacobsen, E. (1991). *Mol. Gen. Genet.*, **225**, 289.

45. Gogarten, J. P., Fichmann, J., Braun, Y., Morgan, L., Styles, P., Taiz, S. L., *et al.* (1992). *Plant Cell*, **4**, 851.

46. Höfgen, R. and Willmitzer, L. (1992). *Plant Sci.*, **87**, 45.

47. Knutzon, D. S., Thompson, G. A., Radke, S. E., Johnson, W. B., Knauf, V. C., and Kridl, J. C. (1992). *Proc. Natl. Acad. Sci. USA*, **89**, 2624.

48. McGurl, B., Pearce, G., Orozco-Cardenas, M., and Ryan, C. A. (1992). *Science*, **255**, 1570.

49. Müller-Röber, B., Sonnewald, U., and Willmotzer, L. (1992). *EMBO J.*, **11**, 1229.

50. Peñarrubia, L., Aguilar, M., Margossian, L., and Fischer, R. L. (1992). *Plant Cell*, **4**, 681.

51. Tieman, D. M., Harriman, R. W., Ramamohan, G., and Handa, A. K. (1992). *Plant Cell*, **4**, 667.

52. Van der Meer, I. M., Stam, M. E., van Tunen, A. J., Mol, J. N. M., and Stuitje, A. R. (1992). *Plant Cell*, **4**, 253.

53. Zhang, H., Scheirer, D. C., Fowle, W. H., and Goodman, H. M. (1992). *Plant Cell*, **4**, 1575.

54. Riesmeier, J., Flügge, U. I., Schulz, B., Heineke, D., Heldt, H. W., Willmitzer, L., *et al.* (1993). *Proc. Natl. Acad. Sci. USA*, **90**, 6160.

55. Elomaa, P., Honkanen, J., Puska, R., Seppanen, P., Helariutta, Y., Mehto, M., *et al.* (1993). *Bio/Technology*, **11**, 508.
56. Hall, L. J., Tucker, G. A., Smith, C. J. S., Watson, C. F., Seymour, G. B., Bundick, Y., *et al.* (1993). *Plant J.*, **3**, 121.
57. Oliver, M. J., Ferguson, D. L., Burke, J. J., and Velten, J. J. (1993). *Mol. Gen. Genet.*, **239**, 425.
58. Salehuzzaman, S. N. I. M., Jacobsen, E., and Visser, R. G. F. (1993). *Plant Mol. Biol.*, **23**, 947.
59. Schmülling, T., Röhrig, H., Pilz, S., Walden, R., and Schell, J. (1993). *Mol. Gen. Genet.*, **237**, 385.
60. Temple, S. J., Knight, T. J., Unkefer, P. J., and Sengupta-Gopalan, C. (1993). *Mol. Gen. Genet.*, **236**, 315.
61. Zrenner, R., Willmitzer, L., and Sonnewald, U. (1993). *Planta*, **190**, 247.
62. Bachem, C. W. B., Speckmann, G.-J., van der Linde, P. C. G., Verheggen, F. T. M., Hunt, M. D., Steffens, J. C., *et al.* (1994). *Bio/Technology*, **12**, 1101.
63. Bejarano, E. R. and Lichtenstein, C. P. (1994). *Plant Mol. Biol.*, **24**, 241.
64. Hajirezaei, M., Sonnewald, U., Viola, R., Carlisle, S., Dennis, D., and Stitt, M. (1994). *Planta*, **192**, 16.
65. Herbers, K., Prat, S., and Willmitzer, L. (1994). *Planta*, **194**, 230.
66. Kossmann, J., Sonnewald, U., and Willmitzer, L. (1994). *Plant J.*, **6**, 637.
67. Kuipers, A. G. J., Soppe, W. J. J., Jacobsen, E., and Visser, R. G. F. (1995). *Mol.Gen Genet.*, **246**, 745.
68. Majeau, N., Arnoldo, M. A., and Coleman, J. R. (1994). *Plant Mol. Biol.*, **25**, 377.
69. Ni, W., Paiva, N. L., and Dixon, R. A. (1994). *Transgenic Res.*, **3**, 120.
70. Riesmeier, J. W., Willmitzer, L., and Frommer, W. B. (1994). *EMBO J.*, **13**, 1.
71. Samac, D. A. and Shah, D. M. (1994). *Plant Mol. Biol.*, **25**, 587.
71a. Van der Krol, A. R., Mur, L. A., de Lange, P., Gerats, G. M., Mol, J. N. M., and Stuitje, A. R. (1990). *Mol. Gen. Genet.*, **220**, 204.
72. Peach, C. and Velten, J. (1991). *Plant Mol. Biol.*, **17**, 49.
73. Van der Krol, A. R., Mol, J. N. M., and Stuitje, A. R. (1988). *Gene*, **72**, 45.
74. Visser, R. G. F. (1989). PhD thesis, State University Groningen, The Netherlands.
75. Kuipers, A. G. J., Soppe, W. J. J., Jacobsen, E., and Visser, R. G. F. (1994). *Plant Mol. Biol.*, **26**, 1759.
76. Visser, R. G. F., Stolte, A., and Jacobsen, E. (1991). *Plant Mol. Biol.*, **17**, 691.
77. Smith, C. J. S., Watson, C. F., Morris, P. C., Bird, C. R., Seymour, G. B., Gray, J. E., *et al.* (1990). *Plant Mol. Biol.*, **14**, 369.
78. Kramer, M., Sanders, R. A., Sheehy, R. E., Melis, M., Kuehn, M., and Hiatt, W. R. (1990). In *Horticultural biotechnology: proceedings of the horticultural biotechnology symposium* (ed. A. B. Bennett and S. D. O'Neill), pp. 347–55. Wiley–Liss, New York, NY.
79. Cech, T. R. (1987). *Science*, **231**, 1532.
80. Cotten, M. (1990). *TIBTECH*, **8**, 174.
81. Steinecke, P., Herget, T., and Schreier, P. H. (1992). *EMBO J.*, **10**, 2007.
82. Edington, B. and Nelson, R. S. (1992). In *Gene regulation: biology of antisense RNA and DNA.* (ed. R. P. Erickson and J. G. Izant), pp. 209–21. Raven Press, New York, NY.
83. Hjalt, T. and Wagner, E. G. H. (1992). *Nucleic Acids Res.*, **20**, 6723.

84. Mlynárová, L., Loonen, A., Heldens, J., Jansen, R. C., Keizer, P., Stiekema, W. J., *et al.* (1994). *Plant Cell*, **6**, 417.
85. Sonnewald, U., Lerchl, J., Zrenner, R., and Frommer, W. (1994). *Plant Cell Environ.*, **17**, 649.
86. Redenbaugh, K., Hiatt, W., Martineau, B., and Emlay, D. (1994). *Trends Food Sci. Technol.*, **5**, 105.
87. Visser, R. G. F. and Jacobsen, E. (1993). *TIBTECH*, **11**, 63.
88. Hovenkamp-Hermelink, J. H. M., Jacobsen, E., Ponstein, A. S., Visser, R. G. F., Vos-Scheperkeuter, G. H., Bijmolt, E. W., *et al.* (1987). *Theor. Appl. Genet.*, **75**, 217.
89. Jacobsen, E., Hovenkamp-Hermelink, J. H. M., Krijgsheld, H. T., Nijdam, H., Pijnacker, L. P., Witholt, B., *et al.* (1989). *Euphytica*, **44**, 43.
90. Bejarano, E. R. and Lichtenstein, C. P. (1992). *TIBTECH*, **10**, 383.
91. Van Blokland, R., De Lange, P., Mol, J. N. M., and Kooter, J. M. (1993). In *Antisense research and application* (ed. B. LeBlue), pp. 125–48. CRC Press, London.
92. Van der Krol, A. R., Mur, L. A., Beld, M., Mol, J. N. M., and Stuitje, A. R. (1990). *Plant Cell*, **2**, 291.
93. Stam, M., Mol, J. N. M., and Kooter, J. M. (1997). *Ann. Bot.*, **79**, 3.
94. Dellaporta, S. L., Wood, J., and Hicks, J. B. (1983). *Plant Mol. Biol. Rep.*, **1**, 19.

Antisense molecules and ribozymes: medical applications

MICHAEL STRAUSS

1. Introduction

The application of antisense oligodeoxynucleotides as tools for specific inhibition of gene expression was an extremely important step in the development of molecular medicine (reviewed in ref. 1, see also Chapters 3 and 11). From the first moment of experimental investigation of the power of these molecules it was obvious that, at some stage, this class of molecules would find its way to clinical application as therapeutic drugs (2). After several years of successful use in cell culture experiments, antisense molecules including synthetic oligos and expressed RNA are increasingly applied in animal experiments. Most of the studies carried out in cell culture and animals so far have been aimed at the investigation of the consequences of the loss of a particular function. The antisense strategy has the advantage over the powerful method of gene knock-outs of allowing transient suppression of gene expression. Thus, it allows one to inactivate, for a short period of time, those functions which would be lethal if inactivated at the gene level (see also Chapter 12 for another approach to the problem of lethality). Since the power of the antisense approach first became obvious, studies focusing on the treatment of diseases have been initiated over the last few years. It seems only a matter of time before this strategy can be applied clinically to the treatment of particular diseases, e.g. infectious diseases like hepatitis B and C, or HIV infection, and even some forms of cancer. This chapter will give a brief overview of the different fields of medical application of antisense molecules, and will provide the reader with protocols for the experimental use of these molecules in cell culture and animal studies.

2. Studies in cell culture

Cell culture is an *in vitro* system compared to whole animals or humans. However, with regard to the actual biological function of antisense molecules, they ever, with regard to the actual biological function of antisense molecules, they must surely represent an *in vivo* situation. This is why studies on the biological function of antisense oligodeoxyribonucleotides and expression constructs have mainly been carried out in cell cultures, and have generated a bulk of data with relevance to the understanding of biological functions and to their use as inhibitors of processes leading to disease. Antisense oligodeoxyribo-nucleotides (Chapter 3) have successfully been applied to the inhibition of various biological functions. After modified oligos, in particular the phospho-thioates (3, 4), became available quite impressive inhibitions of gene function have been obtained. The broadest field of application has probably been, and still is, cancer development. Both oncogenes (5–7) and tumour suppressor gene (8) functions have been inactivated and the effects have been studied. In particular, studies on inactivation of the mutant *Ha-ras* gene product led to the conclusion that antisense oligos could be used as therapeutic drugs to inhibit tumour growth (7, 9). Stimulation of cell division can also be achieved by inhibiting the expression of tumour suppressor genes like *Rb* (8). Another field of early applications was inhibition of HIV gene expression or replication (2, 9, 10).

The application of expression constructs for generating antisense RNA has, however, turned out to be much more complicated. After the publication of the first successful experiments using an antisense construct for the herpes virus *tk* gene (11) many groups have had problems applying the same simple strategy to their gene of interest. It took some time before it was realized that low stability of the antisense RNA and accessibility of the target sequence are important parameters which may limit the application of this strategy. In the last few years, optimized expression constructs have been developed which allow for higher stability (12–14). Selection of a suitable part of the gene to function as an antisense RNA is, however, still a matter of trial and error (see Chapter 7 for protocols to measure hybridization kinetics of antisense RNA:mRNA duplexes, Chapter 5 for analysis of RNA structure, and Chapter 1 for experimental measurement of T_m).

After the concept of using synthetic hammerhead (see Chapter 5) or hair-pin ribozymes was first established (15–17), ribozymes have been designed which are directed against potential cleavage sites in target RNAs of interest. In most cases cleavage has been observed *in vitro* but not in cell culture. In some cases, where an effect in cell culture has been detected, it is not more pronounced than using an antisense oligo suggesting a simple antisense effect of the ribozyme molecule (18). However, finding an efficient ribozyme is still a matter of trying out a large number of candidate target sites and ribozyme structures. The experiences with synthetic ribozymes and expressed antisense RNAs has converged recently in the development of suitable expression cassettes for the expression of ribozymes which are functional in cell culture (12, 14). Concurrently, all four general strategies have also been applied to animals *in vivo*.

2.1 Application of antisense oligodeoxyribonucleotides

The use of synthetic antisense oligodeoxyribonucleotides is the easiest and most broadly applied way of inactivating gene function for a limited period of time. The inhibitory effect is mainly dependent on the perfect base pairing of the oligo to the target RNA. The optimal length of antisense oligos is between 14 and 18 nucleotides but 12-mers may work in some cases because this is the minimal length for an oligo to be absolutely specific for one particular target RNA. However, even 8-mers have been shown to function specifically (20). Besides the length of the oligo, one has to find the optimal target site for its binding within the mRNA. In most cases, the region around the AUG start codon has been chosen as the target sequence. This, most likely, allows for both inhibition of translational initiation by steric hindrance and degradation of the RNA:DNA hybrid by RNase H (21). The latter mechanism would also act if the target sequence lies somewhere else in the mRNA. The cap region (22) and even the 3' untranslated region (23) have also been found to be suitable targets in some cases.

In the early days of the antisense technology, ordinary oligos were used. However, their very short half-life within cells, due to degradation by nucleases, prompted the application of modified, nuclease-resistant molecules. Among the modified oligos, the phosphothioates are the most popular and most inhibitory versions (3, 4, 24). They are efficiently taken up by cells, are not toxic at concentrations which are sufficient to block the function of most mRNAs, and retain the specificity of target binding. The half-life of phospho-thioates in mammalian cells *in vivo* is between two to four days and they are often still detectable after ten days (8). The biological function of the target RNA and its protein product can be inhibited for up to ten days (8). Repeated application can prolong the effect.

For designing an efficient antisense oligo for a particular target sequence a few rules should to be followed:

- no intramolecular base pairing of more than two adjacent nucleotides should be possible
- the G:C content should be 40–60%
- stretches of more than three Gs should be avoided
- if possible, a G or a C should be the last nucleotide at both ends

These rules have been empirically established from our own experience with antisense oligos for about 30 different target RNAs. We normally design our oligos by choosing 15–18 nucleotides around the AUG start codon accord-ing to these rules. The AUG can be in any position within the sequence, ideally it should be in the middle. Confirmation of the suitability of the selected oligo by one of the computer programs (e.g. *OLIGO*) is recommended, particularly as to the presence of self-complementarity, but is not required.

The actual introduction of an oligo into mammalian cells can be achieved by one of the standard methods for gene transfer with little modifications. The easiest way of doing it is the addition of oligos to the growth medium which leads to active uptake by the cells until saturation is reached (see *Protocol 1*). The mechanism of cellular uptake of oligos is not yet clearly established. In contrast to mononucleotides, oligos are rapidly taken up by mammalian cells. There is a suggestion that a cell surface receptor exists for small or even larger pieces of DNA (25). Alternatively, microinjection, lipofection (*Protocol 2*), or electroporation (*Protocol 3*) can be applied. Whereas the first three methods work well only with adherent cells, the latter can also be applied to cells in suspension cultures, in particular to lymphocytes.

Protocol 1.   Uptake of oligos from culture medium according to Strauss *et al.* (8)

Equipment and reagents
- HPLC purified oligo
- Distilled water
- Medium without serum
- Microcentrifuge
- 0.2 μm spin filters (Millipore)
- Sterile pipette tips, plastic tubes

A. *Sterile preparation of oligos*
1. Dissolve dried oligo in distilled water at approx. 1 mg/ml.
2. Spin at full speed through Millipore sterile filter unit in a microcentrifuge.
3. Take an aliquot for OD_{260} measurement and adjust to a suitable concentration (e.g. 1 mg/ml).

B. *Application to cell cultures on plastic*
1. Wash cells at the right density (e.g. 50–70% confluency) with medium without serum, repeat twice.
2. Add medium without serum containing the oligo at a concentration of 5–30 μg/ml (has to be optimized for the respective target mRNA), incubate at 37°C for 1 h.
3. Add serum to the medium at the required concentration (5–10%) depending on the cell type.
4. Incubate cells at 37°C for the required time to see an effect (two to four days) depending on the half-life of the target RNA.

We have compared *Protocol 1* and *Protocol 2* with regard to the extent of down-regulation of the retinoblastoma protein. Whereas *Protocol 1* normally results in reduction of the protein level by 70–85% using 10 μM oligos, *Protocol 2* results in an even stronger reduction of pRb levels (about 90%) using only 0.5 μM oligos (Warthoe and Strauss, unpublished data). Similar efficiencies of uptake have been obtained by others (26, 27). Intracellular uptake seems to be facilitated by lipofection with considerable accumulation of oligos in the nucleus (27). However, it remains to be clarified if the nuclear accumulation is actually responsible for the strong inhibition of protein synthesis.

Protocol 2. Transfer of oligos into cultured cells using cationic liposomes according to Bennett *et al.* (26) as modified by Warthoe and Strauss (unpublished data)

Equipment and reagents
- HPLC purified, sterile oligonucleotide, 20 μM stock solution (see *Protocol 1A*)
- DOTMA (Lipofectin) (Gibco BRL, Life Technologies)

Method

1. Grow cells to 40% confluency in 5 cm culture dishes.

2. Wash 3 × with medium without serum.

3. Add to the cells 2 ml of medium without serum containing 20 μg of lipofectin (= 8 μM).

4. Add oligo to the desired concentration[a] which needs to be optimized.

5. Incubate cells for 4 h at 37°C.

6. Replace the medium by standard culture medium with serum containing the same concentration of oligo as in step 4.

7. Incubate cells for two to three days (depending on the gene product) until the effect of the treatment can be studied.

[a] Concentrations of oligos are usually between 0.1–1 μM.

Protocol 3. Delivery of oligos into mammalian cells by electroporation according to Lukas *et al.* (28, 29)

Equipment and reagents
- HPLC purified, sterile oligo (see *Protocol 1A*)
- Bio-Rad Gene Pulser™ (Bio-Rad, California, USA)
- PBS
- 4 mm sterile electroporation cuvettes (Bio-Rad)

Method

1. Harvest cells and resuspend in cold PBS at a density of 2–4 × 10^7 cells/ml.

2. Mix 50 μl of the cell suspension with 5 μl of oligo to the desired final concentration (up to 40 μM), transfer into a sterile electroporation

Protocol 3. *Continued*

cuvette, and carefully spread the suspension on the bottom of the cuvette to connect the electrodes.

3. Let the cuvette stand on ice for 10 min with occasional shaking.

4. Expose the cells to a single electrical pulse (400 V[a]; 25 μF; infinite resistance).

5. Add 1 ml of pre-warmed complete medium immediately after the pulse, transfer cells into a Petri dish containing medium, and incubate at 37 °C for the required period of time.

[a] This voltage is used for lymphocytes. For fibroblasts and epithelial cells 280–320 V are applied.

2.2 Using synthetic ribozymes

Synthetic ribozymes in the sense of antisense technology are basically oligo-ribonucleotides with catalytic activity for cleavage of target RNAs (15–17). Molecules of this type can be generated by placing the sequence of a hammer-head or hairpin ribozyme in the middle of an antisense sequence. Ribozymes are designed for target sequences which contain the consensus cleavage site NUH with a preference for GUC or CUC. Most mRNAs contain several sites like this. If the flanking sequences of a potential cleavage site allow creation of antisense sequences of six to ten nucleotides on both sides according to the rules given in Section 2.1, ribozymes can be designed. Normally, several different ribozymes for different target sites are tested for their cleavage activity on the target RNA *in vitro* using a purified RNA transcribed *in vitro* (see Chapter 5). However, this does not guarantee for function *in vivo*. Often, biological activity of a ribozyme can not be detected within cells which is most likely due to inaccessibility of the cleavage site within the naturally folded RNA. Thus, finding a functional ribozyme is a matter of trial and error. Computer-aided prediction of the secondary structure of the target RNA can help to some extent (see Chapter 6). Selection of potential loops or single-stranded regions as targets increases the chance of obtaining a ribozyme which functions *in vivo*. In general, a functional ribozyme is superior to an ordinary antisense oligo with regard to the extent of mRNA inactivation owing to its catalytic activity. For delivery of synthetic ribozymes to cells the same methods can be used which were described for the application of antisense oligos (*Protocols 1–3*).

2.3 Using expression constructs

In order to achieve continuous inactivation of the function of a particular gene it is necessary to have the antisense molecule continuously present within the cell. This can be achieved by placing the antisense DNA template of the gene of interest, or part of it, under control of a suitable promoter. Transfection of the resulting construct should lead to expression of the respective antisense RNA. It has indeed been shown that this simple approach can work. For instance, for the herpes thymidine kinase gene where this principle was applied first, the use of the full-length gene in antisense orientation was successful (11). However, in many other cases the full-length gene or cDNA has only marginal effects on the activity of the target RNA. Since antisense RNA probably needs to base pair to the target RNA, the target sequence must be accessible. As the target RNA and the antisense RNA may fold in different ways, the full-length antisense RNA may be disadvantageous to use. According to our experience, it is a good idea to divide the cDNA into fragments of 300–800 nucleotides, clone all fragments into a suitable expression vector, and test them individually for inhibition of gene function. As an example, we have tried to generate antisense RNA genes for the retinoblastoma susceptibility gene (*Rb*). The full-length cDNA was 4.9 kb and was found to be completely inactive as an antisense construct. Most of the fragments were inactive or were inefficient in reducing the level of the Rb protein. The only fragment which was highly efficient in inactivating the function of the gene product was one covering the first 365 nucleotides from the 5′ end (30). Interestingly, the predicted secondary structure of this piece of RNA was very complex with little single-stranded regions (30) making it hard to believe that the predicted structure does indeed apply to the *in vivo* situation.

Whereas the selection of suitable parts of a gene as antisense sequences is largely a matter of trial and error, there are some general rules for choosing suitable expression cassettes and vectors to allow for efficient and continuous expression. Basically, a suitable expression cassette contains the promoter at the 5′ end and a polyadenylation signal at the 3′ end. The source for the latter seems to be less important than its presence in general which should guarantee for polyadenylation of the transcript and thereby for relative stability of the 3′ end. Different types of promoters can be used to drive the expression of an antisense gene *in vivo*. Only transcription from polymerase II promoters allows for capping of the transcripts which provides stability for the 5′ end. *Table 1* summarizes the the three basic systems which are available for expression of antisense RNA.

Several modifications of the basic expression cassettes have been introduced. Embedding the antisense RNA in the structure of a small nuclear RNA seems to provide some stability (12, 31). In addition, stable loop structures can be used to guarantee for an exposed structure of the antisense part of the RNA (14). The expression cassettes, in particular those based on transcription by RNA polymerase II, can be used in combination with various vector systems which are summarized in *Table 2*.

Delivery of antisense constructs for intracellular expression can be done by

Table 1. Properties of various expression systems

Promoter type	Expression levels[a]	Stability	Reference
RNA polymerase II	+/++	High	32
RNA polymerase III	++/+++	Intermediate	12
T7 RNA polymerase	+++	Low	14

[a] Low expression levels (+) are up to 500 RNA molecules per cell, intermediate levels (++) are up to 5000 RNA molecules per cell, high levels are more than 5000 molecules per cell.

Table 2. Transfer methods and vectors for expression of antisense genes

Vector	Transfer efficiency[a]		Stability[b]
	Ex vivo	In vivo	
Plasmid			
Transfection	+	−	T/S
Lipofection	++	+	T
Injection	+++	+	T/S
Viral			
Retrovirus	+++	+	S
Adenovirus	+++	+++	T
Adeno-assoc. virus	+++	+	S

[a] Transfer efficiencies of up to 5% are low (+), 5–20% are intermediate (++), and more than 20% are high (+++).
[b] The stability of the transferred antisense gene and its expression is given as transient (T) if it lasts only for some days, or stable (S) if it is detectable after several weeks.

the same methods which are applied for standard gene transfer. For viral vectors, a recombinant virus has to be constructed. Since it is not in the scope of this article to describe the techniques for generation of viral vectors, only a short summary of the essential steps is given. In the case of retroviral vectors, the recombinant retroviral DNA construct is transfected into a packaging cell line and clones are selected which produce sufficient titres of the viral vector (33). Titres normally do not exceed 10^6. For gene delivery in cell culture one would add 1 ml of this viral supernatant to 10^6 cells and would obtain almost 100% infection for fibroblasts and lower efficiencies for other cell types. Since the titres of retroviral vectors are too low for *in vivo* application, transplantation of the virus-producing cells into the target tissue can be done to allow for continuous production of virus for some days in a relatively small volume (34). If adenoviral vectors are used, the expression cassette is first cloned into a transfer vector where it is flanked by sequences from the adenoviral genome. Upon cotransfection of the transfer vector with viral vector DNA into a helper cell line (293 cells) which provides essential replication functions in *trans*, recombination between both molecules occur with a frequency of about 5% (35). Since adenovirus forms lytic plaques individual plaques have to be screened for recombinants. Adenoviral vectors are produced at titres of about 5×10^8 and can be concentrated by centrifugation up to 10^{13} p.f.u./ml. These titres allow for *in vivo* application in very small volumes. In order to achieve almost 100% infection of cells *in vitro* or *in vivo* adenoviral vectors are applied at a multiplicity of infection of 10–100 (36).

For delivery of antisense constructs *in vitro*, it is normally not essential to achieve 100% efficiency. Several methods are available for transfer of plasmid vectors to cells in culture. Microinjection results in 50–100% efficency, electroporation (see *Protocol 3*) in 40–80% efficiency, lipofection in 10–40%, and calcium phosphate coprecipitation in only 5–20%. The latter two methods are described here. We generally recommend using plasmid preparations with high purity. Purification by resin materials like Qiagen columns is preferred over caesium chloride gradient centrifugation.

Protocol 4. Transfection of antisense expression plasmids by calcium phosphate coprecipitation according to Lieber *et al.* (37)

Equipment and reagents

- Plasmid DNA (1 μg/μl)
- 2 M CaCl$_2$
- 2 × HBS: 50 mM Hepes, 280 mM NaCl, 1.5 mM Na phosphate, equal amounts of mono- and dibasic, pH 6.96
- Sterile distilled water
- Sterile plastic tips
- Micropipette
- Vortex

Method

1. Seed cells in 5 cm Petri dishes or 25 cm^2 tissue culture flasks at a density which results in a 50–70% confluent culture on the next day.

2. Transfer 10 μg DNA into an Eppendorf tube and add distilled water to 220 μl.

3. Add 30 μl of 2 M CaCl$_2$ and mix by vortexing.

4. Add 250 μl of 2 × HBS dropwise with simultaneous vortexing.

5. Transfer the resulting precipitates in 0.5 ml to cell culture vessels which were seeded with cells the day before and contain 4.5 ml of fresh culture medium.

6. Incubate cells with the DNA precipitates overnight.

7. Wash cells with medium without serum and add complete fresh medium.

8. Incubate cells for two to four days if transient effects of antisense function are to be tested, or select stably transformed clones (requires selectable marker on the plasmid).

Protocol 5. Lipofection of plasmid vectors according to the protocol of Life Technologies (38) with modifications

Equipment and reagents
- LipofectAMINE™ (Gibco BRL, LifeTechnologies)
- Plasmid DNA (1 μg/μl)
- Sterile distilled water
- Sterile plastic tips
- Micropipette
- OPTI-MEM (Gibco BRL, Life Technologies)

Method

1. In 6-well tissue culture plates, seed 1–2×10^5 cells in 2 ml of complete medium, and incubate until cells are 60–80% confluent (time depends on the cell type used).

2. Dilute 2 μg of DNA[a] into 100 μl serum-free medium (prefererably OPTI-MEM) and dilute 5 μl of LipofectAMINE[a] into 100 μl serum-free medium separately, combine both solutions, and mix gently.

3. Incubate at room temperature for 30 min to allow complexes to form.

4. Rinse the cells in the meantime once with 4 ml of serum-free medium.

5. Add 0.8 ml of serum-free medium to the DNA–LipofectAMINE complexes and mix gently.

6. Overlay the diluted complex solution onto the rinsed cells and incubate for 2–5 h (depending on the cell type) at 37 °C in a CO_2 incubator.

7. Add 1 ml of complete medium with twofold concentration of serum. If toxicity is a problem with the particular cell type replace medium the next day by complete medium.

8. Assay for antisense function 48–72 h after lipofection.

[a] These amounts of DNA and LipofectAMINE are used for primary epithelial cells and hepatocytes. For most established cell lines 4 μg of DNA and 10 μl of LipofectAMINE are used.

Many gene functions have successfully been inactivated in cell culture by transfection of antisense expression plasmids including the proto-oncogenes *c-myc* (39), *c-myb* (40), *c-fos* (41), *c-abl* (42), and *c-jun* (43). These studies contributed to the elucidation of signalling pathways in mammalian cells (43).

2.4 Selection of target sites for ribozyme cleavage and isolation of ribozymes

Ribozymes are potentially the most efficient antisense molecules with regard to complete inactivation of the target RNA. Expressed ribozymes have been applied successfully to the inactivation of a mutant *Ha-ras* oncogene (44) and BCR–ABL fusions (45) which resulted in repression of the malignant phenotype. Ribozymes targeted to HIV RNA abolished viral replication (46–49). The expression of bovine leukaemia virus can also be inhibited by a ribozyme (50). An anti-lymphocytic choriomeningitis virus ribozyme expressed in tissue culture cells diminished viral RNA levels and led to a reduction in infectious virus yield (51). For efficient treatment of HIV infection the expression of a ribozyme from a retroviral vector could be a clever idea because this might result in successful co-localization of the ribozyme with the target RNA (52). These studies already indicate the potential power of this technique for therapeutic application, particularly to cancer and viral infections.

Creating a ribozyme which is able to cleave target RNA efficiently in cells *in vivo* has been a matter of trial and error so far. The major problem seems to be the actual accessibility of a potential cleavage site. Our laboratory has recently developed a strategy for the selection of accessible cleavage sites in target RNAs (14). It is based on the use of an expression library of 5×10^9 individual ribozymes with random sequences (13 and 15 nt) in their hammerhead-flanking target-binding arms. The library is transcribed *in vitro* and the ribozyme transcripts are incubated with total RNA from cells which contain the target transcript of interest. After cleavage *in vitro*, 3' terminal fragments are purified via oligo(dT), reverse transcribed, and cloned, after PCR using gene-specific primers. By sequencing, the actual cleavage site can be identified. Using primers specific for the flanks of the cleavage site, the respective ribozymes can be reamplified from the library, recloned into the expression vector, and tested for function *in vivo*. Using the message of an overexpressed growth hormone gene as the target, we could indeed show that a ribozyme selected by this protocol can reduce the level of growth hormone to less than 0.2% (14). Since this is the most efficient and reliable procedure to identify suitable target sites in a particular RNA which are accessable *in vivo*, the detailed protocol will be described.

Protocol 6.   Selection of target sites and isolation of functional ribozymes from a library according to Lieber and Strauss (14)

Equipment and reagents

- TKB buffer: 20 mM Tris–HCl pH 7.9, 0.2 mM EDTA, 10 mM 2-mercaptoethanol, 0.1 M KCl, 20% glycerol, 0.5 mM phenylmethylsulfonyl fluoride
- Stop solution: 95% formamide, 20 mM EDTA, 0.05% bromophenol blue, 0.05% xylene cyanol
- Phenol:chloroform (1:1)
- RNasin (Boehringer Mannheim)
- T7 RNA polymerase (New England Biolabs)
- NTP stock solution: containing 10 mM each of ATP, GTP, CTP, UTP

- Oligo(dT) cellulose (PolyATract, Promega)
- NMWL spin filter units, 30 000 (Millipore)
- Superscript II reverse transcriptase (Life Technologies)
- GeneClean (Bio101, LaJolla)
- Terminal deoxynucleotidyl transferase (Life Technologies)
- *Taq* polymerase (Perkin Elmer)
- 10 × PCR buffer: 100 mM Tris–HCl pH 7.9, 500 mM KCl, 20% DMSO, 1% Triton X-100

See Chapter 7 for details on handling RNA.

Protocol 6. *Continued*

A. *Transcription of ribozymes from a library*

1. Mix 2 μl (2 μg) of ribozyme library plasmid DNA with 12.5 μl TKB, add 10 U of RNasin, 2.5 μl of 5 mM NTP stock solution, 2.5 μl 10 mM DTT, 100 U T7 RNA polymerase (Biolabs), and water to a total volume of 25 μl, and incubate at 37°C for 60 min.

2. Add 23 U of DNase I and incubate for 10 min at room temperature.

3. Add an equal volume of phenol:chloroform (1:1), mix and take the aqueous phase, repeat the extraction, and ethanol precipitate.

4. Resuspend RNA in 10 μl of water and run an aliquot on an agarose gel to estimate the amount of RNA. Normally, 5–8 μg RNA/25 μl reaction volume are synthesized.

B. *Cleavage of target RNA by the ribozymes*

1. Mix purified total RNA (1 μg) from cells of interest with 10 μg ribozyme library RNA synthesized by T7 polymerase in 50 mM Tris–HCl pH 7.5, 10 mM $MgCl_2$.

2. Perform heat denaturation by boiling at 95°C for 90 sec and quickly cool on ice.

3. Incubate reaction for 1 h at 37°C.

C. *Identification of cleavage sites by cloning and sequencing*

1. Purify 3′-terminal RNA cleavage products by binding to oligo(dT) cellulose (purified RNA may contain a low percentage of uncleaved gene-specific RNA).

2. Anneal approx. 0.2 μg of purified RNA with 2.5 μM oligo(dT) primer in 10 μl water for 10 min at 70°C.

3. Remove unbound primer by centrifugation through a 30 000 NMWL filter unit.

4. Add 2 μl dNTP stock solution (10 mM), 2 μl reverse transcription buffer, 200 U of superscript II reverse transcriptase, and water to a total volume of 20 μl, and incubate at 37°C for 1 h.

5. Purify cDNA:RNA hybrids by binding to GeneClean and elute with 30 μl of water.

6. Boil the hybrids for 2 min and cool on ice.

7. Add 5 μl dGTP (2 mM), 5 μl tailing buffer, 20 U of terminal deoxynucleotidyl transferase, and water to 50 μl, and incubate at 37°C for 15 min.

8. Take 2.5 μl tailing reaction and add 2 μl dNTP (10 mM), 1.5 mM $MgCl_2$, 10 μl 10 × PCR buffer, 2.5 U of *Taq* polymerase, 15 μM C_{15}-primer, and water to a total volume of 100 μl.

9. Run seven cycles of one-directional PCR with 30 sec 95°C, 30 sec 42°C, 90 sec 72°C.

10. Add 25 μM of the adequate gene-specific downstream primer and carry out 40 cycles of PCR with 60 sec 94°C and 90 sec 72°C.

11. Run reaction products on a gel and cut most prominent band(s) corresponding to the preferred cleavage products.

12. Clone the fragments and sequence them according to standard procedures.

D. *Isolation of ribozymes from a library*

1. Mix 50 ng of plasmids from the ribozyme library with 20 μM of each of the upstream and downstream primers, which are specific for the sequences around the cleavage site of the sequenced fragment, and add 2.5 U *Taq* polymerase, 10 × PCR buffer, and water to give 100 μl total reaction volume.

2. Run PCR for 40 cycles (45 sec 95°C, 45 sec 52°C, 45 sec 72°C).

3. Purify reaction product by spin filtration using NMWL filter units, digest with desired restriction endonuclease, and clone into the proper expression vector.

For expression of ribozyme genes the same vectors as for standard antisense genes can be used (see *Table 2*). Ribozymes selected by the procedure described in *Protocol 6* are highly efficient in inactivating function of the target RNA and even an excess of two- to tenfold over the target RNA can be sufficient to completely abolish the function of a particular gene as demonstrated for overexpressed human growth hormone in stably transformed clones (14).

3. *In vivo* studies in animals

Studies on antisense and ribozyme function in animals serve two purposes. On the one hand, they are useful to analyse the effects of loss of function of gene products at the systemic level and, on the other hand, they are important as pre-clinical studies to test for therapeutic effects. In most experiments carried out so far antisense oligos have been used because systemic delivery of expression constructs still suffers from the relatively low transfer efficiencies. In the case of antisense oligos, direct injection of naked molecules is normally done (*Protocol 7*). Phosphothioate oligos are preferentially used because of their relatively high stability against enzymatic degradation. Several authors have injected oligos into particular regions of the brain, e.g. to inhibit function of angiotensinogen and the AT1 receptor (53), and the D2 dopamine receptor

(54). Biological effects of receptor inactivations could be observed for several days and recovery from the antisense treatment was also detected, in the latter case after two days.

Liver-specific targeting of oligos after intravenous injection can be achieved by coupling of the oligo to an asialoglycoprotein (55). An interesting approach has been taken to study the antiviral activity of oligo (56). The authors treated Peking ducks which were infected with duck hepatitis B virus by injecting them with oligos into the foot pad vein. 20 μg oligo/g body weight were injected once every day over a period of ten days and even four days of treatment had a significant effect (Offensperger, personal communication). This treatment resulted in complete inhibition of surface antigen expression and viral replication indicating that antisense treatment of hepatitis B might be feasible (56). Systemic application in mice and rats can be done by intraperitoneal or intravenous injection of oligos.

Protocol 7. Application of oligos by injection according to Agrawal *et al.* and Lev-Lehman *et al.* (57, 58)

Equipment and reagents
- HPLC purified phosphothioate oligo (prepared as in *Protocol 1*)
- PBS
- Hamilton syringe

Method

1. Dilute oligo stock solution to 1 μg/μl in PBS.

2. Take up in Hamilton syringe and inject 10 μl/g body weight into the peritoneum of three-week-old mice.

3. Alternatively, rats (200–350 g body weight) are injected intravenously with 100 μl at oligo concentrations between 0.1–1 μg/ml using a 1 ml syringe.

Using this protocol, antisense inhibition of acetylcholinesterase gene expression has been obtained which resulted in transient alterations in the haematopoietic system including a 1000-fold reduction of acetylcholinesterase mRNA in lymph nodes and drastic reduction of bone marrow lymphocyte and erythroid fractions (58).

Recent studies have reported that restenosis in rats (59–61) and pigs (62) can be inhibited by targeting the function of cell cycle proteins like MYC (59, 62), cdk2 (60), and PCNA with antisense oligos (61). In all cases, a significant inhibition of proliferation of vascular smooth muscle cells has been observed. A combination of antisense cdk2 and cdc2 oligos delivered by Sendai virus–liposome complexes resulted in an almost complete inhibition of neointima formation (60). After a single intraluminal administration, the phosphothioate oligos persisted for up to two weeks in the medial layer of the vessels

(60). These studies suggest that even a single dose of oligos can have a profound therapeutic effect. Inhibition of restenosis will probably be one of the first clinical applications where antisense therapeutics can prove their value.

Growth of a variety of tumour cell types in nude mice has been successfully inhibited by antisense oligos specific for the mRNA of the transcription factor NF-kB (63). Targeting of the mutant *Ha-ras* mRNA in T24 human bladder carcinoma cells prevented tumour growth in nude mice for up to 14 days and the level of RAS protein was decreased by 90% (64). An antisense *K-ras* retroviral construct has been employed to prevent the growth of human lung cancer cells implanted orthotopically in nude mice (65). Particularly interesting is the first study on inhibition of lymphomas from a cell line with a t(14;18) translocation. After injection of such a cell line into SCID mice and administration of antisense *bcl-2* oligos over a period of 14 days, no lymphomas were detected in treated animals in contrast to the untreated control animals (66). Overexpression of *bcl-2* in these cells prevents apoptosis and block of the *bcl-2* pathway by antisense should drive the cells into apoptosis. Thus, inhibition of *bcl-2* should have a cytotoxic rather than cytostatic effect.

With regard to the pre-clinical assessment of phosphothioate oligo treatment, studies on pharmacokinetics are required which include absorption, distribution, metabolism, and excretion. A recent study in rats using either single injection or chronic delivery via a pump has generated some interesting results which are of general importance (67). After a single injection, the oligo appears most prominently in the liver after 3, 6, and 12 h amounting to about 40% of the dose. A total of 58% was found sequestered in tissues, less than 1% was found circulating in the plasma, and about 15% was found excreted into the urine for a total recovery of 73% (67). The $t_{1/2}$ for distribution out of the plasma was 20–25 min and the $t_{1/2}$ for urinary elimination was 27–41 h. The latter relatively long half-life suggests that administration of an oligo once every other day could be used to maintain a therapeutically effective concentration of an antisense oligo. Whereas phosphothioate oligos are not toxic in mice and rats at concentrations up to 50 mg/kg (10), they were reported to activate complement and to induce haemodynamic changes in monkeys upon rapid intravenous injection (68). However, slow injection seems to reduce this side-effect considerably (68). It is important to know that doses of 450 μg per pregnant mouse (25 g body weight) had no obvious toxic effect on the fetuses (69) suggesting the treatment to be relatively safe for the offspring also.

4. Generation of transgenic mice with antisense/ribozyme constructs

Systemic inactivation of genes can be accomplished by gene knock-out through homologous recombination (70). However, several gene knock-outs have turned out to be lethal in early embryonic development. Hence, it would be extremely important to have an alternative method to accomplish inactivation of gene function only in a particular organ. This could theoretically be achieved by expression of antisense RNA from tissue-specific promoters in transgenic mice. Transgenic mice can routinely be generated by DNA injection into the pronucleus of fertilized eggs and subsequent transfer of the injected eggs into foster mothers (71). Very few applications of antisense expression to transgenic mice have been published so far. Expression of myelin basic protein antisense RNA was found to result in the development of a shiverer behaviour (72). Recently, successful expression and function of ribozymes in transgenic mice has been reported (73, 74). Reduction of β2-microglobulin up to 90% has been achieved in some tissues in one study (73). Pancreas-specific expression of a glucokinase-specific ribozyme was found to result in reduction of glucokinase levels to 30%, accompanied by an impairment of insulin release in response to glucose which might predispose these mice to type II diabetes (74). In both studies, no attempts were made to optimize expression and to select the most suitable target sites for the ribozyme. Thus, it seems likely that selection of optimal target sites and the use of more suitable expression constructs (14) may eventually lead to complete inactivation of gene function in transgenic mice.

5. Clinical application

First clinical phase I trials using antisense strategies have been initiated. These studies are aimed at toxicity testing rather than therapeutic effects. Five patients with relapsed acute myelogeneous leukaemia (AML) received a ten-day continuous intravenous infusion of p53 antisense phosphothioate oligos at doses of 0.05 mg/kg/h. No major toxicity effect was attributable to this treatment (75). Other trials using retroviral expression constructs for antisense RNA to mutant *ras* sequences (76) or ribozymes specific for HIV RNA (77, 78) are in progress. However, significant improvement of expression and delivery systems is still required to achieve therapeutic effects or even a cure for infectious diseases like AIDS or hepatitis.

6. Conclusions

Antisense technology can be used in a more or less predictable manner to inactivate the function of a particular gene. Basically, any gene function of interest can be inactivated either transiently or stably. It seems therefore natural that application of this powerful technology does not remain restricted to basic research on the evaluation of functions of particular gene products, but advances to clinical application. First clinical protocols aiming at short-term treatment of infectious and malignant diseases have been initiated. A later routine use of synthetic oligos requires cheaper methods of production and higher stability, e.g. by encapsulation. The use of antisense or ribozyme expression vectors for 'anti-gene' therapy of infectious diseases and cancer is clearly one of the long-term goals in the development of strategies for gene therapy. However, the success is largely dependent on the improvement of delivery systems. In contrast to 'classical' gene therapy, where transfer efficiencies of 10–20% may be sufficient for the correction of a genetic defect, a 100% efficiency of antisense gene transfer and expression would be required which might be difficult to achieve *in vivo*. Nevertheless, improvement of both the antisense technology and gene delivery will soon lead to a broader clinical application in particular in such diseases as AIDS, hepatitis, and particular forms of cancer.

References

1. Hélène, C. and Toulmé, J. J. (1990). *Biochem. Biophys. Acta*, **1049**, 99.
2. Wickstrom, E. (ed.) (1991). *Prospects for antisense nucleic acid therapy of cancer and AIDS*. Wiley–Liss, New York, NY.
3. Eckstein, F. (1988). *Angew. Chem.*, **22**, 423.
4. Agrawal, S., Goodchild, J., Civeira, M. P., Thornton, A. H., Sarin, P. S., and Zamecnik, P. C. (1989). *Nucleosides Nucleotides*, **8**, 819.
5. Wickstrom, E., Bacon, T. A., Gonzalez, A., Freeman, D. L., Lyman, G. H., and Wickstrom, E. (1988). *Proc. Natl. Acad. Sci. USA*, **85**, 1028.
6. Gewirtz, A. M. and Calabretta, B. (1988). *Science*, **242**, 1303.
7. Daaka, Y. and Wickstrom, E. (1990). *Oncogene Res.*, **5**, 279.
8. Strauss, M., Hering, S., Lieber, A., Herrmann, G., Griffin, B. E., and Arnold, W. (1992). *Oncogene*, **7**, 769.
9. Matsukura, M., Shinozuka, K., Zon, G., Mitsuya, H., Reitz, M., Cohen, J. S., *et al.* (1987). *Proc. Natl. Acad. Sci. USA*, **84**, 7706.
10. Agrawal, S., Goodchild, J., Civeira, M. P., Thornton, A. H., Sarin, P. S., and Zamecnik, P. C. (1988). *Proc. Natl. Acad. Sci. USA*, **85**, 6268.
11. Izant, J. G. and Weintraub, H. (1984). *Cell*, **36**, 1007.
12. Bertrand, E. L., Pictet, R., and Grange, T. (1994). *Nucleic Acids Res.*, **22**, 293.
13. L'Huillier, P. J., Davis, S. R., and Bellamy, A. R. (1992). *EMBO J.*, **11**, 4411.
14. Lieber, A. and Strauss, M. (1995). *Mol. Cell. Biol.*, **15**, 540.
15. Cech, T. R. (1987). *Science*, **236**, 1532.
16. Uhlenbeck, O. C. (1987). *Nature*, **328**, 596.
17. Haseloff, J. and Gerlach, W. L. (1988). *Nature*, **334**, 585.
18. Cotten, M., Schaffner, G., and Birnstiel, M. (1989). *Mol. Cell. Biol.*, **9**, 4479.
20. Fakler, B., Herlitze, S., Amthor, B., Zenner, H. P., and Ruppersberg, J. P. (1994). *J. Biol. Chem.*, **269**, 16187.
21. Walder, R. Y. and Walder, J. A. (1988). *Proc. Natl. Acad. Sci. USA*, **85**, 5011.
22. Bacon, T. A. and Wickstrom, E. (1991). *Oncogene Res.*, **6**, 13.
23. Agrawal, S., Sarin, P. S., Zamecnik, M., and Zamecnik, P. C. (1992). In *Gene regulation: biology of antisense RNA and DNA* (ed. R. P. Erickson and J. G. Izant), pp. 273–83. Raven Press Ltd., New York.
24. Marcus-Secura, C. J., Woerner, A. M., Shinozuka, K., Zon, G., and Quinnan, G. V. (1987). *Nucleic Acids Res.*, **15**, 5749.

25. Loke, S. L., Stein, C., Zhang, X. H., Mori, K., Nakanishi, M., Subhasinghe, C., *et al.* (1989). *Proc. Natl. Acad. Sci. USA*, **86**, 3474.
26. Bennet, C. F., Chiang, M.-Y., Chan, H., Shoemaker, J. E. E., and Mirabelli, C. K. (1992). *Mol. Pharmacol.*, **41**, 1023.
27. Wagner, R. W. (1994). *Nature*, **372**, 333.
28. Lukas, J., Bartek, J., and Strauss, M. (1994). *J. Immunol. Methods*, **170**, 255.
29. Lukas, J., Jadayel, D., Bartkova, J., Strauss, M., and Bartek, J. (1994). *Oncogene*, **9**, 2159.
30. Strauss, M., Hamann, J., Müller, H., Lieber, A., Sandig, V., Bauer, D., *et al.* (1992). In *DNA replication and the cell cycle. 43. Colloquium Mosbach 1992*, pp. 221–8. Springer–Verlag, Berlin, Heidelberg.
31. Izant, J. G. (1992). In *Gene regulation: biology of antisense RNA and DNA* (ed. R. P. Erickson and J. G. Izant), pp. 183–96. Raven Press Ltd., New York.
32. Prochownik, E. V., Kukowska-Latallo, J. F., and Rodgers, C. (1988). *Mol. Cell. Biol.*, **8**, 3683.
33. Grossman, M. and Wilson, J. M. (1993). *Curr. Opin. Genet. Dev.*, **3**, 110.
34. Culver, K. W., Ram, Z., Wallbridge, S., Ishii, H., Oldfield, E. H., and Blaese, R. M. (1992). *Science*, **256**, 1550.
35. Stratford-Perricaudet, L. D., Levrero, M., Chasse, J. F., Perricaudet, M., and Briand, P. (1990). *Hum. Gene Ther.*, **1**, 141.
36. Li, Q. T., Kay, M. A., Finegold, M., Stratford-Perricaudet, L. D., and Woo, S. L. C. (1993). *Hum. Gene Ther.*, **4**, 403.
37. Lieber, A., Sandig, V., Sommer, W., Bähring, S., and Strauss, M. (1993). In *Methods in enzymology* (ed. R. Wu), Vol. 217, pp. 47–66. Academic Press, London.
38. Gibco BRL (Life Technologies, Inc.). LIPOFECTAMINE™ Product Information Sheet.
39. Yokoyama, K. and Imamoto, F. (1987). *Proc. Natl. Acad. Sci. USA*, **84**, 7363.
40. Gerwitz, A. M., Anfossi, G., Venturelli, D., Valpreda, S., Sims, R., and Calabretta, B. (1989). *Science*, **245**, 180.
41. Holt, J. T., Gopal, T. V., Moulton, A. D., and Nienhuis, A. W. (1986). *Proc. Natl. Acad. Sci. USA*, **83**, 4794.
42. Caracciolo, D., Valtieri, M., Venturelli, D., Peschle, C., Gerwitz, A. M., and Calabretta, B. (1989). *Science*, **245**, 1107.
43. Prochownik, E. V. (1992) In *Gene regulation: biology of antisense RNA and DNA* (ed. R. P. Erickson and J. G. Izant), pp. 303–16. Raven Press Ltd., New York.
44. Kashani, M., Funato, T., Florenes, V. A., Fodstad, O., and Scanlon, K. (1994). *Cancer Res.*, **54**, 900.
45. Shore, S. K., Nabissa, P. M., and Reddy, E. P. (1993). *Oncogene*, **8**, 3183.
46. Crisell, P., Thompson, S., and James, W. (1994). *Nucleic Acids Res.*, **21**, 5251.
47. Lisziewicz, J., Sun, D., Smythe, J., Lusso, P., Lori, F., Louie, A., *et al.* (1993). *Proc. Natl. Acad. Sci. USA*, **90**, 8000.
48. Sarver, N., Cantin, E. M., Chang, P. S., Zaia, J. A., Ladne, P. A., Stephens, D. A., *et al.* (1990). *Science*, **247**, 1222.
49. Yu, M., Ojwang, J., Yamada, O., Hampel, A., Rapapport, J., Looney, D., *et al.* (1993). *Proc. Natl. Acad. Sci. USA*, **90**, 6340.
50. Cantor, G. H., McElwain, T. F., Birkebak, T. A., and Palmer, G. H. (1993). *Proc. Natl. Acad. Sci. USA*, **90**, 10932.
51. Xing, Z. and Whitton, J. L. (1993). *J. Virol.*, **67**, 1840.

52. Sullenger, B. A. and Cech, T. R. (1993). *Science*, **262**, 1566.
53. Ogawa, S., Olazabal, U. E., Parhar, I. S., and Pfaff, D. W. (1994). *J. Neurosci.*, **14**, 1766.
54. Gyorko, R., Wielbo, D., and Phillips, M. I. (1993). *Regul. Pept.*, **10**, 167.
55. Lu, X. M., Fishman, A. J., Jyawook, S. L., Hendricks, K., Tompkins, R. G., and Yarmush, M. L. (1994). *J. Nucl. Med.*, **35**, 269.
56. Offensperger, W. B., Offensperger, S., Walter, E., Teubner, K., Igloi, G., Blum, H. E., *et al.* (1994). *EMBO J.*, **12**, 1257.
57. Agrawal, S., Temsamani, J., and Tang, J. Y. (1991). *Proc. Natl. Acad. Sci. USA*, **88**, 7595.
58. Lev-Lehmann, E., Ginsberg, D., Hornreich, G., Ehrlich, G., Meshorer, A., Eckstein, F., *et al.* (1994). *Gene Ther.*, **1**, 127.
59. Bennett, M. R., Anglin, S., McEwan, J. R., Jagoe, R., Newby, A. C., and Evan, G. I. (1994). *J. Clin. Invest.*, **93**, 820.
60. Morishita, R., Gibbons, G. H., Ellison, K. E., Nakajima, M., von der Leyen, H., Zhang, L., *et al.* (1994). *J. Clin. Invest.*, **93**, 1458.
61. Simons, M., Edelman, E. R., and Rosenberg, R. D. (1994). *J. Clin. Invest.*, **93**, 2351.
62. Shi, Y., Fard, A., Galeo, A., Hutchinson, H. G., Vermani, P., Dodge, G. R., *et al.* (1994). *Circulation*, **90**, 944.
63. Higgins, K. A., Perez, J. R., Coleman, T. A., Dorshkind, K., McComas, W. A., Sarmiento, U. M., *et al.* (1993). *Proc. Natl. Acad. Sci. USA*, **90**, 9901.
64. Gray, G. D., Henandez, O. M., Hebel, D., Root, M., Pow-Sang, J. M., and Wickstrom, E. (1993). *Cancer Res.*, **53**, 577.
65. Georges, R. N., Mukhopadhyay, T., Zhang, Y., and Roth, J. A. (1993). *Cancer Res.*, **53**, 1743.
66. Pocock, C., Al-Mahdi, N., Hall, P., Morgan, G., and Cotter, F. (1993). *Blood*, **92** (suppl.), 200a.
67. Iversen, P. L., Mata, J., Tracewell, W. G., and Zon, G. (1994). *Antisense Res. Dev.*, **4**, 43.
68. Galbraith, W. M., Hobson, W. C., Giclas, P. C., Schechter, P. J., and Agrawal, S. (1994). *Antisense Res. Dev.*, **4**, 201.
69. Gaudette, M. F., Hampikian, G., Metelev, V., Agrawal, S., and Crain, W. R. (1994). *Antisense Res. Dev.*, **3**, 391.
70. Morrow, B. and Kucherlapati, R. (1993). *Curr. Opin. Biotechnol.*, **4**, 577.
71. Smithies, O. (1993). *Trends Genet.*, **9**, 112.
72. Katsuki, M., Sato, M., Kimura, M., Yokoyama, M., Koboyashi, K., and Nomura, T. (1988). *Science*, **241**, 593.
73. Larsson, S., Hotchkiss, G., Andäng, M., Nyholm, T., Inzunza, J., Jansson, I., *et al.* (1994). *Nucleic Acids Res.*, **22**, 2242.
74. Efrat, S., Leiser, M., Wu, Y.-J., Fusco-DeMane, D., Emran, O. A., Surana, M., *et al.* (1994). *Proc. Natl. Acad. Sci. USA*, **91**, 2051.
75. Bayever, E., Iversen, P. L., Bishop, M. R., Sharp, J. G., Tewary, H. K., Arneson, M. A., *et al.* (1994). *Antisense Res. Dev.*, **3**, 383.
76. Roth, J. (1993). *Hum. Gene Ther.*, **4**, 383.
77. Leavitt, M. C., Yu, M., Yamada, O., Kraus, G., Looney, D., Poeschla, E., *et al.* (1994). *Hum. Gene Ther.*, **5**, 1115.
78. Wong-Staal, F. (1994). *Hum. Gene Ther.*, **5**, 788.

11

Non-antisense effects of oligodeoxynucleotides

C. A. STEIN and ARTHUR KRIEG

1. Introduction

A conceptual problem in the application of oligodeoxynucleotides (oligos) as antisense, anti-gene, or anti-protein agents is that there is no essential difference in the oligo synthesized for either of these three applications. The designation of an oligo as 'antisense', or 'triplex', or 'aptamer' is actually frequently based more on the intention of the investigator than on any scientifically determined mechanism of action. In principle, any oligo might have any of these activities, depending on the relative values of K_d for different oligo binding sites and on the intracellular localization of the oligo. As an example, a 16 base oligo synthesized to be a triplex-forming oligo was shown to have fortuitous complementarity to an mRNA sequence at 12 non-contiguous bases; this was sufficient to hybridize and activate RNase H, causing degradation of that RNA (1). In addition, many 'antisense' oligos have been found to show non-sequence-specific effects; this may perhaps be due to their polyanionic backbones (reviewed in ref. 2).

The problems of non-sequence specificity have become highlighted by the increasing use of modified oligos, particularly those that are nuclease-resistant (2). Current efforts in this area were made possible by the seminal chemical synthesis of phosphorothioate (PS) oligos by Stec *et al.* (3), on which most of this manuscript will focus. The work of Stec *et al.*, was based in part on a series of observations initially made by Eckstein and his co-workers (for a review, see ref. 4). Eventually, when the synthesis of full-length PS oligos became automated and the oligos more readily available, it was easily shown that PS oligos were quite resistant to the effects of a variety of nucleases, included P1, S1, and snake venom phosphodiesterase (5). In general, degradation was anywhere from 10- to 1000-fold slower for the PS oligo than for its phosphodiester (PO) congener. However, it became clear relatively early that despite the promise of PS oligos, non-sequence-specific effects would be a problem with which experimentalists would have to contend. Cazenave *et al.* (6), first noticed that PS oligos could non-sequence specifically block the translation of rabbit β-globin mRNA into protein in both rabbit reticulocyte lysate and in wheat germ extract. In unpublished observations, Stein noted that the ability of PS oligos to non-sequence specifically inhibit translation was correlated with their ability to bind directly to cell fractions enriched in ribosomes. Sequence-specific effects on globin mRNA translation could also be observed with the reticulocyte lysate system in the presence of PS oligos (6). However, the sequence-specific effect was highly PS oligo concentration-dependent, and rapidly gave way to non-sequence-specific inhibition. Furthermore, Gao *et al.* (7) found that PS oligos, non-sequence specifically, could inhibit the function of RNase H1 and 2, enzymes responsible for cleaving the mRNA strand of an mRNA:oligo duplex. Two factors probably contribute to the increased non-antisense effects of PS oligos. First, they have a reduced T_m for hybridization to complementary mRNA compared to the PO backbone. Secondly, PS oligos have an increased affinity for binding to proteins.

On the basis of these observations, made in cell-free systems under relatively stringent conditions, it would have been predictable that PS oligos would exhibit profound non-sequence specificity in tissue culture. Indeed, as discussed below, this is precisely what occurred (examples of known non-antisense effects of PO and PS oligos are listed in *Table 1*). Furthermore,

Table 1. Non-sequence-specific effects of phosphorothioate oligos

Oligo	Target/effect
SdCn	rsCD4/blocks binding of rsgp120; prevents HIV-1 induced syncytia formation
SdCn, SdT2G4T2	v3 loop of rsgp120
SdCn	HIV-1 reverse transcriptase/inhibition of activity
SdCn	DNA polymerases/inhibition of activity
SdT28	*Taq* polymerase/inhibition of activity
SdCn and others	PKC, multiple isoforms/competitive inhibitor of PKC β1 phosphorylation of substrate
SdCn, SdTn, G-quartet,	bFGF/blocks binding of bFGF to cell surface receptors; inhibits tritiated
others	thymidine uptake in NIH 3T3 cells
SdCn, SdTn	PDGF/inhibits PDGF stimulation of uPA activity
SdCn, SdTn	aFGF; FGF-4; VEGF
SdCn, G-quartet, others	Extracellular matrix derived from NIH 3T3 cells
Multiple	Fibronectin/blocks binding to cell surface receptors
Multiple	Laminin/blocks binding to sulfatide
Multiple	Sp1/non-specific induction of activity *in vivo*
SdC28, SdT28	Vacuolar H+—ATPase/inhibition of proton pumping activity
G-quartet or G-rich oligos	Inhibition of IFN-γ binding and/or secretion
CpG motif	Lymphocyte activation
Any S-oligos, sequence selectivity	Some inhibition of tyrosine kinases

it has become clear that in addition to their non-sequence-specific effects, PS and PO oligos also have sequence-specific or sequence-selective effects that are not due to an antisense mechanism of action. It is now clear that some published 'antisense' experimental systems in fact result from such non-antisense effects. The purpose of this chapter is to point out the known non-antisense activities of PO and PS oligos so that the experimenter who wishes to perform antisense experiments can anticipate undesired effects and reduce the probability of drawing mistaken conclusions. Of course, it is quite likely that other non-antisense effects of oligos remain to be decribed, so this chapter will also present the types of controls that should be employed.

2. Non-antisense effects of oligodeoxynucleotides

2.1 Anti-HIV effects of PS oligos

In any discussion of the non-sequence-specific effects of PS oligos, it is appropriate to examine the effects of these compounds on HIV replication. Examination of these effects provides a paradigm for those that may be observed in other systems. Furthermore, the effects of PS oligos on HIV replication have been relatively well studied and are better understood than most other systems, although many points still remain to be clarified.

Many experiments seem to suggest that PS oligos can inhibit HIV-1 replication in a sequence-specific manner (8). However, PS oligos may also inhibit HIV-1 replication by a mechanism that has been termed 'non-sequence-specific'. This term is something of a misnomer, because it implies that all PS oligos of identical length, regardless of sequence, will have an approximately equal inhibitory effect on HIV-1 replication in a given test system. That this is not the case will be demonstrated later.

The finding that PS oligos can non-sequence specifically inhibit HIV-1 replication was first made by Matsukura *et al.* (9). These observations were made in the context of experiments that were, in fact, designed to look for the sequence-specific effects of PS oligos. The cell line used, ATH8 cells, is a T lymphocyte line which is sensitive to the cytopathic effects of the virus. A series of PS oligos were synthesized which were complementary to the 5' region of the HIV-1 rev mRNA, and as controls, a series of cytidine (SdC) and adenosine (SdA) PS homopolymers, of chain lengths 5, 14, 18, 21, and 28 were also evaluated. All PS oligos, antisense as well as controls, blocked the cytopathic effect of the virus. The most effective construct was SdC28, where virtually complete blockade could be seen at a concentration of 3 μM. The ability to block the cytopathic effect was highly dependent on length, as for a 14-mer, a 10 μM concentration was required, and even 25 μM was ineffective for a 5-mer. The cytoprotective effect appeared to plateau at an oligo length of about 18–21 bases. The homoadenosine construct was somewhat less active than the homocytidine construct, perhaps, as has been suggested (10), due to depurination caused by acid treatment during the synthesis with the resulting generation of abasic sites. A large number of PS homopolymers and mixed base sequences (11), and even a phosphoroselenoate oligo (SedC28) (12) have been evaluated in the *de novo* infection system. All of these constructs appear to have a similar ability to block the cytopathic effect of HIV-1 in ATH8 cells. The non-sequence-specific effects of PS oligos on HIV replication were to a large extent confirmed by subsequent studies, especially those of Zelphati *et al.* (13), who used acutely infected CEM cells.

However, it was also demonstrated that the inhibition of the cytopathic effect of HIV-1 could not be due to the induction of interferon by the PS oligos. This question had been raised because it had long been known that sulfurized analogues of poly-r(I:C) were potent interferon inducers. In fact, addition of 1000 units of recombinant α or γ interferon to the assay system in no way altered the cytopathic effect of the virus.

Similar results to those obtained in (8), were also obtained by Agrawal *et al.* (14). They used Molt 3 and H9 cells, which, in contrast to the ATH8 cells, are tolerant to the virus. 20-mer PS homopolymers were constructed that were complementary to the HIV-1 mRNA splice donor and splice acceptor sites, and maximally inhibited HIV-1 p24 protein production and HIV-1 induced syncytial formation at concentrations of 20–100 mg/ml. SdC20, SdT20, and SdG20 were all thought to be about equally effective, whereas, similar to the observations in (8), SdA20 was less effective. Because of the way in which this assay system was employed, i.e. initially only a small number of cells were infected, the non-sequence-specific effects of the PS oligos on viral replication were accentuated.

Additional non-sequence-specific effects as measured by diminished p24 protein production were observed by Balotta *et al.* (15), who targeted sequence-specific 27- and 28-mers to the HIV-1 vpr mRNA in primary human macrophages from several donors. Some non-complementary controls were as active as the antisense constructs, but unfortunately, several of the antisense constructs and controls contained a G-quartet. The presence of the G-quartet makes interpretation of data extremely difficult. Indeed, even the octamer SdT2G4T2 (16) which forms a parallel-stranded tetrameric guanosine quartet structure, is a relatively potent inhibitor of HIV replication (concentration at 50% inhibition (IC_{50}) = 0.5 μM). In this case, the presence of tetramers appears to account for all of the antiviral activity, as heat-induced dissociation significantly vitiates it. Significant inhibition of Friend murine leukaemia virus replication in tissue culture was observed with PS GT-rich oligos (17).

Modification of a PS oligo by a 5′ terminal cholesteryl moiety markedly increases its anti-HIV activity (18, 19). This effect is particularly pronounced for shorter oligos; for example, the anti-HIV activity of a cholesteryl 10-mer

(Chol SdC10) was approximately equivalent to that of SdC28. Chol-oligos effectively blocked HIV-1 induced syncytia formation, a process which occurs exclusively at the level of the cell surface.

Several potential mechanisms of action have been suggested which may account for the non-sequence-specific ability of PS oligos to inhibit HIV-1 replication. The first mechanism suggested, although it is by no means proven, involves inhibition of the HIV-1 reverse transcriptase (RT) (20). For this mechanism to be successful, naked PS oligo must, after internalization in lymphoid cells, be able to penetrate the endosomal membrane, and attain an appropriate concentration in the cytoplasm. However, it has never been demonstrated in any cell type that PS oligos can penetrate the endosomal membrane, and their cytoplasmic concentrations are equally unknown in any cell type.

Despite all of these problems, Majumdar *et al.* (20) showed that SdC28 and related oligos could interfere with the *in vitro* DNA polymerase activity of HIV-1 RT. Both PS and PO oligos were competitive inhibitors of the binding of the template primer r(A)810-d(T)14 to the free enzyme. The K_i for OdC28 was 560 nM, and the K_i for SdC28 was 2.8 nM. The value of K_d was approximately equal to the value of K_i, which tends to confirm that the mechanism was simple competitive. Furthermore, a Hill plot of the inhibition by SdC28 gave a line with a linear slope = 1, suggesting a reaction order of unity. Maury *et al.* (21) confirmed many of these results, and also demonstrated uncompetitive inhibition of RT by the annealed template primer poly(rI) SdC14 as the concentration increased from 0.25 µM to 1 µM. In general, these PS duplexes have a higher primer activity (i.e. incorporation of dCTP into newly synthesized DNA) than when the PO congener is used, but reasons for this increase currently remain unclear. The HIV-1 RT-associated RNase H activity was also sensitive to SdC28, but the DNA-dependent DNA polymerase activity was not (22). 5' cholesteryl modified PS oligos have also been studied as HIV-1 RT inhibitors, and are more potent than the unmodified PS oligos (19).

Calf α< pig liver γ (20), and human DNA polymerases (γ > α > β) are all inhibited by SdC28, and by shorter chain oligos as well (7). The herpes simplex virus-2 DNA polymerase was also inhibited, but was more sensitive to PS oligos than were the human polymerases. The function of the HIV-1 integrase, which is encoded by the *pol* gene, is also affected by PS oligos. The integrase is a protein whose C terminal end (residues 211–284) contains some 22% basic amino acids (arginine and lysine), and whose activity is compromised by the polyanion suramin (23). However, as is the case with suramin, it is unclear if the attainable nuclear concentration of PS oligos is sufficient to inhibit this protein in intact cells.

Another way in which PS oligos can non-sequence specifically inhibit HIV-1 replication is by blocking the binding of the viral envelope glycoprotein gp120 to its cell surface receptor, CD4 (10). The ability of PS oligos to do this, is highly length-dependent, with SdC28 > SdC15 >> SdC5. Mapping studies

(24) have revealed that there are most likely two distinct polyanion binding sites on recombinant CD4. They both occur on the N terminal D1 domain: one loop of this domain, known as C'C'', is analogous to the immunoglobulin CDR2 loop and is involved in gp120 binding, perhaps through residues 41–56 (25, 26), PS oligos probably bind at or near this site. An additional oligo binding site appears to be on the FG (CDR3-like loop), which is also found on the D1 domain. This was demonstrated by the fact that PS oligos could block the binding of the mAb L71.1.1, a CDR3-like loop-specific antibody. These data are consistent with that of Parish *et al.* (27), who suggested that a secondary polyanion binding site, distinct from the primary gp120 binding site, exists on CD4. It is possible that the length dependence of the binding, which increases dramatically at a PS oligo length of 15–18 bases, may be due to the requirement that PS oligos be long enough to bridge the two binding sites. Computer modelling has shown that this can occur only with an oligo of some 13–15 bases or greater. It should be kept in mind that PS oligos are just one of a large number of polyanions which can competitively bind to CD4: these polyanions include pentosan polysulfate, suramin, reactive red 120, and trypan blue.

Non-specific interactions also occur between PS oligos and the v3 loop of the viral envelope glycoprotein, gp120 (28). The binding constant for PS oligos showed a discrete increase in gp120 binding for PS oligos >12-mer in length. However, similar to what was observed for CD4, there was no further increment in binding constant beyond an 18-mer. In their ability to bind to the v3 loop, PS oligos once again resemble other charged organic polyanions, such as pentosan polysulfate and dextran sulfate. Other experiments have shown that PS oligos are weak non-specific competitors of the binding of gp120 to neural glycolipids, including sulfatide (C. Stein and N. Latov, unpublished data). In addition, Wyatt *et al.* (16) have demonstrated that the tetrameric form of SdT2G4T2 (termed a 'G tetrad') also appears to bind to the v3 loop, whereas the PO congener and the control sequence SdTGTGTG bind very weakly, if at all. The K_d value for tetramer binding to gp120 appears to be < 1 μM. Although not specifically stated in (16), SdT2G4T2 will undoubtedly also bind to the D1 domain of CD4, as described above. Indeed, the ability of PS oligos to bind non-sequence specifically to many proteins required for HIV replication may account in part for their anti-HIV activity. In summary, aside from possible antisense effects against HIV, PS oligos have both sequence-specific and sequence non-specific activities. In practice, it is extremely difficult to dissect the observed biological effects into sequence-specific versus non-sequence-specific components. Such difficulties undoubtedly exist in other systems as well. It seems increasingly likely that many 'antisense' effects observed with PS oligos are in reality a complex amalgam of sequence-specific and non-sequence-specific effects. The experimenter must always keep this possibility in the forefront, as failure to do so may result in erroneous conclusions being drawn.

2.2 Protein kinase C

One cellular protein of extreme current interest whose interactions with PS oligos has recently been studied, is protein kinase C (PKC). A general discussion of the biochemistry of this major intracellular signal transducer is beyond the scope of this article, save to relate that some of its isoforms are Ca^{2+}-, phospholipid-, and ATP-dependent serine–threonine kinases. In *in vitro* experiments (29) in the presence of lipid cofactor (usually phosphatidylserine), PS oligos, in a length- and concentration-dependent manner, will block the PKC β1 isoform phosphorylation of substrate (an octapeptide fragment of the epidermal growth factor receptor). This inhibitory activity is linear competitive with respect to substrate, but not competitive with ATP. The K_i is 2 μM, and the K_d value is 5.4 μM; the fact that they are almost identical argues that the inhibition is truly competitive. Isoform-specific inhibition was not observed. Precisely the same results, with an almost identical value of K_i, was observed for pentosan polysulfate. Strikingly, suramin, which is also an inhibitor of PKC β1 phosphorylating activity, is a competitive inhibitor with respect to ATP, and a mixed competitive–non-competitive inhibitor with respect to substrate. In the absence of lipid cofactor, all three polyanions (SdC28, pentosan polysulfate, and suramin) were stimulators of PKC β1-induced substrate phosphorylation, although the mechanism for this remains uncertain. In the absence of substrate, SdC28 will induce PKC β1 autophosphorylation. PKC δ, which is not Ca^{2+}-dependent, is affected in a complex manner by PS oligo concentration, and the presence of cofactors and phorbol esters. However, because of the relatively high value of IC_{50} for PKC β1 inhibition by SdC28 (5 μM), and the uncertainty surrounding the true value of intracytoplasmic concentration of any PS oligos even after prolonged incubation, it is unclear if PKC inhibition actually occurs *in vivo*.

2.3 Growth factors

One of the major naturally occurring classes of charged polyanions are glycosaminoglycans (GAGs), of which heparin is a notable example. Other examples of sulfated GAGs include heparan, dermatan, and chondroitin sulfates. These materials bind and sequester a variety of growth factors to the basement membrane, and in some cases, are responsible for the presentation of the growth factor (e.g. bFGF) to its appropriate cell surface receptor. Thus, GAGs are intimately involved in cellular growth, and in important growth supporting activities, such as angiogenesis. Many synthetic polyanions, including pentosan polysulfate (30) and suramin (28) can bind heparin-binding growth factors, and block their binding to cell surface receptors. The ability of these compounds (particularly suramin), to do this, has been taken advantage of by several investigators (11, 31) who have examined the activity of suramin as an antineoplastic agent. Without implying that inhibition of growth factor binding to the cell surface is the only mechanism of action for this agent, it appears to be clear that significant clinical activity has been observed in hormone refractory metastatic prostate cancer, and in chemotherapy refractory indolent lymphoma.

Like suramin and heparin, PS oligos are also capable of interacting with heparin-binding growth factors, and can block binding of them to their cell surface receptors. Guvakova *et al.* (32) have studied the interactions of PS and PO oligos with heparin-binding growth factors. They showed that oligos bound not only to basic fibroblast growth factor (bFGF), but to other members of the FGF family including aFGF, Kaposi's growth factor (FGF-4), platelet-derived growth factor (PDGF), and vascular endothelial growth factor (VEGF), but not to EGF (epidermal growth factor), which is non-heparin binding. PS, but not PO oligos, could block the binding of $[^{125}I]$bFGF and $[^{125}I]$PDGF to NIH 3T3 cells. The value of IC_{50} for SdC28 was approximately 5 μM, and was highly chain length-dependent. Furthermore, these authors also examined the effects of 18-mer PS oligos of different sequences on $[^{125}I]$bFGF binding to low and high affinity bFGF binding sites on both NIH 3T3 fibroblasts and DU145 prostate cancer cells.

Inhibition of binding to both classes of sites by the PS oligos were observed, and sequence-selective inhibition was found for the high affinity sites. In general, inhibition of binding was increased when the PS oligo contained four consecutive Gs (the G-quartet motif). These non-antisense sequence-selective effects of the G-quartet are of great concern since two of the most widely used 'antisense' oligos, to c-Myc and to c-Myb, both contain G-quartets, while their sense controls do not.

PS homopolymers of cytidine and thymidine, like heparin, also release bFGF bound to low affinity receptors in extracellular matrix. Like suramin and heparin, they are potent inhibitors of heparanase (I. Vlodavsky and C. Stein, unpublished results), and inhibit bFGF-induced DNA synthesis in 3T3 fibroblasts. All of these effects are concentration- and length-dependent, and are probably to some extent sequence-selective as well. Further, SdC28 blocks the PDGF-induced increase in urokinase-type plasminogen activator (uPA) activity in smooth muscle cells, and is also a potent inhibitor of smooth muscle migration (E. L. Rabbani and C. A. Stein, unpublished results).

2.4 Extracellular matrix elements

We have recently shown (32a), that PS oligos can adhere avidly to extracellular matrix (ECM) derived from confluent cultures of NIH 3T3 cells. Matrix elements that bind PS oligos include fibronectin and laminin, but not vitronectin. PS oligos appear to block the binding of fluoresceinated fibronectin to its cell surface receptor, whereas, again by analogy with suramin, PS oligos block the binding of laminin to sulfatide. These characteristics of PS oligo behaviour may underlie their ability to block the adhesion and spreading of trypsinized cells to plastic substrata. This effect, as might be predicted, is markedly vitiated when the plates are pre-coated with either fibronectin or laminin. Interestingly, PS oligos containing the G-quartet are more potent inhibitors of adhesion than those which lack it. These observations suggest an alternative explanation for the findings of Watson and Pons (33), who found that an antisense c-myb oligo, which contained a G-quartet, was a more potent blocker of cellular adhesion than the sense control.

The sequence-selective protein binding properties of PS oligos are clearly affected by the presence of a G-quartet, and oligos bearing this motif may influence cellular growth and proliferation in striking, 'sequence-independent' ways (34). We (C. A. Stein and Z. Khaled, unpublished observations) have determined that the ability of PS oligos to bind to certain proteins in a non-sequence-specific manner may actually be enhanced as the number of contiguous guanosine residues is increased beyond four. This is associated with the increased ability of these oligos to precipitate proteinaceous material from complete media. These effects appear to plateau with a stretch of seven to nine contiguous guanosine residues.

2.5 Effects of PS oligodeoxynucleotides on transcription factor activities

Perez *et al.* have reported that aside from possible antisense effects, PS oligos induce increased binding of the transcription factor Sp1 to its target in a non-sequence-specific manner (35). This induction was quite rapid and apparent within 30 minutes. Sp1 activity was increased in human and murine cell lines and in primary cells and was induced *in vivo* as well as *in vitro*. The central role of Sp1 in the transcriptional regulation of multiple genes raises concerns that phosphorothioate oligos may have non-antisense effects due to their activation of this or other nuclear binding proteins. In this context, it is noteworthy that the sequence-specific binding of some transcription factors to double-stranded oligos is lost if the oligos are PS modified (36). On the other hand, Bielinska *et al.* have reported that double-stranded PS oligos containing the binding site for NF-KB or octamer transcription factors specifically compete with the function of the native sites (36a).

2.6 PS and chimeric oligodeoxynucleotides can specifically or non-specifically inhibit protein tyrosine kinases and polynucleotide kinase

Besides their effects on PKC (see above), Bergan *et al.* have recently identified an oligo sequence that specifically inhibits the p210bcr–abl tyrosine kinase, but not other kinases (37). This inhibition was seen with PS but not PO oligos. We have found that all 20-mer PS oligos inhibit the *src* family kinases lyn, fyn, and lck, as well as the EGF receptor tyrosine kinase (37a). However, shorter PS oligos, such as 6-mers, showed sequence-specific inhibition. Inhibition was not seen with the 'reverse' oligo, i.e. the same six bases in the opposite order (37a). Sequence-specific inhibition was also seen with 20-mer chimeric oligos in which the first two and the last five internucleotide linkages were PS modified in an attempt to confer stability while reducing non-specific effects. PO oligos did not inhibit any tested kinases. We have also demonstrated length-dependent inhibition of polynucleotide kinase by PS but not by PO or chimeric PS–PO oligos (37b). Since these inhibitory effects could be seen at nanomolar concentrations, this raises yet another possible type of non-antisense effects of PS oligos, that may not be detected with routine controls.

2.7 PO and PS oligodeoxynucleotides containing G-rich regions block interferon-gamma binding and/or secretion

Ramanathan *et al.* have reported that PO oligos containing G-rich regions (including a G-quartet) block the binding of IFN-γ to its cellular receptor, but a control sense oligo does not (38). These effects, which were seen at standard oligo concentrations used in 'antisense' experiments, resulted in decreased induction of immune and adhesion proteins in response to IFN-γ treatment of cells. It is worth noting that the oligos causing these effects had been synthesized as triplex oligos, and that the results were initially interpreted as resulting from a triplex mechanism of action. In addition to blocking the binding of IFN-γ, G-rich PS oligos block the *in vitro* production of IFN-γ by lymphocytes activated with a variety of mitogens (39).

2.8 Identification of a mitogenic CpG motif in PO and PS oligodeoxynucleotides

2.8.1 The CpG motif

In the course of investigating the lymphocyte stimulatory effects of two antisense oligos specific for endogenous retroviral sequences, 3D and 3M (40–42) (*Table 2*), we found that two out of 24 'control' PO oligos (including various scrambled, sense, and mismatch controls for a panel of 'antisense' oligos) also mediated B cell activation, while the other 'controls' had no effect. Like the 'antisense' oligos, these 'control' oligos (1 and 2, *Table 2*) induced splenic B cells to incorporate [^{3}H]uridine and to secrete immunoglobulin (*Table 2*).

The mechanism of this B cell activation by the 'control' oligos did not appear to involve antisense effects:

(a) Comparison to vertebrate DNA sequences listed in GenBank showed no greater homology than that seen with non-stimulatory oligos.

(b) Oligos 1 and 2 showed no hybridization to Northern blots.

(c) Changes in the oligo sequence did not necessarily eliminate the stimulation.

Table 2. Immune stimulatory oligo nucleotide sequences

ODN	Sequence (5' to 3')[b]	Stimulation index[a]	
		[³H]uridine[c]	IgM production[d]
1	GCTAGAC<u>CG</u>TTAG<u>CG</u>T	6.1 ± 0.8	17.9 ± 3.6
1a	......T........	1.2 ± 0.2	1.7 ± 0.5
1b	......Z........	1.2 ± 0.1	1.8 ± 0.0
1c	............Z..	10.3 ± 4.4	9.5 ± 1.8
1d	....A..........	5.5 ± 1.0	ND
1e	.....T..A......	4.3 ± 1.8	ND
1f	..AT......GAGC.	13.0 ± 2.3	18.3 ± 7.5
2	ATGGAAGGTCCAGC<u>G</u>TTCTC	2.9 ± 0.2	13.6 ± 2.0
2a	..<u>C</u>..CTC..<u>G</u>.........	7.7 ± 0.8	24.2 ± 3.2
2b	..Z..CTC.ZG..Z......	1.6 ± 0.5	2.8 ± 2.2
2c	..Z..CTC..<u>G</u>.........	3.1 ± 0.6	7.3 ± 1.4
2d	..<u>C</u>..CTC..<u>G</u>.....Z..	7.4 ± 1.4	27.7 ± 5.4
2e	...........A......	5.6 ± 2.0	ND
3D	GAGAA<u>CG</u>CTGGACCTTCCAT	4.9 ± 0.5	19.9 ± 3.6
3Da	.......<u>C</u>..........	6.6 ± 1.5	33.9 ± 6.8
3Db	.......<u>C</u>......<u>G</u>..	10.1 ± 2.8	25.4 ± 0.8
3Dc	...C.A............	1.0 ± 0.1	1.2 ± 0.5
3Dd	.....Z............	1.2 ± 0.2	1.0 ± 0.4
3De	.............Z......	4.4 ± 1.2	18.8 ± 4.4
3Df	......A..........	1.6 ± 0.1	7.7 ± 0.4
3Dg	........CC.G.ACTG..	6.1 ± 1.5	18.6 ± 1.5
3M	TCCATGT<u>CG</u>GTCCTGATGCT	4.1 ± 0.2	23.2 ± 4.9
3Ma	......CT............	0.9 ± 0.1	1.8 ± 0.5
3Mb	......Z............	1.3 ± 0.3	1.5 ± 0.6
3Mc	..........Z........	5.4 ± 1.5	8.5 ± 2.6
3Md	......A..T..........	15.1 ± 10	ND
3Me	................C..A.	3.6 ± 0.2	14.2 ± 5.2
4	TCAACGTT	6.1 ± 1.4	19.2 ± 5.2
4a	....GC..	1.1 ± 0.2	1.5 ± 1.1
4b	...G<u>C</u>GC.	4.5 ± 0.2	9.6 ± 3.4
4c	...T<u>CG</u>A.	2.7 ± 1.0	ND
4d	C.......	3.9 ± 1.4	ND
4e	—.......	1.3 ± 0.2	1.1 ± 0.5
4f	C	1.2 ± 0.2	ND
4g	..TT..AA	1.3 ± 0.2	ND
4h	—.....CT	1.4 ± 0.3	ND

[a] Stimulation indexes are the mean standard deviation compared to wells cultured with no added oligo. ND = not done.

[b] CpG dinucleotides are underlined. Dots indicate identity; dashes indicate deletions. Oligos 3D and 3M have been previously reported (40). **Z** indicates 5-methylcytosine.

[c] Effects of oligos on B cell RNA synthesis (as a measure of cell activation) were assessed essentially as described (40).

[d] Resting B cells were isolated and cultured in 30 μM oligo or 20 μg/ml LPS for 48 h, then used to determine the number of B cells actively secreting IgM by ELIspot assay (42a).

The fact that the two 'control' oligos caused B cell activation similar to that of the two 'antisense' oligos raised the possibility that perhaps all four oligos were stimulating B cells through some non-antisense mechanism involving a sequence motif that was absent in all of the other non-stimulatory control oligos. In comparing these sequences and those of multiple other oligos with base substitutions, we eventually realized that all of the stimulatory oligos contained CpG dinucleotides that were in a different sequence context from the non-stimulatory controls, suggesting the possible existence of a mitogenic CpG motif.

To test the hypothesis that the CpG motif was responsible for the stimulation, we switched either the CpG or other bases in the oligos to eliminate or to increase the number of CpG dinucleotides present. Eliminating the CpG invariably abolished stimulation by oligos (42a) (*Table 2*; compare oligo 1 to 1a, 3D to 3Dc, and 3M to 3Ma). (Note: the second CpG in oligo 1 is too close to the 3′ end to effect stimulation—see below.) Changes in the oligo sequences that did not affect the CpG or the immediate flanking bases did not affect the level of stimulation (compare oligo 1 to 1f, 3D to 3Dg, 3M to 3Me). On the other hand, increasing the number of CpG motifs generally increased the level of B cell stimulation (compare oligo 2 to 2a, or 3D to 3Da and 3Db). In general, B cell stimulation by these oligos was not competed by non-stimulatory phosphodiester oligos, even at a fivefold molar excess. CpG oligos caused no detectable activation of any T cell population.

We synthesized over 300 oligos ranging in length from 5–42 bases that contained methylated, unmethylated, or no CpG dinucleotides in different sequence contexts, and examined their *in vitro* effects on spleen cells (representative sequences are listed in *Table 2*). From these studies we conclude that the bases flanking the CpG dinucleotide play an important role in determining the B cell activation induced by an oligo. The optimal stimulatory motif consists of a CpG flanked by two 5′ purines (preferably a GpA dinucleotide) and two 3′ pyrimidines (preferably a TpC or TpT dinucleotide). Mutations of oligos to bring the CpG motif closer to this ideal, improved stimulation (e.g. compare oligo 2 to 2e, 3M to 3Md, *Table 2*) while mutations that disturbed the motif reduced stimulation (e.g. compare oligo 3D to 3Df, 4 to 4b, 4c, and 4d, *Table 2*).

We also tested the B cell mitogenicity of oligos in which cytosines in CpG motifs or elsewhere were replaced by 5-methylcytosine. Oligos containing methylated CpG motifs caused no mitogenic effect (*Table 2*; oligo 1b, 2b, 2c, 3Dd, and 3Mb). As a control for possible anti-proliferative effects of methylcytosines, oligos in which other cytosines were methylated retained their stimulatory properties (oligo 1c, 2d, 3De, and 3Mc). These data suggest that to avoid this type of non-antisense effect, it may be desirable to use oligos in which cytosines in CpGs are replaced by 5-methylcytosine.

Based on these studies we have determined the formula of the consensus optimal mitogenic CpG motif to be:

$$R_1R_2CGY_1Y_2$$

where R_1 is a purine (mild preference for G), R_2 is a purine (strong preference for A) or thymidine, Y_1 is a pyrimidine (strong preference for T), and Y_2 is a pyrimidine (42a). Oligos with unmethylated CpG dinucleotides in partial matches to this motif give variable stimulation.

CpG oligos can cause stimulation with either a PO or PS backbone, but our studies to date indicate that the constraints of sequence context are more rigid for PS oligos than for PO. For example, we have noted that the sequence of oligo 3D, which is an excellent match of the consensus CpG element, is extremely mitogenic when synthesized as a PS. However, oligo 3M, which has a pyrimidine at position R_2 and a purine at position Y_1, has no mitogenic effect as a PS despite good stimulation as a PO (*Table 2* and data not shown). When oligo 3Md was made as a PS, it caused very strong B cell stimulation.

Addition of a CpG to the 5' end of an oligo or addition of a CpG in an unfavourable sequence context could actually reduce B cell proliferation induced by a stimulatory CpG oligo. There is insufficient space to show all the oligo sequences that led to this conclusion, but the role of flanking sequences is the reason that the second CpG near the 3' end of oligo 1 can be mutated without loss of stimulation (*Table 2*, oligo 1c, 1f).

The optimal *minimal* sequence that we could identify which would stimulate B cells was TCAACGTT, which contained a six base 'palindrome,' AACGTT (*Table 2*; several other 8-mer oligos containing palindromes with the formula shown above were also stimulatory). In contrast to the longer oligos (15-mers or longer), short oligos required a perfect palindrome to be stimulatory. In longer oligos, the palindrome could be disrupted (as long as the general formula above was maintained) with no substantial change in stimulation.

2.8.2 Biological effects of CpG oligos

To understand the mechanism whereby CpG oligos induce B cell activation and Ig secretion, the time course of this effect was studied. When the cells were pulsed at the same time as oligo addition and harvested just 4 h later, there was already a twofold increase in [^{3}H]uridine incorporation (*Figure 1*). Stimulation peaked at 48 h and then decreased (*Figure 1*). After 24 h, there was no more detectable intact oligo (43, data not shown), perhaps accounting for the subsequent fall in stimulation.

Cell cycle analysis was used to obtain additional information regarding the proportion of B cells activated by CpG oligos. Optimal concentrations of the B cell mitogen lipopolysaccharide (LPS) induced about 80% of B cells to enter the cell cycle. Remarkably, CpG oligos induced cycling in > 95% of B cells (*Figure 2*) (42a), indicating that essentially all B cells become activated.

Certain B cell lines such as WEHI-231 are induced to undergo growth

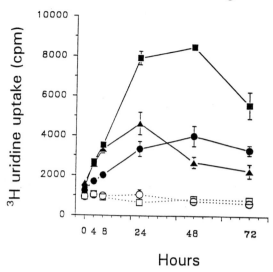

Figure 1. The kinetics of oligo stimulated [³H]uridine uptake in DBA/2 B cells are shown. Cells were cultured with either no additive (open circle), or the following oligos from *Table 2*: oligo 1 (solid square, CpG); 4 (solid triangle, CpG); 3D (solid circle, CpG), or 4a (open box, non-CpG) at 20 μM. Time on the *x* axis is the number of hours cells were cultured with the oligo prior to pulsing with 1 μCi [³H]uridine. Cells were pulsed for 4 h and then harvested.

arrest and/or apoptosis in response to cross-linking of their antigen receptor by anti-IgM (44). WEHI-231 cells are rescued from growth arrest by certain stimuli such as LPS and by the CD40 ligand (45). Like LPS, CpG oligos rescued WEHI-231 cells from growth arrest (45a).

Investigations on mycobacterial DNA sequences have demonstrated that PO oligos which contain certain palindrome sequences can activate natural killer cells (46, 47). The sequences that had this effect all had a CpG motif conforming to the rules outlined above, suggesting that the mechanism is the same. We have also found that CpG oligos induce natural killer activity in both human and mouse cells, and are more potent than interferon γ (47a). This activation does not require a palindrome, as long as the oligo is close to the consensus motif.

To determine whether CpG oligos can cause *in vivo* immune stimulation, DBA/2 mice were injected with several PS CpG or non-CpG oligos IP at a dose of 33 mg/kg (approximately 500 μg/mouse). Pharmacokinetic studies in mice indicate that this dose of PS oligos gives levels of approximately 10 μg/g in spleen tissue (within the effective concentration range determined from our *in vitro* studies) for longer than 24 h (48). Spleen cells from the mice were examined 24 h after oligo injection for expression of B cell surface activation markers and for their spontaneous proliferation using [³H]thymidine. Expres-

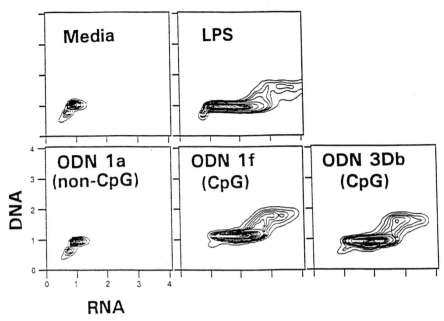

Figure 2. 2×10^6 B cells were cultured for 48 h in 2 ml tissue culture medium alone, or with 30 μg/ml lipopolysaccharide (LPS), or with the indicated phosphorothioate modified oligo at 1 μM. Cell cycle analysis was performed as described (29). The y axis gives staining with propidium iodide (DNA) and the x axis with acridine orange (RNA).

sion of activation markers was significantly increased in B cells from mice injected with CpG oligos, but not from mice injected with saline or non-CpG oligos (42a) (*Figure 3*). Spontaneous [³H]thymidine incorporation was increased by two- to sixfold in spleen cells from mice injected with stimulatory oligos compared to saline or non-CpG oligo injected mice.

There are several published reports (including three authored by one or both of the authors of this chapter) in which oligos were found to induce lymphocyte activation or protect lymphocytes against apoptosis or growth inhibition (40–42, 46, 47, 49–58). In all of these reports, the effects seen can be explained by the existence of a CpG motif in the positive oligos but not in the control oligos. Thus, the effects seen in some reports of 'antisense' oligos may not be due to an antisense mechanism of action.

It is not yet clear whether CpG oligos also have effects on other cell types. We believe that the most important lesson from these studies is that both PO and PS oligos can have potent non-antisense effects that would not necessarily be detected by a few control oligos. It seems almost certain that in coming years, other potent sequence motifs will be identified in other systems. The greater the experimenters index of suspicion, and the more types of controls performed (see below), the less the risk of misinterpreting experimental results.

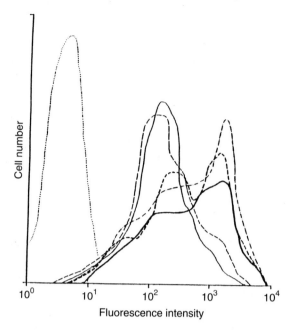

Figure 3. Mice were injected IP with 0.25 ml of sterile PBS or 500 μg of the indicated phosphorothioate oligo dissolved in PBS to examine *in vivo* immune activation by a CpG oligo. 24 h later, spleen cells were harvested, washed, and stained for three colour flow cytometry using phycoerythrin-conjugated 6B2 to gate on B cells in conjunction with biotin-conjugated anti-Ly-6A/E (Pharmingen, San Diego, CA) (a marker for B cell activation—other activation markers such as class II major histocompatibility complex protein expression showed similar increases). The dotted line is staining of isotype matched control ab; light solid line, PBS injected mouse; long dashed line, control oligo 1a (non-CpG) injected mouse; heavy solid line, oligo 1 (CpG); dot-dash line, oligo 3Db (CpG); short dashed line, oligo 2a. Two mice were studied for each condition and analysed individually. The experiment is one of three performed with similar results.

2.8.3 Mechanism of action of the CpG motif

Unlike antigens that trigger B cells through their surface immunoglobulin receptor, CpG oligos did not induce any detectable Ca^{2+} flux, changes in protein tyrosine phosphorylation, or phosphatidylinositol-3-kinase activation. To detect possible cell membrane proteins with binding specificity for CpG oligos, we performed flow cytometry measuring the surface binding, uptake, and egress of FITC oligos with or without a CpG motif. These studies showed no difference between the oligos suggesting that specificity does not lie at the level of cell binding or uptake. CpG oligos appeared to require cell uptake for their activity, since oligos covalently linked to a solid Teflon support were non-stimulatory, as were biotinylated oligos linked to either avidin beads or avidin-coated Petri dishes (data not shown). CpG oligos conjugated to either

FITC or biotin retained full mitogenic properties, indicating no apparent steric hindrance.

The optimal CpG motif, GACGTT/C, matches the DNA binding site for the cyclic AMP response element binding protein and related transcription factors of the CREB/ATF family. 11 members of this family of bZip proteins have been cloned to date, some of which appear to activate, and some inhibit, transcription. Since CREB/ATF proteins appear to regulate transcription of many immune (and other) genes, they are attractive candidates for mediators of CpG oligo effects. Further studies will be required to determine whether immune activation by CpG oligos can be explained in whole or part through effects on CREB/ATF proteins.

3. Immune stimulation by nucleic acids

Aside from the CpG motif, other nucleic acid structures have also been reported to have immune effects, the molecular mechanisms of which are obscure.

3.1 Polynucleotides

Several polynucleotides have been extensively evaluated as biological response modifiers. Perhaps the best example is poly r(I:C) which is a potent inducer of interferon (IFN) production, as well as a macrophage activator, and inducer of natural killer activity (59–62). It appears that this murine natural killer activation primarily is due to induction of IFN-β secretion (63). This activation is specific for the ribose sugar since deoxyribose was ineffective. Its potent *in vitro* anti-tumour activity led to several clinical trials using poly r(I:C) complexed with poly-L-lysine and carboxymethyl cellulose (to reduce degradation by RNase). Unfortunately, toxic side-effects have thus far prevented poly r(I:C) from becoming a useful therapeutic agent.

Several observations suggest that certain DNA structures may also have the potential to activate lymphocytes. For example, Bell *et al.* (64), reported that nucleosomal protein–DNA complexes (but not naked DNA) in spleen cell supernatants caused B cell proliferation and immunoglobulin secretion. In other cases, naked DNA has also been reported to have immune effects. Tokunaga *et al.* (65) have reported that dG•dC induces IFN-γ and natural killer activity, and Messina *et al.* (66) have reported that 260–800 bp fragments of poly (dG)•(dC) and poly (dG•dC) are mitogenic for B cells.

3.2 Nucleotides

Guanine ribonucleosides substituted at the C8 position with either a bromine or a thiol group are B cell mitogens and promote B cell differentiation and immunoglobulin secretion (67, 68). 8-mercaptoguanosine and 8-bromoguanosine also can substitute for the cytokine requirement for the generation of cytotoxic T lymphocytes (67), augment murine natural killer activity (69), and synergize with IL-2 in inducing murine lymphokine activated killer cells (70). At least some of these immune augmenting activities appear to be due to the induction of IFN (70). Macrophage cytotoxicity and cytokine secretion is also increased.

The mechanism of action of these compounds is not yet clear. Stimulation appears to require cell uptake, and is associated with the presence of high affinity and low affinity classes of binding sites (71). By preparing numerous structural analogues, Goodman and Goodman have demonstrated that one region of this ribonucleoside appears to bind to a protein mediating mito-genicity, while a second region of the molecule appears to interact with a protein that mediates adjuvanticity (72).

Recently, a 5′ triphosphorylated thymidine produced by a mycobacterium was found to be mitogenic for a subset of human T cells (73). The mechanism through which this works has not yet been elucidated.

4. Problems in the interpretation of data derived from *in vitro* and *in vivo* use of oligodeoxynucleotides

This heading takes its name from an editorial published in 1994 (74). Unfortunately, it has been a tendency of some researchers to view oligos as biologicals, and to use them as such in experiments. However, it is perhaps more accurate and useful to view oligos instead as chemicals. Oligos are charged polyanions; as noted above, they can and do interact with many proteins which serve a variety of biochemical roles. This implies that the biological effects observed after treatment of cells with 'antisense' PS oligos, and to a lesser extent, PO oligos, actually result from a complex mix of sequence-specific (antisense or non-antisense) plus non-sequence-specific effects. For this reason, it may frequently be difficult, if not impossible, to make definitive claims about the mechanism in a given system.

Considering all of their non-antisense effects, it is probably wise to avoid the use of oligos containing a G-quartet for antisense experiments. If a G-quartet containing oligo is used, it is critical that the control oligos retain the G-quartet, in the same relative position as in the antisense construct. If a multi-base mismatched oligo is used as a control, the mismatches should not be placed in the middle of the G-quartet. Likewise, if an 'antisense' oligo contains a CpG dinucleotide, it may be prudent to test a control oligo in which that cytosine is replaced by a 5-methylcytosine to verify that the supposed 'antisense' activity does not disappear. In our experience, PS oligos with length > 15-mer and concentration > 10 μM will frequently produce non-antisense effects in tissue culture, especially if they contain G-quartets or CpG dinucleotides. Given all of these observations, it becomes difficult to understand how antisense effects can be demonstrated using an antisense PS oligo targeted to c-myb (which contains a G-quartet) in an assay which measures arterial intimal smooth muscle proliferation after balloon catheter injury (75). Indeed, Burgess *et al.* (76), have found that the anti-proliferative activity of both c-myb and c-myc oligos containing G-quartets on smooth muscle cells is caused by a non-antisense mechanism, perhaps related to those described above. Many of the observations made on PS oligos apply to phosphorodithioate (PS2) oligos as well (77).

In order to reduce the risk of encountering such non-antisense effects, we have suggested several general rules for the use of control oligos (74):

1. The experimenter must demonstrate a decrease in target protein. There are some exceptions to this rule; in primary cells, oligo uptake may be heterogeneous, and in a subset of cells, quite minimal (78). Thus, target protein levels may change in a subset of cells, but overall the change may be minimal. A control protein with a half-life similar to the target protein should also be examined. For example, it is not reasonable to claim specificity if myb protein levels, as compared to β-actin levels, are decreased, as the two have very different half-lives. Furthermore, measurement of mRNA levels by RT-PCR is less desirable than measurement of protein levels subsequent to oligo treatment.

2. Because of the large number of non-sequence-specific effects generated by oligos, a large number of control oligos must be used. Unfortunately, at this time it is uncertain what that number is. Four general classes of controls have been proposed for 'antisense' oligos (74) including:

(a) Sense control. This is a good control for the oligo backbone and self-complementarity, but does not maintain composition.

(b) Scrambled control. This is a good control for composition, but is unlikely to detect the presence of active motifs, e.g. the G-quartet. A control which is composed of a mixture of oligos containing each base at each position is a good control for the backbone, but is likewise irrelevant for the control of active structural motifs. A 'reverse' oligo, i.e. the identical sequence with the opposite polarity, has been suggested by K. Schlingensiepen (Max Planck Institute, Göttingen), and might be a good control in some circumstances.

(c) Mismatched control. This control can demonstrate target hybridization specificity, but the results may be easily open to misinterpretation depending on where the mismatches are made. For example, if the mismatch happened to disrupt a G-quartet or CpG motif, the results may be thought wrongly to be due to an antisense mechanism of action.

(d) Mismatched target control (i.e. using cells with a mutant gene, deleted gene, etc.). The absence of non-sequence-specific effects as shown by this control might be encouraging, but the control cells may be different in unknown ways from the target cells in critical parameters such as internalization and compartmentation.

3. Finally, it would not be inappropriate to determine the extent to which an oligo targeted to a specific mRNA is capable of binding to the very protein produced as a translation product of that mRNA. This is especially true if that protein was a secreted autocrine or paracrine effector protein, and the oligo is present in the culture media. Clearly, on the basis of the above discussions, should this interaction occur, unpredictable non-specific effects will also likely occur.

As the knowledge base on oligos continues to increase and their non-antisense properties become better understood, our ability to make maximal use of these fascinating substances will also dramatically increase. One such use may, surprisingly, develop into a treatment for glaucoma. We (F. Supek, J. Tonkinson, M. Wax, and C. A. Stein, manuscript in preparation) have determined that PS oligos can inhibit the function of the proton vacuolar ATPase (H^+–ATPase). This enzyme and related molecules are the proteins primarily responsible for acidification of the endosome. However, in the ciliary body of the eye, an active form appears to be found in the cell membrane, where it is associated with the production of aqueous humour. When the corneas of experimental rabbits are treated with an aqueous solution of SdC28, intra-ocular pressure is markedly decreased, and may remain so for hours. The PS oligo appears to be non-toxic to the cornea, unlike pentosan polysulfate, which does not inhibit the H^+–ATPase, and does not lower intraocular pressure. An extensive confirmatory trial of these observations is currently in progress.

The possible therapeutic application of non-antisense effects of PS oligos are also pointed out by a close examination of the oligo ISIS 2922, an 'antisense' oligo with remarkably potent antiviral effects against cytomegalovirus (CMV). ISIS 2922 is reported to have encouraging *in vivo* efficacy in early clinical trials against CMV retinitis. The sequence of this oligo is quite interesting because it has two atypical CpG motifs: there is a GCG at both the extreme 5′ and 3′ ends (79). Although not fully recognized at the time, published studies on ISIS 2922 demonstrate that its antiviral effect can not be due to 'antisense' since internal mismatch control oligos lost very little antiviral activity despite a severe drop in the T_m. On the other hand, deletion of a single base from one of the CpG motifs caused a 40% drop in antiviral efficacy, and deletion of a single base from both of the CpG motifs abolished antiviral effect despite little change in the T_m for hybridization to the supposed mRNA target (79). It seems likely that it will be possible to develop even better antiviral oligos once the essential role of CpG motifs in this effect is recognized.

Finally, it should be emphasized that the difficulties encountered with PS oligos do not rule out their use as antisense agents. Certainly, if the PS oligo length is sufficiently short, the concentration sufficiently low, the sequence appropriately chosen, and the proper control oligos synthesized and tested, sequence-specific inhibition of genetic expression can probably be demonstrated in certain cell lines. On the other hand, if investigators do not consider the fundamental properties of these fascinating compounds when they plan and interpret their experiments, the literature and those reading it will only become increasingly confused.

Acknowledgements

C. A. S. is the Irving Associate Professor of Medicine, Columbia University. His research is partially supported by R29 60639. A. M. K. is a Pfizer Scholar whose work has been supported by grants from the Lupus Foundation of America, The Arthritis Foundation, The Veterans Administration, and NIH R29-AR42556–01.

References

1. Praseuth, D., Guieysse, A.-L., Itkes, A. V., and Helene, C. (1993). *Antisense Res. Dev.*, **3**, 33.
2. Stein, C. A. and Cheng, Y.-C. (1993). *Science*, **261**, 1004.
3. Steck, W. J., Zon, G., Egan, W., and Stec, B. (1984). *J. Am. Chem. Soc.*, **106**, 6077.
4. Eckstein, F. (1985). *Annu. Rev. Biochem.*, **54**, 367.
5. Stein, C. A., Subasinghe, C., Shinozuka, K., and Cohen, J. (1988). *Nucleic Acids Res.*, **16**, 3209.
6. Cazenave, C., Stein, C. A., Loreau, N., Thuong, N. T., Neckers, L., Subasinghe, C., *et al.* (1989). *Nucleic Acids Res.*, **17**, 4255.
7. Gao., W.-Y., *et al.* (1992). *Mol. Pharmacol.*, **41**, 223.
8. Matsukura, M., Shinozuka, K., Zon, G., Mitsuya, H., Reitz, M., Cohen, J. S., *et al.* (1987). *Proc. Natl. Acad. Sci. USA*, **84**, 7706.
9. Matsukura, M., Zon, G., Shinozuka, K., Robert-Guroff, M., Stein, C. A., Mitsuya, H., *et al.* (1989). *Proc. Natl. Acad. Sci. USA*, **86**, 4244.
10. Stein, C. A., Tonkinson, J., and Yakubov, L. (1991). *Pharmacol. Ther.*, **52**, 365.
11. Stein, C. A., Matsukura, M., Subasinghe, C., Cohen, J., and Broder, S. (1989). *AIDS Res. Hum. Retroviruses*, **5**, 639.
12. Mori, K., Boiziau, C., Cazenave, C., Matsukura, M., Subasinghe, C., Cohen, J., *et al.* (1989). *Nucleic Acids Res.*, **17**, 8207.
13. Zelphati, O., Imbach, J.-L., Signoret, N., Zon, G., Rayner, B., and Leserman, L. (1994). *Nucleic Acids Res.*, **22**, 4307.
14. Agrawal, S., Goodchild, J., Civeira, M. P., Thornton, A. H., Sarin, P. S., and Zemecnik, P. C. (1988). *Proc. Natl. Acad. Sci. USA*, **85**, 7079.
15. Balotta, C., Lusso, P., Crowley, R., Gallo, R. C., and Franchini, G. (1993). *J. Virol.*, 4409.
16. Wyatt, J. R., Vickers, T. A., Roberson, J. L., Buckheit, R. W. Jr., Klimkait, T., DeBaets, E., *et al.* (1994). *Proc. Natl. Acad. Sci. USA*, **91**, 1356.
17. Ojwang, J., Okleberry, K. M., Marshall, H. B., Vu, H. M., Huffman, J. H., and Rando, R. F. (1994). *Antiviral Res.*, **25**, 27.
18. Letsinger, R. L., Zhang, G., Sun, D. K., Ikeuchi, T., and Sarin, P. S. (1989). *Proc. Natl. Acad. Sci. USA*, **86**, 6553.

19. Stein, C. A., Pal, R., Hoke, G., Mumbauer, S., Kinstler, O., and Letsinger, R. (1991). *Biochemistry*, **30**, 2439.

20. Majumdar, C., Stein, C. A., Cohen, J. S., Broder, S., and Wilson, S. (1989). *Biochemistry*, **28**, 1340.

21. Maury, G., El Alaoui, A., Morvan, F., Müller, B., Imbach, J.-L., and Goody, R. S. (1992). *Biochem. Biophys. Res. Commun.*, **186**, 1249.

22. Y.-C. Cheng. Personal communication.

23. Carteau, S., Mouscadet, J. F., Goulaouic, H., Subra, F., and Auclair, C. (1993). *Arch. Biochem. Biophys.*, **305**, 606.

24. Yakubov, L., Khaled, Z., Zhang, L.-M., Truneh, A., Vlassov, V., and Stein, C. A. (1993). *J. Biol. Chem.*, **268**, 18818.

25. Clayton, L., Hussey, R., Steinbrich, R., Ramachandran, H., Husain, Y., and Reinherz, E. (1988). *Nature*, **335**, 363.

26. Arthos, J., Deen, K. C., Chaikin, M., Fornwald, J., Sathe, G., Sattentau, Q., *et al.* (1989). *Cell*, **57**, 469.

27. Parish, C., Low, L., Warren, H., and Cunningham, A. (1990). *J. Immunol.*, **145**, 1188.

28. Stein, C. A., Cleary, A., Yakubov, L., and Lederman, S. (1993). *Antisense Res. Dev.*, **3**, 19.

29. Darzynkiewicz, Z., Traganos, F., Sharpless, T., and Melamed, M. R. (1976). *Proc. Natl. Acad. Sci. USA*, **78**, 2881.

30. Wellstein, A., Zugmaier, G., Califano, J., Dern, F., Paik, S., and Lippman, M. (1991). *J. Natl. Cancer Inst.*, **83**, 716.

31. Eisenberger, M., Reyno, L., Jodrell, D., Sinibaldi, V., Tkaczuk, T., Sridhara, R., *et al.* (1993). *J. Natl. Cancer Inst.*, **85**, 611.

32. Guvakova, M., Yakubov, L., Vlodavsky, I., Tonkinson, J., and Stein, C. A. (1995). *J. Biol. Chem.*, **270**, 2620.

32a. Khaled, Z., Benimetskaya, L., Kahn, T., Zeltzer, R., Sharma, H., Narayanan, R., and Stein, C. A. (1996). *Nucleic Acids Res.*, **24**, 737.

33. Watson, P. H., Pon, R. T., and Shiu, R. P. (1992). *Exp. Cell Res.*, **202**, 391.

34. Yaswen, P., Stampfer, M., Ghosh, K., and Cohen, J. (1993). *Antisense Res Dev.*, **3**, 67.

35. Perez, J. R., Li, Y., Stein, C. A., Majumder, S., Van Oorschot, A., and Narayanan, R. (1994). *Proc. Natl. Acad. Sci. USA*, **91**, 5957.

36. Brown, D. A., Kang, S.-H., Gryaznov, S. M., DeDionisio, L., Heidenreich, O., Sullivan, S., *et al.* (1994). *J. Biol. Chem.*, **269**, 26801.

36a. Bielinska, A., Shivdasani, R. A., Zhang, L., and Nabel, G. J. (1990). *Science*, **250**, 997.

37. Bergan, R., Connel, Y., Fahmy, B., Kyle, E., and Neckers, L. (1994). *Nucleic Acids Res.*, **22**, 2154.

37a. Krieg, A. M., Matson, S., Cheng, K., Fisher, E., Koretzky, G. A., and Koland, J. G. (1997). *Antisense Res. Dev.*, **7**, 115.

37b. Teasdale, R., Matson, S. J., Fisher, E., and Krieg, A. M. (1994). *Antisense Res. Dev.*, **4**, 295.

38. Ramanathan, M., Lantz, M., MacGregor, R. D., Garovoy, M. R., and Hunt, C. A. (1994). *J. Biol. Chem.*, **269**, 24564.

39. Pisetsky, D. S. and Halpern, M. D. (1994). *Arthritis Rheum.*, **37**, S308.

40. Krieg, A. M., Gause, W. C., Gourley, M. F., and Steinberg, A. D. (1989). *J. Immunol.*, **143**, 2448.

41. Krieg, A., Tonkinson, J., Matson, S., Zhao, Q., Saxon, M., Zhang, L.-M., *et al.* (1993). *Proc. Natl. Acad. Sci. USA*, **90**, 1048.
42. Mojcik, C., Gourley, M. F., Klinman, D. M., Krieg, A. M., Gmelig-Meyling, F., and Steinberg, A. D. (1993). *Clin. Immunol. Immunopathol.*, **67**, 130.
42a. Krieg, A. M., Yi, A., Matson, S., Waldschmidt, T. J., Bishop, G. A., Teasdale, R., *et al.* (1995). *Nature*, **374**, 546.
43. Matson, S. and Krieg, A. M. (1992). *Antisense Res. Dev.*, **2**, 325.
44. Jakway, J. P., Unsinger, W. R., Gold, M. R., Mishell, R. I., and DeFranco, A. L. (1986). *J. Immunol.*, **137**, 2225.
45a. Yi, A. -K., Hornbeck, P., Lafrenz, D. E., and Krieg, A. M. (1996). *J. Immunol.*, **157**, 4918.
45. Tsubata, T., Wu, J., and Honjo, T. (1993). *Nature*, **364**, 645.
46. Yamamoto, S., Yamamoto, T., Kataoka, T., Kuramoto, E., Yano, O., and Tokunaga, T. (1992). *J. Immunol.*, **148**, 4072.
47. Kuramoto, E., Yano, O., Kimura, Y., Baba, M., Makino, T., Yamamoto, S., *et al.* (1992). *Jpn. J. Cancer Res.*, **83**, 1128.
47a. Ballas, Z. K., Rasmussen, W. L., and Krieg, A. M. (1996). *J. Immunol.*, **157**, 1840.
48. Agrawal, S., Temsaman, J., and Tang, Y. (1991). *Proc. Natl. Acad. Sci. USA*, **88**, 7595.
49. Hatzfeld, J., Li, M.-L., Brown, E. L., Sookdeo, H., Levesque, J.-P., O'Toole, T., *et al.* (1991). *J. Exp. Med.*, **174**, 925.
50. Shi, Y., Glynn, J. M., Guilbert, L. J., Cotter, T. G., Bissonnette, R. P., and Green, D. R. (1992). *Science*, **257**, 212.
51. Tanaka, T., Chu, C. C., and Paul, W. E. (1992). *J. Exp. Med.*, **175**, 597.
52. Branda, R. F., Moore, A. L., Mathews, L., McCormack, J. J., and Zon, G. (1993). *Biochem. Pharmacol.*, **45**, 2037.
53. McIntyre, K. W., Lombard-Gillooly, K., Perez, J. R., Kunsch, C., Sarmiento, U. M., Larigan, J. D., *et al.* (1993). *Antisense Res. Dev.*, **3**, 309.
54. Pisetsky, D. S. and Reich, C. F. (1993). *Life Sci.*, **54**, 101.
55. Fischer, G., Kent, S. C., Joseph, L., Green, R., and Scott, D. W. (1994). *J. Exp. Med.*, **179**, 221.
56. Scheuermann, R. H., Racila, E., Tucker, T., Yefenof, E., Street, N. E., Vitetta, E. S., *et al.* (1994). *Proc. Natl. Acad. Sci. USA*, **91**, 4048.
57. Sarmiento, U. M., Perez, J. R., Becker, J. M., and Narayanan, R. (1994). *Antisense Res. Dev.*, **4**, 99.
58. Li, M.-L., Cardoso, A. A., Sansilvestri, P., Hatzfeld, A., Brown, E. L., Sookdeo, H., *et al.* (1994). *Leukemia*, **8**, 441.
59. Talmadge, J. E., Adams, J., Phillips, H., Collins, M., Lenz, B., Schneider, M., *et al.* (1985). *Cancer Res.*, **45**, 1058.
60. Wiltrout, R. H., Salup, R. R., Twilley, T. A., and Talmadge, J. E. (1985). *J. Biol. Resp. Mod.*, **4**, 512.
61. Krown, S. E. (1986). *Semin. Oncol.*, **13**, 207.
62. Ewel, C. H., Urba, S. J., Kopp, W. C., Smith II, J. W., Steis, R. G., Rossio, J. L., *et al.* (1992). *Cancer Res.*, **52**, 3005.
63. Ishikawa, R. and Biron, C. A. (1993). *J. Immunol.*, **150**, 3713.
64. Bell, D. A., Morrison, B., and VandenBygaart, P. (1990). *J. Clin. Invest.*, **85**, 1487.
65. Tokunaga, S., Yamamoto, S., and Namba, K. (1988). *Jpn. J. Cancer Res.*, **79**, 682.
66. Messina, J. P., Gilkeson, G. S., and Pisetsky, D. S. (1993). *Cell. Immunol.*, **147**, 148.
67. Feldbush, T. L. and Ballas, Z. K. (1985). *J. Immunol.*, **134**, 3204.

68. Goodman, M. C. (1991). *Immunopharmacology*, **21**, 51.
69. Koo, G. C., Jewell, M. E., Manyak, C. L., Sigal, N. H., and Wicker, L. S. (1988). *J. Immunol.*, **140**, 3249.
70. Thompson, R. A. and Ballas, Z. K. (1990). *J. Immunol.*, **145**, 3524.
71. Goodman, M. G. (1990). *J. Biol. Chem.*, **265**, 22467.
72. Goodman, M. G. and Goodman, J. H. (1994). *J. Immunol.*, **153**, 4081.
73. Constant, P., Davodeau, F., Peyrat, M. A., Poquet, Y., Puzo, G., Bonneville, M., *et al.* (1994). *Science*, **264**, 267.
74. Stein, C. and Krieg, A. M. (1994). *Antisense Res. Dev.*, **4**, 309.
75. Simons, M., Edelman, E., DeKeyser, J., Langer, R., and Rosenberg, R. (1992). *Nature*, **359**, 67.
76. T. Burgess. Personal communication.
77. Tonkinson, J. L., Guvakova, M., Khaled, Z., Lee, J., Yakubov, L., Marshall, W. S., *et al.* (1994). *Antisense Res Dev.*, **4**, 269.
78. Zhao, Q., Waldschmidt, T., Fisher, E., Herrera, C. J., and Krieg, A. M. (1994). *Blood*, **84**, 3660.
79. Azad, R. F., Driver, V. B., Tanaka, K., Crooke, R. M., and Anderson, K. P. (1993). *Anti. Agents Chemo.*, **37**, 1945.

12

Antisense rescue

JEFFREY T. HOLT

1. Introduction

Genetic manipulation of cells has provided a powerful approach to study gene function and the regulation of gene expression. Although genetic manipulation includes overexpression and 'underexpression' strategies, methods which inhibit gene function are particularly important because they probe the usual function of a gene product within an appropriately expressing cell. Antibody injection (1), antisense methods (2–6), and gene targeting (7) can be used for gene inhibition; and each approach has its strengths and weaknesses.

Antisense strategies can provide convincing and informative data when appropriate controls are included and experimental artefacts are avoided. Transfection of antisense plasmids generally requires the isolation of stable transformants which can produce artefactual results as a consequence of selection bias or inducer effects. Selection bias occurs when constitutive expression of an antisense gene inhibits viability. An example from our experience were studies of a constitutive anti-*fos* RNA regulated by the Rous Sarcoma virus LTR. Transfection with this construct resulted in a marked decrease in the numbers of stable transformants, while transfection of an equivalent construct, but carrying a defective promoter, produced the expected number of transformants. Because the few transformants which survive selection bias represent survivors which obscure the normal antisense phenotype, we prefer the use of inducible promoters to regulate antisense RNA expression so the antisense is not active during clonal selection. Although the use of inducible promoters diminishes the worrying problems of selection bias, one must control for unexpected effects of inducers on cell growth or other phenotypic parameters.

In order to verify that gene silencing is antisense-mediated, and not due to non-specific artefacts, we have developed an antisense rescue method. Here, the target mRNA expressed from the endogenous alleles is inactivated by expression of an antisense gene. Concomitantly, expression of this mRNA is replaced by an antisense-resistant sense gene, cotransfected into the cell. Antisense-resistant genes are constructed by deleting the sequence that serves

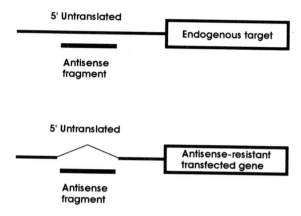

Figure 1. Illustration of cotransfection strategy for antisense inhibition of an endogenous target gene and complementation with antisense-resistant deletion mutant. The antisense sequence, shown as a short black bar, is directed against the 5′ untranslated region of the endogenous target gene, which is deleted from the antisense-resistant deletion mutant.

as the target for the antisense RNA, for example a portion of the 5′ untranslated region, as illustrated in *Figure 1*.

This method is a quick alternative to 'classical gene targeting'. It does not require homologous recombination to replace/disrupt both endogenous alleles. Thus antisense rescue allows a newly introduced sense gene, carrying a defined mutation, to substitute for the expression of endogenous alleles. Using inducible promoters for both the antisense and new sense genes even potentially lethal disruptions can be studied. The method can be restricted to specific cell or tissue types etc. by using the appropriate promoters. For example, while knocking-out expression of a target gene in all cell types with antisense, the sense gene may be re-expressed in a specified cell type using a specific promoter. The most valuable application of this method may be in gene therapy, where a 'gene transplant' is used to cure a dominant genetic disorder. Expression of the mutant gene could be suppressed by antisense and replaced by expression of an antisense-resistant wild-type gene.

The antisense rescue method described here provides a novel way to dissect protein function. We have employed this method to define both specific and generalized functions of the *c-fos* gene (8).

2. The antisense rescue approach

Below protocols for the application of the rescue approach are provided for cultured mammalian cells. With appropriate modifications, this method can also be used for other experimental systems.

In the system described below, a hormone-inducible promoter, mouse mammary tumour virus (MMTV) is used to drive expression of both the anti-sense and the substitute sense gene. These are induced by the steroid dexamethasone. The antisense gene vector carries the neomycin-resistance marker, *nptII*, to select for G418-resistant transfected cells. Cotransfection is readily obtained by using a ninefold excess of the vector carrying the antisense-resistant sense gene. Thus a second selectable marker is not required.

In our studies of the human *c-fos* gene (8), we constructed the antisense-resistant gene by deleting ~ 100 bp of the 5′ proximal untranslated region, thus eliminating the target sequence for the 84 bp antisense *c-fos* construct (2). Murine fibroblasts expressing both the human antisense and the anti-sense-resistant genes were obtained by the cotransfection strategy described in *Protocol 1*. Unless otherwise specified, standard procedures for tissue culture should be followed, for example as described in '*Animal cell culture: a practical approach*' (second edition).

Protocol 1.   Cotransfection of mouse fibroblasts with inducible antisense vector and antisense-resistant target genes

Equipment and reagents

- 2 × Hepes: 1.6 g NaCl, 0.074 g KCl, 0.027g $Na_2HPO_4.2H_2O$, 0.2 g dextrose, 1 g Hepes in 100 ml of distilled water, pH to 7.05 with 0.5 N NaOH, and filter
- Trypsin/EDTA (Sigma)
- Cell culture media: Dulbecco's modified Eagle media (DMEM) with 10% calf serum (Hyclone)
- Tissue culture hood

- Phosphate-buffered saline (PBS) pH 7.4 without calcium or magnesium
- Glycerol solution: 45 ml glycerol, 11.5 ml distilled H_2O, 15 ml 2 × Hepes
- G418: 100 mg/ml in distilled water (1000 × stock)
- CO_2 incubator
- pH meter

A. *Preparation of cells for transfection*

1. Grow mouse fibroblast cells in 100 mm tissue culture dishes.

2. Trypsinize cells the evening before, use 400 000 cells per plate. Plating an optimal cell number is essential and can be achieved by either careful cell counts or plating cells at several cell densities and choosing the correct set of plates.

3. Examine plates, perform transfection when cells are ~ 75–80% confluent.

B. *Cotransfection*

1. Mix 18 μg of antisense-resistant expression vector DNA (without selectable marker) and 2 μg of antisense expression vector containing a G418 selectable marker in less than 450 μl.

2. Add water to a volume of 450 μl.

3. Add 50 μl 2.5 M $CaCl_2$.

Protocol 1. *Continued*

4. Add the DNA mixture to 500 µl of 2 × Hepes. It is very important to mix in this order.

5. Allow to sit at room temperature for 30 min.

6. Add the mixture to cell plates and media and incubate at 37°C for 4 h.

7. Aspirate media and add 3 ml of glycerol solution. Quickly aspirate glycerol and wash cells with 5 ml PBS. Caution: more than 5 min exposure to glycerol will damage cells.

8. Add 10 ml fresh media to plates and incubate at 37°C for 12 h.

9. Wash cells two or three times with 5 ml PBS.

10. Add 1 ml trypsin solution, incubate at 37°C until cells detach.

C. *G418 selection*

1. Plate 200 000 cells per 100 mm dish in fresh media.

2. Add 1000 × G418 solution to plates the following morning to a final concentration of 100 µg/ml, i.e. approx. 48 h after transfection.

3. Change media every two days until clones are readily seen at seven to ten days.

The use of elevated serum concentrations (20%) increases the number of clones expressing anti-fos RNA (2, 8), presumably because anti-fos RNA produces severe growth inhibition of BalbC/3T3 cells when cells are grown logarithmically in low serum. The MMTV promoter is somewhat leaky and this low level expression has heightened effects at low serum concentrations.

Once clones have been identified, it is important to determine whether anti-sense inhibition is successful and/or that the antisense-resistant gene is expressing the target gene product. In our system, this is demonstrated following induction of expression from the dexamethasone-inducible promoters driving these genes. To demonstrate a switch in production from the endogenous *c-fos* protein to the one encoded by the introduced *c-fos* gene, stable transfectant cell lines are quiesced in media with 0.5% serum for 48 h, pre-treated for 1 h with either ethanol vehicle (1 µl of 100% ethanol/ml) or 1 µM dexamethasone, and then restimulated with media containing 10% serum for 30 min in the presence of [^{35}S]methionine as described in *Protocol 2*. Immunoprecipitation assays should show an inhibition of the endogenous gene with a corresponding replacement by the antisense-resistant transfected gene. In these experiments we could distinguish between the endogenous and the antisense-resistant genes because the protein sizes are different. Alternatively, one could use species-specific antibodies (e.g. an antibody directed against a human gene product expressed in a mouse cell) or employ an epitope tag.

Protocol 2. Immunprecipitation to demonstrate antisense inhibition and antisense rescue

Equipment and reagents

- Antibody (Ab) buffer: to make 100 ml mix, 2 ml of 1 M Tris pH 7.5, 1 ml of 5 M NaCl, 5 ml of 10% NP-40, 5 ml of 10% deoxycholate, 2.5 ml of 20% SDS, 0.5 ml of aprotinin (Sigma A-6012), 200 ml of 0.5 M EDTA—add 0.04 g iodoacetamide/25 ml of Ab buffer just prior to use
- Staph-A washing buffer: 1% BSA, 0.01% NaN$_3$
- [^{35}S]methionine: 100 Ci/mmol
- 10 × PBS pH 7.4, without calcium and magnesium
- 25% TCA (w/v)

- Radioimmunoprecipitation assay (RIPA) buffer: to make 100 ml, mix 1 ml of 1 M Tris 7.4, 3 ml of 5 M NaCl, 10 ml of 10% NP-40, 10 ml of 10% deoxycholate, 0.5 ml of 20% SDS, and 0.5 ml of aprotinin (Sigma A-6012)
- Protein sample buffer: 80 mM Tris–HCl pH 6.8, 10% glycerol, 5% SDS, 1% bromophenol blue
- ScintiVerse II (Fisher)
- Scintillation counter
- Sonicator
- Vacuum flask with glass filter

A. *Methionine depletion*

1. Wash plates twice with room temperature PBS.

2. Add 2 ml DMEM (minus methionine) to each 100 mm plate.

3. Incubate at 37°C for 30 min.

4. Add [^{35}S]methionine to plates to give 200–350 μCi/ml.

5. Incubate at 37°C for 1 h, rocking plates every 5 min.

6. Aspirate the radioactive supernatant and dispose of it with radioactive liquid waste.

7. Wash the plates with 5 ml cold 1 × PBS.

8. Aspirate the PBS and store plates at –70°C.

B. *Cell lysis and sonication*

1. Keep samples on ice. Add 800 μl of Ab buffer.

2. Scrape cells from the plate surface using a rubber policeman.

3. Transfer to a 15 ml conical tube.

4. Rinse the plate with 200 μl of freshly prepared Ab buffer. Transfer to a tube containing the rest of the sample.

5. Immerse the sonicator just below the sample surface, then sonicate at a low setting to avoid foaming. Small bubbles rising from the bottom of the tube indicate that sonication is complete.

C. *TCA precipitation*

1. Set-up three glass tubes per sample in an ice bucket and then add: 0.5 ml distilled water, 5.0 μl of sample, and 1.0 ml cold 25% TCA.

2. Gently swirl tubes and incubate on ice for a minimum of 30 min.

Protocol 2. *Continued*

3. Set-up vacuum flask, rubber stopper with filter, chimney, and clamp.
4. Pour contents of glass tube over the filter. Refill the glass tube with cold 10% TCA and pour over filter. Repeat twice.
5. Rinse the sides of the chimney with acetone and let filter dry.
6. Place filter into scintillation vial and add 5–10 ml ScintiVerse II and count to determine methionine incorporation.

D. *Pre-clearing samples for immunoprecipitation*
1. Combine lysate and Ab buffer in an Eppendorf tube (1.5 ml total volume).
2. Incubate samples with 5 μl of pre-immune serum for 2 h with gentle rocking at 4°C (e.g. use goat pre-immune serum if the anti-mouse *c-fos* antibody was prepared in goat), supplied by Vector.
3. While samples are pre-clearing, prepare freshly washed Staph-A as follows: mark tube to indicate initial volume of Staph-A, then centrifuge at 4°C for 2 min at 2000 *g* to spin down the Staph-A, and aspirate the supernatant. Add Staph-A washing buffer up to the initial volume and resuspend pellet. Repeat wash a total of five times and incubate Staph-A on ice for at least 30 min.
4. Add 80 μl of freshly washed Staph-A to samples and incubate samples for 30 min with gentle rocking at 4°C.
5. Centrifuge samples in microcentrifuge at 4°C for 10 min.
6. Transfer the pre-cleared supernatant to fresh tubes.

E. *Immunoprecipitation*
1. Add 5 μl of specific antibody to the pre-cleared supernatant and incubate overnight at 4°C with gentle rocking.
2. Add 60 μl of the Staph-A suspension previously prepared.
3. Incubate at 4°C with gentle rocking for 30 min and then pellet in microcentrifuge for 2–3 min.
4. Add 500 μl of cold RIPA buffer to the pellets and vortex to break up pellet completely. (This is very important!)
5. Centrifuge and aspirate supernatant.
6. Repeat wash with RIPA buffer for a total of five times.
7. Carefully remove all supernatant and add 60 μl of freshly prepared sample buffer (1 ml sample buffer plus 40 μl β-mercaptoethanol).
8. Vortex to break up pellet completely. (This is very important!)
9. Heat samples at 95°C for 3 min and spin in a microcentrifuge for 5 min.

10. Transfer the supernatant to a fresh tube, avoiding the pellet.

11. Load the samples on freshly prepared 10% SDS–polyacrylamide gels and subject to electrophoresis at 250 V.

12. Fix gel in 30% methanol, 10% glacial acetic acid for 60–75 min with shaking at room temperature.

13. Pour off fixative and incubate with pure glacial acetic acid for 5 min at room temperature with shaking.

14. Pour off glacial acetic acid and incubate gels in 20% PPO (in glacial acetic acid) for 1–1.5 h.

15. Return PPO to brown bottle (to be reused) and wash gel under a stream of tap-water for at least 30 min.

16. Dry gel down for 60 min at 75°C and expose film.

Protein levels of exogenous and endogenous target genes can also be qualitatively determined by cell immunofluorescence (9). A method for performing immunofluorescence on cultured cells is presented in *Protocol 3*. This also requires the use of an antibody that can distinguish the endogenous gene product from the antisense-resistant gene product.

Protocol 3.   Cell immunofluorescence

Equipment and reagents

- 3% formaldehyde: dilution of Dowex deionized concentrated formaldehyde (37%) in PBS
- 50 mM glycine
- 0.5% Triton X-100
- FITC antibody (Jackson labs)
- Polyvinyl alcohol mix (Sigma)

Method

1. Split cells into 6-well plates.

2. The next day, wash cells twice with 1 ml PBS.

3. Fix cells in 3% formaldehyde in PBS for 20 min at room temperature.

4. Wash cells twice with 1 ml PBS.

5. Quench cells with 50 mM glycine in PBS at room temperature for 10 min.

6. Wash cells twice with 1 ml PBS.

7. Permeabilize cells with 0.5% Triton X-100 in PBS at room temperature for 5 min.

8. Wash cells twice with 1 ml PBS.

9. Block plates at 4°C for 30 min with, e.g. goat 5% serum if the FITC anti-rabbit IgG antibody was raised in goat.

Protocol 3. *Continued*

10. Wash cells five times with 1 ml PBS.

11. Add antibody specifically recognizing the target gene product (e.g. raised in rabbit as above) to the wells, at six different dilutions: (1:100, 1:200, 1:500, 1:700, 1:1000, 1:2000).

12. Incubate at 4°C overnight.

13. Wash immunofluorescence plates five times with 1 ml PBS.

14. Add the FITC antibody, diluted 1:1000 in PBS containing 5% serum as specified in step 9. Addition of this non-specific serum may have to be varied to obtain the optimal signal to background ratio.

15. Incubate at 4°C for 2.5 h.

16. Wash four times with 1 ml PBS.

17. Mount cells with polyvinyl alcohol mix, avoid air bubbles.

18. View under fluorescent microscope. Keep plates in dark when not viewing to preserve immunofluorescent signal as long as possible.

Acknowledgements

This work was supported by NIH grant CA51735.

References

1. Kovary, K. and Bravo, R. (1991). *Mol. Cell. Biol.*, **11**, 4466.
2. Holt, J. T., Gopal, T. V., Moulton, A. D., and Nienhuis, A. W. (1986). *Proc. Natl. Acad. Sci. USA*, **83**, 4794.
3. Kerr, L. D., Holt, J. T., and Matrisian, L. M. (1988). *Science*, **242**, 1424.
4. Kim, S.-J., Denhez, F., Kim, K. Y., Holt, J. T., Sporn, M. B., and Roberts, A. B. (1989). *J. Biol. Chem.*, **264**, 19373.
5. Nishikura, K. and Murray, J. M. (1987). *Mol. Cell. Biol.*, **7**, 639.
6. Robinson-Benion, C., Kamata, N., and Holt, J. T. (1991). *Antisense Res. Dev.*, **1**, 21.
7. Johnson, R. S., Spiegelman, B. M., and Papaioannou, V. (1992). *Cell*, **71**, 577.
8. Holt, J. T. (1993). *Mol. Cell. Biol.*, **13**, 3821.
9. Jotte, R. M., Kamata, N., and Holt, J. T. (1994). *J. Biol. Chem.*, **269**, 16383.

13

Shotgun antisense mutagenesis

TIMOTHY P. SPANN, DEBRA A. BROCK, and
RICHARD H. GOMER

1. Introduction

1.1 Overview of shotgun antisense

The use of antisense RNA and DNA to block gene expression has proven to be a valuable technique for the examination of gene function in a number of different systems. Typically, a known cDNA fragment is directionally cloned into an expression vector so that cells transfected with this construct will synthesize antisense RNA transcripts. The antisense transcript will cause the loss or reduction of the corresponding gene product. The phenotypes of transformed cells are then assayed and possible functions of the gene are inferred. If, instead of transforming cells with one antisense construct, a population of cells is transformed with an antisense cDNA library, a complex pool of transformed cells is produced. This type of protocol (shotgun antisense) will yield many different clones; in each clone, the expression of a different gene product will be blocked by an antisense transcript. Mutants generated by shotgun antisense can be screened in the same manner as mutants generated by any other means. However, when a mutant is selected for further study, PCR can be used to immediately isolate the antisense cDNA, which can then be directly cloned and sequenced. This allows one to examine the sequence of the protein encoded by the repressed gene a few days after identifying a mutant with an interesting phenotype (*Figure 1*). To confirm that the mutant phenotype was caused by the antisense cDNA construct, the PCR product isolated from the mutant of interest is cloned into the antisense vector and used to transform cells. A matching phenotype indicates that the cDNA caused the phenotype. Alternatively, the cDNA can be used in a gene disruption construct, which in turn can be used to make a genomic knock-out of the corresponding gene. We have developed a shotgun antisense procedure for the simple eukaryote *Dictyostelium discoideum* (1).

1.2 The need for large quantities of cDNA

A critical step in making a shotgun antisense library is generating a large amount of representative cDNA. Most mutagenesis applications require the

Shotgun Antisense Mutagenesis

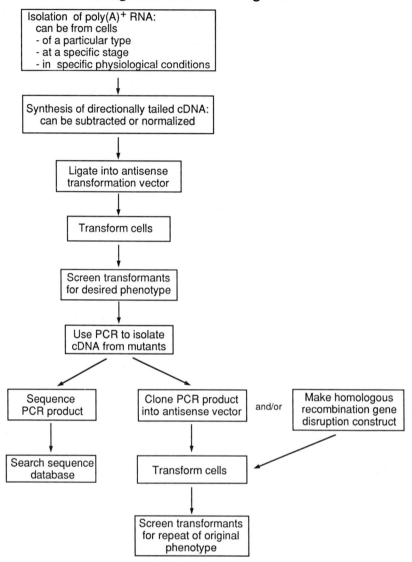

Figure 1. Outline of the shotgun antisense procedure. Shotgun antisense is a way of mutating cells so that the 'mutated' gene's cDNA can be isolated and sequenced immediately after selection of a mutant. This section describes the specialized protocol necessary for making the tailed cDNA pool used for the construction of the shotgun antisense library. Conventional techniques can then be used to transform cells, isolate transformants with desired phenotypes, and isolate the antisense cDNA by PCR from the chosen transformant.

screening of a large number of mutants. In antisense mutagenesis, the number of mutants obtained will be dependent on the transformation efficiency of the vector in which the cDNA is cloned. Unfortunately, the selection of an adequate vector is not obvious, which often requires that the antisense library be cloned into a few different vectors and thus requires a large amount of antisense cDNA.

A typical cDNA synthesis reaction will yield 40–60 ng of cDNA from 2 μg of mRNA. To construct a library, cDNA must next be digested with restriction enzymes and ligated into a vector (resulting in significant loss of cDNA). However, transformation of eukaryotic cells needs microgram amounts of DNA, and thus the cDNA library must be amplified. Phage vectors are normally used to amplify cDNA libraries because these vectors can transform bacteria with very high efficiency allowing a representative library to be made from small quantities of cDNA.

For antisense transformation, plasmids rather than phage must be used. However, it is difficult to make cDNA libraries in plasmids because the relatively low transformation efficiency of plasmids into bacteria causes plasmid libraries to contain many fewer independent clones than phage libraries. In addition, amplification of a library as plasmids in *E. coli* can lead to a reduction in the diversity of the cDNA represented in the library, since plasmids containing some inserts propagate more slowly than others.

In this chapter we describe how to prepare microgram amounts of cDNA by amplification with PCR rather than by amplification by propagation in bacteria, and how to construct antisense libraries from the amplified cDNA by ligation into an equally large amount of antisense vector.

2. Isolation of mRNA

The first step in making cDNA is obtaining mRNA from the cells of interest. We have prepared poly(A)$^+$ RNA in a number of ways from *Dictyostelium* cells, and have found that the Fast Track mRNA Isolation Kit (Invitrogen) reliably produces good yields of high quality poly(A)$^+$ RNA. Similar commercially available products most likely work as well. An advantage of these kits is that they allow the researcher to avoid working with DEPC (diethyl pyrocarbonate) and excessive amounts of phenol:chloroform. As an aside, we found that if techniques involving LiCl are used to prepare the RNA, it is important to use sterile filtered LiCl rather than autoclaved LiCl to avoid interference with the oligo(dT) binding steps.

3. Synthesis of cDNA

To synthesize cDNA of known orientation, we used the SuperScript Plasmid System (Gibco BRL). The kit follows standard cDNA synthesis protocols.

The first cDNA strand is made by a reverse transcriptase primed by an oligo(dT) primer with a single-stranded version of the *Not*I recognition site at the 5' end. The second strand is synthesized by DNA polymerase I in the presence of RNase H and ligase. A *Sal*I adapter is then blunt-end ligated to the cDNA, and the cDNA is digested with *Not*I. The resulting *Not*I–*Sal*I fragments are cloned into a vector which has been digested with *Not*I and *Sal*I.

We modified the kit by using an oligonucleotide (CGGGATCCGGAATCC–CTTTTTTTTTTTTTTTT) with a single-stranded *Bam*HI recognition site 5' of the (dT) run as the primer for the first stand synthesis reaction. We made this change because *Bam*HI is a convenient site in our antisense expression vectors. A second modification was to digest the cDNA with *Bam*HI instead of *Not*I. Use an excess of enzyme. For the estimated 10–20 ng of cDNA, we used 60 U of *Bam*HI in a 50 µl reaction which was incubated for 2 h at 37 °C. Following phenol:chloroform extraction and ethanol precipitation, the cDNA was ready for gel filtration to size select and/or ligate into the vector of choice. As a result of altering the primer used in the cDNA synthesis reaction, our cDNA was incompatible with the commercially prepared vector accompanying the kit, so we prepared our own vectors into which we ligated the cDNA.

4. Vector preparation for directional cloning

The next step is to ligate the cDNA into a plasmid (not necessarily the antisense plasmid) to allow PCR amplification of the cDNA. For the ligation of cDNA into the vector to be successful, the vector must be completely digested by both restriction enzymes so that very little uncut or single cut vector is present. We accomplished this by gel purifying the vector after each restriction digest as suggested in Sambrook *et al.* (2). Digest the vector with the restriction enzyme which is the least efficient (*Sal*I in our case) to ensure that it has cleaved the plasmid, since the cut and the supercoiled plasmid are readily separated by gel purification. We used the GeneClean Kit (Bio101) to purify DNA from the agarose gel. Use a large well so that the gel is not overloaded and the bands can be well resolved, and visualize the DNA with long wavelength UV through a glass plate to reduce nicking. When we tested our *Bam*HI/*Sal*I cut vectors, by self-ligating and bacterial transformation, we found that a vector treated in this manner transformed *E. coli* ~ 0.05% as efficiently as an equal amount of uncut vector.

5. Ligation of cDNA into pBluescript

The last step before PCR amplification is the ligation of the cDNA into a vector. Mix 10 ng of cDNA, 40 ng of pBluescript, 4 µl of 5× BRL ligase buffer, and 1 U of T4 ligase in a 20 µl reaction, and incubate for 3 h at 21 °C. This ligation will serve as the template for the polymerase chain reaction.

6. PCR amplification of the cDNA

The ligation described above provides enough template for hundreds of polymerase chain reactions which would result in milligrams of cDNA. To conveniently clone this cDNA, the primers must be chosen to allow the PCR product to be digested with the appropriate restriction enzymes, *Bam*HI and *Sal*I in the protocol we describe. The distance of the restriction enzyme site from the end of the fragment affects the efficiency of cutting for many enzymes. For *Bam*HI a few nucleotides is sufficient; however, *Sal*I needs to be 20 bp from the end. To amplify the cDNA from pBluescript, we used the T3 primer, CAATTAACCCTCACTAAAGG, and the T7 primer, TAATACGACTCAC-TATAGGG. The primers are stored in aliquots at 1 mg/ml at –70°C. The PCR procedure is given in *Protocol 1*.

Protocol 1. PCR amplification of cDNA

Equipment and reagents

- Thermocycler
- PCR reagent kit (e.g. Perkin Elmer)
- Primers (as described above)

Method

1. Make the template by diluting 1 µl of the cDNA/vector ligation into 75 µl of water.

2. Set-up the PCR reaction as follows:

   - Template   1 µl
   - 10 × buffer   10 µl
   - T7 primer   1 µl
   - T3 primer   1 µl
   - H₂O   69.5 µl

3. Heat at 95°C for 5 min.

4. Put on ice for 3 min, add 0.5 µl *Taq* polymerase, mix, and quickly spin.

5. Place the tubes in the thermocycler and heat to 80°C; when hot add 8 µl of a solution containing 125 mM of each nucleotide.

6. Run 30 cycles of (1 min 94°C; 1 min 44°C; 2 min 72°C).

7. After the last cycle, incubate for 8 min at 72°C.

---

7. Gel purification of PCR product

PCR products ranging in size from 200 to 10000 bp are purified from primers by a 1% agarose TAE gel. Load one reaction on a 2 cm wide lane. The samples should not be exposed to UV light: cut off a marker lane, stain it with ethidium bromide, determine which portion of the gel to save, and then cut out the appropriate piece of the gel. Purify the DNA using the GeneClean Kit.

8. Restriction enzyme digestion and second gel purification of PCR product

Digest the PCR product (typically 10 μg) with 200–300 U of restriction enzyme (in our case *Bam*HI and *Sal*I) in 100 μl reactions. Isolate fragments ranging in size from 200 bp to 10 kb as described above.

9. Ligation of amplified cDNA with antisense vector

9.1 Choice of antisense vector

A general consideration for antisense vector selection is that the vector sequences flanking the cDNA must be known, so that PCR can be used to isolate the antisense cDNA from mutant transformants. Furthermore, care must be taken when designing the oligonucleotide primers so that when a mutant generated by shotgun antisense is selected and the antisense cDNA is isolated by PCR, the PCR products can be cleaved with appropriate restriction enzymes to facilitate the cloning of cDNA.

9.2 Ligation

Ligate the amplified cDNA from Section 8 into the appropriate transformation or transfection vector. Prepare the vector as described in Section 4. Given the large amounts of cDNA that can thus be synthesized, an antisense library can be constructed without expansion in bacteria, or alternatively, smaller amounts of cDNA and vector can be used in the ligation and the ligated product used to transform bacteria. In the latter case, grow the transformed cells in a litre of media for eight hours and purify the plasmid DNA by conventional means.

10. Use of shotgun antisense library

Use the library prepared in Section 9 to transform cells with the same method used for conventional (single antisense DNA species) antisense. Screen for transformants with the desired phenotypes. Prepare DNA from the transformants (often simple protocols such as boiling the cells will work) and isolate the antisense cDNA by PCR.

Acknowledgements

We thank Diane Hatton for the preparation of the manuscript. R. H. G. is an assistant investigator of the Howard Hughes Medical Institute.

References

1. Spann, T. P., Brock, D. A., Lindsey, D. F., Wood, S. A., and Gomer, R. H. (1996). Mutagenesis and gene identification in *Dictyostelium* by shotgun antisense. *Proc. Natl. Acad. Sci. USA*, **93**, 5003.
2. Sambrook, J., Fritsch, E. F., and Maniatis, T. (ed.) (1989). *Molecular cloning: a laboratory manual*, 2nd edn. Cold Spring Harbor Laboratory Press. Cold Spring Harbor, NY.

Detection of sense:antisense duplexes by structure-specific anti-RNA antibodies

NOÉMI LUKÁCS

1. Introduction

It is generally assumed that inhibition of gene expression by antisense RNA involves the formation of sense:antisense RNA hybrids (for review see refs 1–3). The formation of such hybrids is well established in prokaryotic systems and has also been shown experimentally in a few eukaryotic systems (4–6). In most eukaryotic organisms, however, attempts to detect sense:antisense RNA complexes have failed, although the biological effects of the antisense RNA were obvious. Since in several cases the simultaneous synthesis of both RNA strands has been confirmed by nuclear run-on transcription assays, and since reduced steady state levels of sense RNA have been detected, it has been proposed that sense:antisense hybrids are very rapidly degraded *in vivo* (7–9). In addition to the classical duplex formation/degradation mechanism other antisense control mechanisms may also exist, and have been discussed (1–3, 7–9).

This chapter outlines how double-stranded RNA (dsRNA)-specific monoclonal antibodies could be used to analyse possible antisense mechanisms. The anti-dsRNA antibodies have been successfully used to detect the spontaneous formation of RNA duplexes during cotranscription of sense and antisense RNAs *in vitro* (10). The other future application procedures suggested here, are aimed at the development of analytical tools to detect, characterize, and enrich double-stranded RNAs *in vivo* and/or *in vitro*.

2. dsRNA-specific antibodies and their suggested applications in antisense research

The monoclonal antibodies J2 (IgG2a) and K2 (IgM), the use of which is described below, specifically recognize the structure of double-stranded RNA molecules and do not cross-react with DNA, ssRNA, or RNA:DNA hybrids

(11). Binding of these antibodies to dsRNA has been found to be independent of the sequence and nucleotide composition of the dsRNAs tested, provided that an RNA helix of at least about 40–50 bp are present (12). It has not yet been unequivocally explained why the large epitope size is necessary for antigen recognition. The most probable explanation seems to be the necessity of bivalent antibody binding to form stable antigen–antibody complexes.

Double-helical regions of 40–50 bp are usually not present in ssRNAs, therefore the antibodies can be used as structural probes to detect intermolecular RNA:RNA hybrids that are expected when sense:antisense RNA complexes form. Since the detection principle lacks sequence specificity, the use of correct negative controls is extremely important. For example, if the control without antisense RNA already shows high dsRNA-specific signals, the immunological methods will not give reliable results with respect to the sense:antisense duplexes of interest. Samples obtained by *in vitro* transcription usually do not contain high amounts of dsRNA, but extracts made, e.g. from yeast or from sugar beet do, because of the presence of dsRNA viruses in nearly all strains, cultivars, or individual plants of these species (13, 14).

We suggest the following fields of application in antisense research:

(a) The efficiency of duplex formation after cotranscription or mixing of (partially) complementary RNA molecules can be determined by ELISA (enzyme linked immunosorbent assay) or by dsRNA immunoblotting (11). The convenience of the immunological methods allows for a large variety of RNA structures, concentrations, and experimental conditions to be quickly tested. In addition, by combining temperature-gradient gel electrophoresis (see Chapter 1) and immunoblotting the stability and structure of the complexes can be analysed, and coexisting duplexes and alternative RNA structures can also be detected without prior purification (14).

(b) The presence and the localization of RNA duplexes *in vivo* may be detected by immunofluorescence or by immunogold labelling (15). Since the antibodies will react with dsRNAs of any sequence, only a strict correlation between dsRNA-specific labelling with antibodies and the synthesis of a given antisense RNA may be taken as an indication for the formation of sense:antisense RNA hybrids.

(c) The concentration of natural or experimentally-induced antisense RNA complexes is expected to be low. We have developed an immunoaffinity chromatographic procedure which can be used to enrich for dsRNA from nucleic acid extracts as well as from unfractionated aqueous extracts (16). The method has successfully been used to purify viral dsRNAs from tobacco and an endogenous high molecular weight dsRNA from french beans. The application of this method in antisense RNA research may lead to the identification of naturally occurring duplexes, and may also allow one to answer the question of whether and how widely natural antisense regulation occurs in eukaryotes.

3. Analysis of duplex-forming capacity of antisense sequences *in vitro*

In Chapter 1 computer programs for predicting the secondary structure and stability of RNA molecules are presented. These programs help to identify sequence regions which may be good targets for antisensing and to estimate the probability of duplex formation. However, especially in the case of larger RNA molecules, kinetic processes often play an important role in folding. Several metastable RNA structures can arise and coexist and, in addition, proteins present in the system can also interact with the RNA and thereby alter its structural features (1). All these processes make it extremely difficult to predict satisfactorily the efficiency of duplex formation. Direct experimental analysis of the products arising during cotranscription of complementary RNA molecules *in vitro* may substantially improve the reliability of predictions.

A quick and convenient way to detect and quantify the RNA duplexes is the immunochemical analysis of samples withdrawn directly from the aqueous transcription mixture. Using anti-dsRNA monoclonal antibodies in sandwich ELISA (or in competition ELISA) (12) about 1 ng dsRNA can be detected in a sample of 50–100 μl, which may contain up to 25–50 μg of other nucleic acids and also proteins (11). This sensitivity has been achieved with conventional chromogenic substrates, hence with fluorogenic substrates a ten times higher sensitivity would be expected.

3.1 Sandwich ELISA

In sandwich ELISA the dsRNA-specific J2 (IgG2a) antibody is immobilized via protein A bridges on the surface of a microtitre plate. After incubation with the transcription mixture dsRNA bound to J2 is detected by the anti-dsRNA antibody K2 (IgM) and an appropriate IgM-specific secondary antibody. The method and its application for the detection of infection by RNA viruses in plants have been described (11, 14); a protocol proposed for antisense research is given below (*Protocol 1*).

Protocol 1. Sandwich ELISA

Equipment and reagents

- Protein A (RepliGen): 4 μg/ml in PBS
- Purified J2 antibody or J2 hybridoma supernatant: commercially available from Phytotest GmbH, Hilden, FRG
- K2 hybridoma supernatant (Phytotest GmbH, Hilden, FRG)
- Alkaline phosphatase-conjugated F(ab')$_2$—fragment of goat anti-mouse IgM (μ) from Jackson ImmunoResearch, USA; diluted 1:5000 in PBS containing 2% bovine serum albumin (BSA)
- *p*-Nitrophenyl phosphate (NPP) 1 mg/ml in 500 mM NaHCO$_3$ pH 9.5, plus 0.5 mM MgCl$_2$
- Purified viral dsRNA, e.g. dsCARNA5 (17) as a standard
- PBS (phosphate-buffered saline): 10 mM sodium phosphate pH 7.0, and 150 mM NaCl
- Polyvinyl chloride microtitre plates (Micro-test III, Falcon)
- Microplate Reader (MR 5000, Dynatech)

Protocol 1. *Continued*

Method

1. Pre-coat plates with 400 ng/well protein A at 4°C overnight.

2. Saturate free binding sites with 2% BSA in PBS (37°C, 60 min).

3. Wash plate three times with PBS plus 0.5% (v/v) Tween 20.

4. Add 100 μl/well J2 hybridoma supernatant diluted 1:3 with PBS containing 2% BSA. (Alternatively: add 100–200 ng/well purified J2 in PBS plus 2% BSA.)

5. Incubate at 37°C for 60 min, then wash three times with PBS plus 0.5% Tween 20.

6. Add an appropriate amount of the *in vitro* translated sense and antisense RNA. To obtain the calibration curve for quantitative estimation of the duplexes add dsRNA standards at 0.5–7.5 ng/well.

7. Incubate and wash as described in step 5.

8. Add 100 μl K2 supernatant (diluted 1:2) per well, incubate, and wash as above.

9. Add 100 μl/well alkaline phosphatase-conjugated anti-IgM antibody, incubate, and wash as above.

10. Add 100 μl/well NPP solution and measure the OD values with the Microplate Reader at 410 nm.

3.2 dsRNA immunoblotting

The ELISA methods are well suited for quantitative measurements, but their sensitivity is low, and they do not allow for the characterization of nucleic acid structures which arise during cotranscription and lead to a positive reaction with the specific antibodies. The latter goal at about 20 times higher sensitivity of detection (i.e. 50 pg dsRNA/gel band), can be achieved by dsRNA immunoblotting (11). In this procedure the nucleic acids present in the sample are first separated by polyacrylamide gel electrophoresis (PAGE) under nondenaturing conditions, whereby the electrophoretic mobility of the molecules is determined by their size and by the structure which they adopt under the given conditions. After a subsequent electrophoretic transfer to a positively charged membrane, RNA duplexes can be detected by using dsRNA-specific antibodies.

Immunological detection of membrane bound duplexes is also possible in a dot blot assay, i.e. without prior electrophoretic separation, and may thus substitute for the more time-consuming sandwich ELISA. This method has been used for the comparison of duplex-forming capacity of different antisense RNA sequences during cotranscription with potato spindle tuber viroid RNA (10). However, when immunoblotting is combined with conventional or temperature-gradient gel electrophoresis (TGGE), additional structural information about the sense:antisense complexes can be obtained, and the thermal stability may be calculated (see Chapter 1 and the literature cited therein).

Protocol 2. dsRNA immunoblotting

Equipment and reagents

- Slab gel with 5% polyacrylamide in 1 × TBE (89 mM Tris, 89 mM boric acid, pH 8.3, 2.5 mM EDTA)
- Biodyne B membrane (Pall Bio-Support membrane, Dreieich, FRG)
- Blocking reagent (Boehringer Mannheim)
- J2 hybridoma supernatant (Phytotest GmbH, Hilden, FRG)
- PBS (see *Protocol 1*)
- Trans-Blot SD semi-dry transfer cell (Bio-Rad)

- Alkaline phosphatase-conjugated F(ab')$_2$ fragment of goat anti-mouse IgG(H + L): Jackson ImmunoResearch, USA
- Substrate solution: 175 μg/ml 5-bromo-4-chloro-3-indolyl phosphate (BCIP) and 340 μg/ml nitroblue tetrazolium salt (NBT) in 100 mM Tris–HCl pH 9.5, 100 mM NaCl, 5 mM MgCl$_2$
- Power supply and slab gel apparatus
- Rocking platform

Method

1. Load the sample and electrophorese at room temperature in a conventional slab gel using 1 × TBE as electrophoresis buffer.

2. Transfer nucleic acids from the gel onto the positively charged Biodyne B membrane in the semi-dry transfer cell according to the manufacturer's instructions at 0.5 A, 30 min.

3. Saturate free binding sites on the membrane by incubating in 0.5% blocking reagent plus 50 μg/ml sonicated herring sperm DNA in PBS at 4°C overnight.

4. Incubate the membrane with J2 supernatant on a rocking platform at room temperature for 2 h.

5. Wash membrane twice with PBS plus 0.1% (v/v) Triton X-100 (room temperature, 2 × 30 min).

6. Incubate with the alkaline phosphatase-conjugated secondary antibody, diluted 1:3000 in PBS plus 2% BSA (room temperature, 60 min).

7. Repeat step 5.

8. Stain in the dark with the substrate solution. The coloured bands appear usually within 30 min.

Note. It is essential to use polyacrylamide and not agarose gels prior to immunoblotting, because agarose dramatically reduces the sensitivity of dsRNA detection.

When aiming at the characterization of the stability of the duplexes, TGGE should be used instead of conventional PAGE. A TGGE protocol is described in Chapter 1. If necessary, the conditions can be optimized by varying the temperature and the urea and buffer concentrations. To allow for subsequent blotting, the gels should be cast on the hydrophobic surface of the supporting film. For electrophoretic transfer and for immunological detection of duplexes, the same conditions are used as in *Protocol 2*, steps 2–8. As described in Chapter 1, in certain cases TGGE combined with silver staining of the nucleic acids can also be used to detect sense:antisense hybrids without the need for immunoblotting. This is possible, because dsRNAs can be identified on the basis of their very characteristic transition curves. However, since the banding pattern on the gels is often very complex and since the extraction and precipitation of nucleic acids prior to TGGE can artificially promote double-strand formation, the combination of TGGE with the immunological detection of duplexes is recommended, to solve at least the former problem.

Both the sandwich ELISA and the immunoblotting procedures have been proposed for the analysis of duplex-forming capability of different antisense RNA sequences after cotranscription or mixing with the corresponding sense RNA *in vitro*. An important additional application of these methods is to analyse unfractionated nucleic acid extracts. Using these methods for extracts from higher plants, we have found that in several plant species and cultivars, dsRNA can be immunologically detected even if there is no indication of virus infection (14, E. Düren, R. Lück, and N. Lukács, unpublished data). The concentration of dsRNAs is usually so low that they are just detectable on immunoblots as faint bands which reproducibly appear in extracts made from the same cultivar (14). The origin and physiological role of these dsRNAs is not yet known.

4. Detection of dsRNA by immunofluorescence *in situ*

It would greatly help the understanding of antisense mechanisms if the existence of sense:antisense RNA hybrids in eukaryotic cells could be proven and their localization *in situ* could be determined. Although it is not yet known

Figure 1. dsRNA in virus-infected groundnut leaves is detected by immunofluorescence. Chlorotic regions of leaves infected by groundnut rosette virus (GRV) (a and b) and normal leaf tissue from non-infected plants (c) were fixed with 3% paraformaldehyde in 0.1 M Na cacodylate pH 7.3 at room temperature for 2 h. After washing and dehydration the samples were embedded in Lowicryl K4M (Plano, Marburg, FRG) according to the manufacturer's instructions. Semi-thin sections were incubated with the anti-dsRNA J2 antibody (a and c), or with a control antibody (b). Bound antibodies were visualized by incubation with biotinylated goat anti-mouse IgG follwed by incubation with FITC-conjugated streptavidin (both reagents from Dianova, Hamburg, FRG). (a) A representative section of GRV-infected tissue with very strong dsRNA-specific immunofluorescence signals in the cytoplasm and the chloroplasts of mesophyll cells. (b) No fluorescent label is seen after incubating with the control antibody. The result demonstrates the presence of—presumably viral—dsRNA in infected leaves. (c) A slight immunofluorescence after J2 incubation is observed in uninfected tissue as well. The bar represents 30 μm. E, epidermis: P, palisade parenchyme: S, spongy parenchyma: V, vascular bundle. Figure courtesy of Gabriele Tauscher (15).

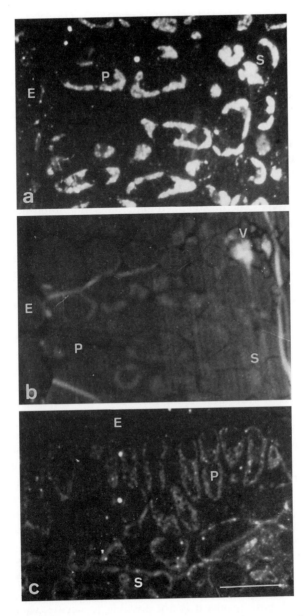

whether the sensitivity necessary for the detection of sense:antisense duplexes can be reached with anti-dsRNA antibodies, the use of such antibodies clearly offers a new experimental approach worthwhile testing experimentally. Earlier data from the literature as well as our studies on the reactivity of monoclonal anti-dsRNA antibodies with uninfected cells and with cells infected by ssRNA viruses, indicate that infected cells can show strong immunofluorescent signals which are absent in non-infected animal or plant cells (15, 18, 19). Although, as illustrated in *Figure 1*, the difference in fluorescence intensity is obvious between infected and non-infected cells (*Figure 1a* and *c*), the non-infected cells also show some fluorescence after incubation with the anti-dsRNA antibody (*Figure 1c*). This fluorescence is weak, but is clearly higher than the background observed after incubation with a non-specific control antibody (compare *Figure 1b* and *c*). The bright fluorescent labelling of infected cells in *Figure 1a* correlates well with the finding of high levels of dsRNA in an aqueous extract of the same plant (11). Since no fixation step is involved in the preparation of aqueous extracts, this provides circumstantial evidence that native dsRNA structures are being detected by immunofluorescence microscopy after fixation with formaldehyde. Unfortunately, a universal method for maintaining native dsRNA structures in fixed material is not established.

dsRNA-specific fluorescence signals have also been observed in the cytoplasm of some non-infected animal cell lines such as mosquito or Vero cells (20, 21). However, fluorescence labelling by anti-dsRNA antibodies is not common to all eukaryotic cells and several widely used cell lines, e.g. baby hamster kidney (BHK) and HeLa cells are dsRNA negative (18, 20, 22). At present it is not possible to decide whether in the dsRNA positive cells the antibodies bind to an intermolecular hybrid of two complementary RNAs or whether they react with base paired regions of highly structured single-stranded RNA molecules.

Whatever RNA structures the antibodies react with, the reactivity pattern is selective and reproducible. This is demonstrated in *Figure 2*, where the immunofluorescence pattern of primary spermatocyte nuclei from *Drosophila hydei* is shown. As in other developing germ cells, the spermatocytes have a very high transcriptional activity and the transcription products are not immediately used but stored in the nucleus (23). The Y chromosome of wild-type *D. hydei* unfolds to form a set of at least five different lampbrush loops each of which contains different families of repetitive DNA and all of which are intensively transcribed (24). Although transcripts are found in all morphologically distinct parts of the spermatocyte nuclei (24), labelling with the anti-dsRNA antibody was only found in the pseudonucleolus (Ps) and along the threads (Th) and no fluorescence was seen, for example in the nucleolus organizer (N) or on tubular ribbons (Tr). The recognition of Ps and Th transcripts may indicate the formation of long RNA helixes either by intermolecular interactions or by intramolecular base pairing of exceptionally long self-complementary regions of the RNA.

On the basis of these results it is not possible to decide whether, and in which systems, it will be possible to detect sense:antisense duplexes *in situ* by using dsRNA-specific antibodies but the obvious next steps are:

(a) To use cell lines or organisms which are originally completely dsRNA negative as the biological system of choice for analysing possible sense:antisense interactions.

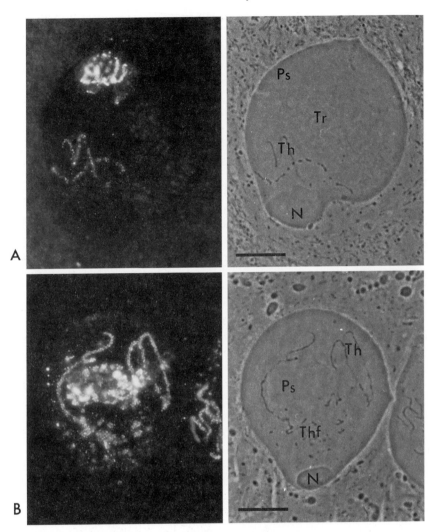

Figure 2. The pseudonucleolus and the threads in spermatocyte nuclei of *D. hydei* are labelled by the J2 antibody. Squash preparations from the testes were fixed first in 96% ethanol (2 min) and then in 3.7% unbuffered formaldehyde solution (5 min) (23). The slides were stored overnight in PBS and then incubated with the dsRNA-specific J2 antibody. Bound antibody was visualized by using FITC labelled goat anti-mouse IgG (H + L). Fluorescent staining of the lampbrush loops pseudonucleolus (Ps) and threads (Th) is shown on the left, the corresponding phase-contrast micrographs are on the right. (A) The wild-type nucleus shown exhibits five lampbrush loops (24) of which Ps, Th, and Tr (tubular ribbons) are clearly visible in phase-contrast. (B) The nucleus of the mutant contains only the Ps and Th; the latter are partly fragmented (Thf) (24). In both types of nuclei only the Ps and Th are labelled by J2. The nucleolus organizer (N) does not show any labelling. The bar represents 5 μm. (Figure courtesy of Dr K. H. Glätzer, Institut für Genetik, Düsseldorf.)

(b) To use the K2 antibody which is an IgM and gives higher sensitivity.

(c) To verify the colocalization of double-stranded RNA and a specific anti-sense RNA by *in situ* hybridization.

5. Enrichment of dsRNA by immunoaffinity chromatography

Affinity chromatography on a dsRNA-specific antibody matrix should result in the enrichment of sense:antisense duplexes and may thus allow for the isolation of duplexes at sufficiently high concentrations for detection. It is possible to choose between several alternative chromatographic procedures, but it should always be taken into account that the extraction procedure must be quick and should lead neither to strand separation within duplexes nor to

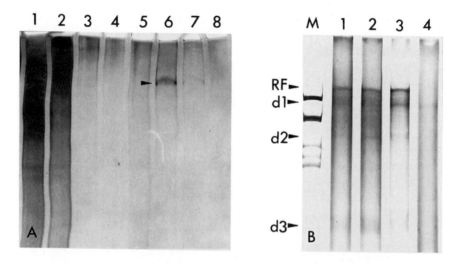

Figure 3. Immunoaffinity chromatography is used to purify dsRNA from different cell samples. (A) Purification of L-A dsRNA from yeast total nucleic acid extract. 1 ml extract containing 13.6 mg nucleic acids was diluted 1:10 with PBS and loaded onto 1.5 ml J2–protein A affinity matrix. Chromatography was as described in *Protocol 4*. Fractions were analysed by electrophoresis in 5% polyacrylamide 1 × TBE gels (see *Protocol 2*) followed by silver staining. Lane 1 shows the total nucleic acid extract, lane 2 is the flow-through, lanes 3–5 represent the different washes, lanes 6–8 are the eluates. The position of L-A dsRNA is shown by the arrow. (B) Purification of viral dsRNAs from aqueous extacts of TMV-infected tobacco. 3 g leaves were shock frozen, homogenized, and extracted with 10 ml extraction buffer (see text). Chromatography using 300 µl J2–protein A matrix was as described in *Protocol 4*. The eluate was treated with different nucleases and then analysed by electrophoresis in 5% polyacrylamide 1 × TBE gels and silver staining. Lane 1 represents the untreated eluate; the samples in lanes 2–4 were digested with DNase I, nuclease S1, and RNase III, respectively. M: reovirus dsRNAs as marker (1.2–3.9 kb). According to their electrophoretic mobility and nuclease sensitivity the major bands in the eluate represent the TMV dsRNAs RF and dsRNA1–3 (designated as d1-d3 in *Figure 3B*) described in the literature (29).

duplex formation as an *in vitro* artefact. The latter may happen during nucleic acid extraction when the deproteinized nucleic acids reach high concentrations after precipitation, especially if the experimental conditions (i.e. low ionic strength, high temperature) promote denaturation of intrachain structures. To purify the dsRNA species shown in *Figure 3B* from plants infected with tobacco mosaic virus (TMV) we made aqueous extracts from shock frozen leaves using 3 ml extraction buffer (0.1 M sodium phosphate pH 7.2, 1 mM phenylmethylsulfonyl fluoride, 5 mM dithiothreitol, 1.5% (w/v) bentonite, and 1.0% (v/v) Tween 20) per gram leaves (16).

To prepare a good immunoaffinity matrix it is absolutely necessary to immobilize the anti-dsRNA antibodies via protein A or through their carbohydrate moieties, because direct coupling to the CNBr-activated matrix leads to losses both in activity and in specificity.

Protocol 3. Preparation of immunoaffinity matrix by coupling anti-dsRNA IgG antibodies covalently to protein A–Sepharose

Reagents

- Protein A–Sepharose[a]: protein A coupled covalently to CNBr-activated Sepharose 4B (Pharmacia-LKB) according to the manufacturer's instructions (the affinity matrix usually has a protein A concentration of 1.0–1.2 mg/ml)
- Purified J2 (IgG2a) antibody at 1–2 mg antibody/ml in PBS (see *Protocol 1*)
- Dimethyl-suberimidate dihydrochloride (Janssen, Belgium) 5 mg/ml in 0.1 M borate buffer pH 8.0
- Citrate buffer 0.1 M pH 4.5

Method

1. Wash protein A–Sepharose twice with 5 vol. PBS in a batch procedure.
2. After centrifugation of the affinity matrix add 1–2 mg purified J2 to 1 ml gel pellet in PBS and incubate for 15 min at room temperature.
3. Separate unbound J2 and matrix by centrifugation, and wash the gel three times with 5–10 vol. of PBS. The pellet after the third wash usually contains 1.2–1.7 mg J2 /ml bound reversibly to protein A.
4. Incubate the gel in 0.1 M borate buffer pH 8.0 for 10 min.
5. Discard the supernatant and incubate gel with the dimethyl-suberimidate solution for 30 min to covalently couple the Fc part of the IgG to protein A (25).
6. Remove reagent by washing twice with PBS.
7. Remove unbound J2 by incubation in 0.1 M citrate buffer pH 4.5.
8. Equilibrate gel matrix with PBS and store in PBS containing 0.2% NaN$_3$ at 4°C.

[a] Other matrices such as protein A-agarose (Bio-Rad) may also be used instead of protein A–Sepharose.

An affinity matrix of similar quality was obtained when antibodies were coupled to Affi-Gel Hz (Bio-Rad) according to the manufacturer's instructions. Here the hydroxyl groups of the carbohydrate moieties on the Fc part of the antibody are first oxidized to aldehydes by sodium periodate and then coupled to the gel matrix derivatized with hydrazides (26). After incubation with 1.7 mg antibody /ml of gel, we obtained a matrix having an antibody concentration of 1.1 mg/ml. The dsRNA binding capacity of both kinds of matrices was determined by using purified L-A dsRNA from yeast (13). The J2–protein A column had a capacity of 4–25 μg dsRNA/ml gel, the J2–Affi-Gel Hz a capacity of 30–100 μg dsRNA/ml (16).

Protocol 4. Immunoaffinity chromatography

Equipment and reagents

- Immunoaffinity matrix with covalently coupled anti-dsRNA antibody
- Mobicol tubes (v = 1.3 ml) (Mobitec, Göttingen, FRG): the lower end of the tubes is closed with a 2 mm thick glass filter; an adapter for syringes is fitted at the upper end
- Rotating shaker

- PBS (see *Protocol 1*)
- 10 mM potassium phosphate pH 7.2 supplemented with 0.3% or 5% sodium deoxycholate (DOC)
- tRNA from *E. coli* (Boehringer Mannheim): 1 mg/ml in TE buffer
- TNE: 50 mM Tris–HCl pH 8.0, 100 mM NaCl, 10 mM EDTA

Method

1. Add 300 μl affinity matrix to 10 ml aqueous extract (or to a crude nucleic acid extract isolated under non-denaturing conditions) and rotate the tube end-over-end in a shaker for 30 min at room temperature.

2. Centrifuge at low revolution, e.g. at 500 r.p.m. for 3 min. Discard supernatant, add 10 ml PBS to the pellet, and rotate for further 15 min.

3. Fill the slurry into a Mobicol tube and wash the column three times with 500 μl PBS and once with 500 μl 0.3% DOC in 10 mM K phosphate pH 7.2. (The use of a low salt buffer is necessary because, at high salt concentrations and also at pH values under 7.0, DOC precipitates in the column.)

4. Elute dsRNA by incubation with 500 μl 5% DOC in 10 mM K phosphate for 15 min. Repeat this step three times.

5. Combine the three eluate fractions, dilute them 1:1 with distilled water, and adjust the concentration to 0.25 M ammonium chloride. Add tRNA to a final concentration of 2.4 μg/ml and precipitate nucleic acids with 3 vol. of ethanol.

6. The column can be reused after washing twice with 500 μl 10 mM phosphate buffer pH 7.2 and equilibrating with 10 ml PBS.

As shown in *Figure 3*, dsRNA is specifically bound to and elutes from the affinity column using the procedure described in *Protocol 4*. (dsRNA can also be efficiently eluted with 1% SDS, but in that case the column can not be reused.) To verify that the nucleic acids eluted were indeed dsRNAs, the fractions were treated with DNase I, nuclease S1, and with RNase III (*Figure 3B*). No changes were observed after DNase I digestion. Treatment with the single-strand-specific nuclease S1 left the discrete nucleic acid bands unaltered, but resulted in the removal of a smear which could represent a heterogeneous mixture of single-stranded contaminants or partially double-stranded RNA. No discrete bands were observed after incubation with the dsRNA-specific RNase III. The results show that naturally occurring dsRNAs are greatly enriched by immunoaffinity chromatography on J2 antibody matrix.

6. Potential applications of dsRNA-specific antibodies for *in vivo* studies

Antibodies and antibody fragments have already been used for 'intracellular immunization', i.e. to influence the activity of their antigens simply by forming antigen–antibody complexes inside the cell (27, 28). Similarly, dsRNA-specific antibodies might also be used to alter, *in vivo*, the state of equilibria in which sense:antisense RNA duplexes are involved. Two possible methods to introduce the antibodies into living cells or organisms have been tested by our group (22). First, animal cells (NRK cells) cultured *in vitro* were electroporated in the presence of 3 mg/ml J2 antibody in PBS containing 10 mM $MgCl_2$. To reach high antibody concentrations inside the cells, the pulses had to be conducted at high electric field and capacity. Electric pulses at 700 V/cm and 960 μF resulted in an average uptake of about 70 fg antibody per cell, i.e. in an intracellular antibody concentration of about 10^{-8} M. Unfortunately only 30% of the cells survived these harsh conditions, however, more than 80% of the surviving cells contained fluorescently labelled antibody 8 h after electroporation. No deleterious effects of the antibodies on cellular morphology or metabolism were observed, but as a result of cell division and antibody decomposition *in vivo* only 40% of the cells showed antibody-specific fluorescence 36 h later (22).

These observations suggest that introduction of dsRNA-specific antibodies by electroporation with the aim of stabilizing and/or protecting sense:antisense RNA hybrids may be useful if short-term studies (8–24 h) are to be carried out. For long-term studies and for optimal targeting of the antibodies to different organelles inside the cell, it will be necessary to produce transgenic lines which express suitable antibody constructs. J2 antibody constructs have already been introduced into tobacco, and several plants containing high concentrations of assembled and active antibody in the apoplast have been characterized in detail (22). At present we are working on the intracytoplasmic expression of active antibody fragments which may then be targeted to other cell compartments. As soon as these transgenic plants have been successfully constructed, they will be made available for research purposes.

Acknowledgements

I am indebted to Dr K. H. Glätzer, Heinrich-Heine-Universität Düsseldorf for generously contributing the previously unpublished photographs for *Figure 2*. I would like to thank G. Tauscher and A. Detke for the results shown in *Figure 1* and *Figure 3*, respectively. I also thank P. Symmons, Deutsche Dynal GmbH, Hamburg for improving the English. This work was supported by grant Lu 386/1–1 and -2 from the Deutsche Forschungsgemeinschaft.

References

1. Eguchi, Y., Itoh, T., and Tomizawa, J. (1991). *Annu. Rev. Biochem.*, **60**, 631.
2. Takayama, K. M. and Inouye, M. (1990). *Crit. Rev. Biochem. Mol. Biol.*, **25**, 155.
3. Murray, J. A. H. and Crockett, N. (1992). In *Antisense RNA and DNA* (ed. J. A. H. Murray), pp. 1–50. Wiley-Liss, New York.
4. Kim, S. K. and Wold, B. J. (1985). *Cell*, **42**, 129.
5. Yokoyama, K. and Imamoto, F. (1987). *Proc. Natl. Acad. Sci. USA*, **84**, 7363.
6. Knecht, D. A. and Loomis, W. F. (1987). *Science*, **236**, 1081.
7. Hélene, C. and Toulmé, J.-J. (1990). *Biochim. Biophys. Acta*, **1049**, 99.
8. Bass, B. L. (1992). In *Antisense RNA and DNA* (ed. J. A. H. Murray), pp. 159–74. Wiley–Liss, New York.
9. Nellen, W. and Lichtenstein, C. (1993). *Trends Biochem. Sci.*, **18**, 419.
10. Matousek, J., Trnená, L., Rakousky, S., and Riesner, D. (1994). *J. Phytopathol.*, **141**, 10.
11. Schönborn, J., Oberstraβ, J., Breyel, E., Tittgen, J., Schumacher, J., and Lukács, N. (1991). *Nucleic Acids Res.*, **19**, 2993.
12. Oberstraβ, J., Richter, A., Fischer, R., and Lukács, N. (1997). Submitted.
13. Wickner, R. B. (1993). *J. Biol. Chem.*, **268**, 3797.
14. Lukács, N. (1994). *J. Virol. Methods*, **47**, 255.
15. Tauscher, G. (1989). Diploma thesis, Heinrich-Heine-Universität Düsseldorf.
16. Detke, A. (1989). Diploma thesis, Heinrich-Heine-Universität Düsseldorf.
17. Kaper, J. M. and Tousignant, M. E. (1978). *Virology*, **85**, 323.
18. Stollar, V. and Stollar, B. D. (1970). *Virology*, **42**, 276.
19. Lin, N. S. and Langenberg, W. G. (1985). *Virology*, **142**, 291.
20. Stollar, B. D. and Stollar, R. K. V. (1979). *Science*, **200**, 1381.
21. Mac Donald, R. D. (1980). *Can. J. Microbiol.*, **26**, 256.
22. Richter, A. (1995). PhD thesis, Heinrich-Heine-Universität Düsseldorf.
23. Glätzer, K. H. and Meyer, G. F. (1981). *Biol. Cell.*, **41**, 165.
24. Trapitz, P., Glätzer, K. H., and Bünemann, H. (1992). *Mol. Gen. Genet.*, **235**, 221.
25. Gersten, D. M. and Marchalonis, J. J. (1978). *J. Immunol. Methods*, **24**, 305.

26. Prisyazhnoy, V. S., Fusek, M., and Alakhov, Y. B. (1988). *J. Chromatogr.*, **424**, 243.
27. Carlson, J. R. (1988). *Mol. Cell. Biol.*, **8**, 2638.
28. Biocca, S., Neuberger, M. S., and Cattaneo, A. (1990). *EMBO J.*, **9**, 101.
29. Zelcer, A., Weaber, K. F., Balázs, E., and Zaitlin, M. (1981). *Virology*, **113**, 417.

A1

List of suppliers

Amersham
Amersham International plc., Lincoln Place, Green End, Aylesbury, Buckinghamshire HP20 2TP, UK.
Amersham Corporation, 2636 South Clearbrook Drive, Arlington Heights, IL 60005, USA.

Anderman
Anderman and Co. Ltd., 145 London Road, Kingston-Upon-Thames, Surrey KT17 7NH, UK.

Beckman Instruments
Beckman Instruments UK Ltd., Oakley Court, Kingsmead Business Park, London Road, High Wycombe, Bucks HP11 1J4, UK.
Beckman Instruments Inc., PO Box 3100, 2500 Harbor Boulevard, Fullerton, CA 92634, USA.

Becton Dickinson
Becton Dickinson and Co., Between Towns Road, Cowley, Oxford OX4 3LY, UK.
Becton Dickinson and Co., 2 Bridgewater Lane, Lincoln Park, NJ 07035, USA.

Bio
Bio 101 Inc., c/o Statech Scientific Ltd, 61–63 Dudley Street, Luton, Bedfordshire LU2 0HP, UK.
Bio 101 Inc., PO Box 2284, La Jolla, CA 92038–2284, USA.
Bioblock Scientific, BP 111, F-67403, Illkirch, Cedex, France.

Bio-Rad Laboratories
Bio-Rad Laboratories Ltd., Bio-Rad House, Maylands Avenue, Hemel Hempstead HP2 7TD, UK.
Bio-Rad Laboratories, Division Headquarters, 3300 Regatta Boulevard, Richmond, CA 94804, USA.

Boehringer Mannheim
Boehringer Mannheim UK (Diagnostics and Biochemicals) Ltd, Bell Lane, Lewes, East Sussex BN17 1LG, UK.
Boehringer Mannheim Corporation, Biochemical Products, 9115 Hague Road, P.O. Box 504 Indianapolis, IN 46250–0414, USA.
Boehringer Mannheim Biochemica, GmbH, Sandhofer Str. 116, Postfach 310120 D-6800 Ma 31, Germany.

British Drug Houses (BDH) Ltd, Poole, Dorset, UK.

Clontech Laboratories, Inc., 4030 Fabian Way, Palo Alto, CA 94303-4607, USA.

Cruachem Ltd, Todd Campus, Acre Road, Glasgow G20 0UA, UK.

Difco Laboratories

Difco Laboratories Ltd., P.O. Box 14B, Central Avenue, West Molesey, Surrey KT8 2SE, UK.

Difco Laboratories, P.O. Box 331058, Detroit, MI 48232–7058, USA.

Du Pont

Dupont (UK) Ltd., Industrial Products Division, Wedgwood Way, Stevenage, Herts, SG1 4Q, UK.

Du Pont Co. (Biotechnology Systems Division), P.O. Box 80024, Wilmington, DE 19880–002, USA.

European Collection of Animal Cell Culture, Division of Biologics, PHLS Centre for Applied Microbiology and Research, Porton Down, Salisbury, Wilts SP4 0JG, UK.

Falcon (Falcon is a registered trademark of Becton Dickinson and Co.).

Fisher Scientific Co., 711 Forbest Avenue, Pittsburgh, PA 15219–4785, USA.

Flow Laboratories, Woodcock Hill, Harefield Road, Rickmansworth, Herts. WD3 1PQ, UK.

Fluka

Fluka-Chemie AG, CH-9470, Buchs, Switzerland.

Fluka Chemicals Ltd., The Old Brickyard, New Road, Gillingham, Dorset SP8 4JL, UK.

Gibco BRL

Gibco BRL (Life Technologies Ltd.), Trident House, Renfrew Road, Paisley PA3 4EF, UK.

Gibco BRL (Life Technologies Inc.), 3175 Staler Road, Grand Island, NY 14072–0068, USA.

Glen Research Inc, 44901 Falcon Place, Sterling, VA 22170, USA.

Arnold R. Horwell, 73 Maygrove Road, West Hampstead, London NW6 2BP, UK.

Hybaid

Hybaid Ltd., 111–113 Waldegrave Road, Teddington, Middlesex TW11 8LL, UK.

Hybaid, National Labnet Corporation, P.O. Box 841, Woodbridge, NJ. 07095, USA.

HyClone Laboratories 1725 South HyClone Road, Logan, UT 84321, USA.

International Biotechnologies Inc., 25 Science Park, New Haven, Connecticut 06535, USA.

Invitrogen Corporation

Invitrogen Corporation 3985 B Sorrenton Valley Building, San Diego, CA. 92121, USA.

Invitrogen Corporation c/o British Biotechnology Products Ltd., 4–10 The Quadrant, Barton Lane, Abingdon, OX14 3YS, UK.

Kodak: Eastman Fine Chemicals 343 State Street, Rochester, NY, USA.

Life Technologies Inc., 8451 Helgerman Court, Gaithersburg, MN 20877, USA.

Merck

Merck Industries Inc., 5 Skyline Drive, Nawthorne, NY 10532, USA.

Merck, Frankfurter Strasse, 250, Postfach 4119, D-64293, Germany.

Millipore

Millipore (UK) Ltd., The Boulevard, Blackmoor Lane, Watford, Herts WD1 8YW, UK.

Millipore Corp./Biosearch, P.O. Box 255, 80 Ashby Road, Bedford, MA 01730, USA.

New England Biolabs (NBL)

New England Biolabs (NBL), 32 Tozer Road, Beverley, MA 01915–5510, USA.

New England Biolabs (NBL), c/o CP Labs Ltd., P.O. Box 22, Bishops Stortford, Herts CM23 3DH, UK.

Nikon Corporation, Fuji Building, 2–3 Marunouchi 3-chome, Chiyoda-ku, Tokyo, Japan.

Perkin-Elmer

Perkin-Elmer Ltd., Maxwell Road, Beaconsfield, Bucks. HP9 1QA, UK.

Perkin-Elmer Ltd., Post Office Lane, Beaconsfield, Bucks, HP9 1QA, UK.

Perkin-Elmer-Cetus (The Perkin-Elmer Corporation), 761 Main Avenue, Norwalk, CT 0689, USA.

Perspective Biosystem, University Park, MIT, 38 Sidney Street, Cambridge, MA 02139, USA.

Pharmacia, 23 Grosvenor Road, St. Albans, AL1 3AW.

Pharmacia Biotech Europe Procordia EuroCentre, Rue de la Fuse-e 62, B-1130 Brussels, Belgium.

Pharmacia Biosystems

Pharmacia Biosystems Ltd. (Biotechnology Division), Davy Avenue, Knowlhill, Milton Keynes MK5 8PH, UK.

Pharmacia LKB Biotechnology AB, Björngatan 30, S-75182 Uppsala, Sweden.

Promega

Promega Ltd., Delta House, Enterprise Road, Chilworth Research Centre, Southampton, UK.

Promega Corporation, 2800 Woods Hollow Road, Madison, WI 53711–5399, USA.

Qiagen

Qiagen Inc., c/o Hybaid, 111–113 Waldegrave Road, Teddington, Middlesex, TW11 8LL, UK.

Qiagen Inc., 9259 Eton Avenue, Chatsworth, CA 91311, USA.

Schleicher and Schuell

Schleicher and Schuell Inc., Keene, NH 03431A, USA.

Index

Index